Telecommunication Breakdown

Telecommunication Breakdown

Concepts of Communication Transmitted via Software-Defined Radio

C. Richard Johnson, Jr.

School of Electrical and Computer Engineering
Cornell University

and

William A. Sethares

Department of Electrical and Computer Engineering
University of Wisconsin—Madison

PEARSON

Prentice
Hall

Upper Saddle River, New Jersey 07458

Library of Congress Cataloging-in-Publication Data on file

Vice President and Editorial Director, ECS: *Marcia J. Horton*
Publisher: *Tom Robbins*
Vice President and Director of Production and Manufacturing, ESM: *David W. Riccardi*
Executive Managing Editor: *Vince O'Brien*
Managing Editor: *David A. George*
Production Editor: *Scott Disanno*
Director of Creative Services: *Paul Belfanti*
Creative Director: *Carole Anson*
Art Director: *Jayne Conte*
Cover Designer: *Bruce Kenselaar*
Art Editor: *Greg Dulles*
Manufacturing Manager: *Trudy Pisciotti*
Manufacturing Buyer: *Lisa McDowell*
Marketing Manager: *Holly Stark*

 © 2004 by Pearson Prentice Hall
Pearson Education, Inc.
Upper Saddle River, New Jersey 07458

Pearson Prentice Hall® is a trademark of Pearson Education, Inc.

MATLAB is a registered trademark of The MathWorks, Inc., 3 Apple Hill Drive, Natick, MA 01760-2098.

The author and publisher of this book have used their best efforts in preparing this book. These efforts include the development, reserach, and testing of the theories and programs to determine their effectiveness. The author and publisher make no warranty of any kind, expressed or implied, with regard to these programs or the documentation contained in this book. The author and publisher shall not be liable in any event for incidental or consequential damages in connection with, or arising out of, the furnishing, performance, or use of these programs.

Printed in the United States of America

10 9 8 7 6 5 4 3 2

ISBN 0-13-143047-5

Pearson Education Ltd., *London*
Pearson Education Australia Pty. Ltd., *Sydney*
Pearson Education Singapore, Pte. Ltd.
Pearson Education North Asia Ltd., *Hong Kong*
Pearson Education Canada, Inc., *Toronto*
Pearson Educación de Mexico, S.A. de C.V.
Pearson Education–Japan, *Tokyo*
Pearson Education Malaysia, Pte. Ltd.
Pearson Education, Inc., *Upper Saddle River, New Jersey*

Contents

To the Instructor

...though it's OK for the student to listen in.

Telecommunication Breakdown helps the reader build a complete digital radio that includes each part of a typical digital communication system. Chapter by chapter, the reader creates a MATLAB™ realization of the various pieces of the system, exploring the key ideas along the way. In the final chapter, the reader "puts it all together" to build a fully functional receiver, though it will not operate in real time. *Telecommunication Breakdown* explores telecommunication systems from a very particular point of view: the construction of a workable receiver. This viewpoint provides a sense of continuity to the study of communication systems.

The three steps in the creation of a working digital radio are the following:

1. building the pieces,
2. assessing the performance of the pieces, and
3. integrating the pieces.

In order to accomplish this in a single semester, we have had to strip away some topics that are commonly covered in an introductory course and emphasize some topics that are often covered only superficially. We have chosen not to present an encyclopedic catalog of every method that can be used to implement each function of the receiver. For example, we focus on frequency division multiplexing rather than time or code division methods, and we concentrate on pulse amplitude modulation rather than quadrature modulation or frequency shift keying. On the other hand, some topics (such as synchronization) loom large in digital receivers, and we have devoted a correspondingly greater space to these. Our belief is that it is better to learn one complete system from start to finish, than to half-learn the properties of many.

Our approach to building the components of the digital radio is consistent throughout *Telecommunication Breakdown*. For many of the tasks, we define a "performance" function and an algorithm that optimizes this function. This approach provides a unified framework for deriving the AGC, clock recovery, carrier recovery, and equalization algorithms. Fortunately, this can be accomplished using only the mathematical tools that an electrical engineer (at the level of a college junior) is likely to have, and *Telecommunication Breakdown* requires no more than knowledge of calculus and Fourier transforms. Any of the comprehensive calculus books by Thomas would provide an adequate background along with an understanding of signals and systems such as might be taught using *DSP First* or any of the fine texts cited for further reading in Section 3.8.

Telecommunication Breakdown emphasizes two ways of assessing the behavior of the components of the communication system: by studying the performance functions, and through the use of experiment. The algorithms embodied in the various components can be derived without making assumptions about details of the constituent signals (such as Gaussian noise). The use of probability is limited to naive ideas such as the notion of an average of a collection of numbers, rather than requiring the machinery of stochastic processes. By removing the advanced probability prerequisite from *Telecommunication Breakdown* it is possible to place it earlier in the curriculum.

The integration phase of the receiver design is accomplished in Chapters 9 and 15. Since any real digital radio operates in a highly complex environment, analytical models cannot hope to approach the "real" situation. Common practice is to build a simulation and to run a series of experiments. *Telecommunication Breakdown* provides a set of guidelines (in Chapter 15) for a series of tests to verify the operation of the receiver. The final project challenges the digital radio that the student has built by adding noises and imperfections of all kinds: additive noise, multipath disturbances, phase jitter, frequency inaccuracies, clock errors, etc. A successful design can operate even in the presence of such distortions.

It should be clear that these choices distinguish *Telecommunication Breakdown* from other, more encyclopedic texts. We believe that this "hands-on" method makes *Telecommunication Breakdown* ideal for use as a learning tool, though it is less comprehensive than a reference book. In addition, the instructor may find that the order of presentation of topics is different from that used by other books. Section 1.3 provides an overview of the flow of topics, and our reasons for structuring the course as we have.

How We've Used *Telecommunication Breakdown*

Though this is a first edition, the authors have taught from (various versions of) this text for a number of years. We have explored several different ways to fit coverage of digital radio into a "standard" electrical engineering senior elective sequence.

Perhaps the simplest way is via a "stand-alone" course, one semester long, in which the student works through the chapters and ends with the final project as outlined in Chapter 15. Students who have graduated tell us that when they get to the workplace, where software-defined digital radio is increasingly important, the preparation of this course has been invaluable. Combined with a rigorous course in probability, other students have reported that they are well prepared for the typical introductory graduate-level class in communications offered at research universities.

At both Cornell and the University of Wisconsin (the home institutions of the authors), there is a two-semester sequence in communications available for advanced undergraduates. We have integrated the text into this curriculum in three ways:

1. Teach from a traditional text for the first semester and use *Telecommunication Breakdown* in the second.
2. Teach from *Telecommunication Breakdown* in the first semester and use a traditional text in the second.
3. Teach from *Telecommunication Breakdown* in the first semester and teach a project-oriented extension in the second.

All three work well. When following the first approach, students often comment that by reading *Telecommunication Breakdown* they "finally understand what they had been doing the previous semester." Because there is no probability prerequisite for *Telecommunication Breakdown*, the second approach can be moved earlier in the curriculum. Of course, we encourage students to take probability at the same time. In the third approach, the students were asked to create an extension of the basic pulse amplitude modulation (PAM) digital radio to quadrature amplitude modulation (QAM), to use more advanced equalization techniques, etc. Some of these extensions are available on the enclosed CD.

Contextual Readings

We believe that the increasing market penetration of broadband communications is the driving force behind the continuing (re)design of "radios" (wireless communications devices). Digital devices continue to penetrate the market formerly occupied by analog (for instance, digital television is slated to replace analog television in the U.S. in 2006) and the area of digital and software-defined radio is regularly reported in the mass media. Accordingly, it is easy for the instructor to emphasize the social and economic aspects of the "wireless revolution."

We provide a list of articles appearing in the popular press (in the year just prior to publication of *Telecommunication Breakdown*), and this is available on the CD. For example, articles from this list discuss how local municipalities are investing in wireless Internet connections in order to attract businesses, governmental interests in the efficient use of the electromagnetic spectrum, consumer demand for broadband access to the Internet, wireless infrastructure, etc. The impacts of digital "radios" are vast, and it is an exciting time to get involved. While *Telecommunication Breakdown* focuses on technological aspects of the radio design, almost all of the mass media articles emphasize the economic, political, and social aspects. We believe that this can also add an important dimension to the student's education.

Some Extras

The CD-ROM included with the book contains extra material of interest, especially to the instructor. First, we have assembled a complete collection of slides (in .pdf format) that may help in lesson planning. The final project is available in two complete forms, one that exploits the block coding of Chapter 14 and one that does not. In addition, there are a large number of "received signals" on the CD that can be used for assignments and for the project. An extra chapter called *A Digital Quadrature Amplitude Modulation (QAM) Radio* (and a corresponding set of .pdf lecture slides) is on the CD, and this extends the software-defined radio from pulse amplitude modulation to QAM. Finally, all the MATLAB code that is presented in the text is available on the CD-ROM. Once these are added to the MATLAB path, they can be used for assignments and for further exploration. See the readme file for up-to-date information and a detailed list of the exact contents of the CD.

Mathematical Prerequisites

- G.B. Thomas and R.L. Finney, *Calculus and Analytic Geometry*, 8th edition, Addison-Wesley.
- J.H. McClellan, R.W. Schafer, and M.A. Yoder, *DSP First: A Multimedia Approach* Prentice Hall, 1998.

Acknowledgments

Applied Signal Technology, Aware, Jai Balkrishnan, Ann Bell, Rick Brown, Raul Casas, Wonzoo Chung, Tom Endres, Fox Digital, Matt Gaubatz, John Gubner, Jarvis Haupt, Andy Klein, Brian Evans, Betty Johnson, Mike Larimore, Sean Leventhal, Lucent Technologies, Rick Martin, National Science Foundation, NxtWave Communications (now ATI), Katie Orlicki, Adam Pierce, Tom Robbins, Brian Sadler, Phil Schniter, Johnson Smith, John Treichler, John Walsh, Evans Wetmore, Doug Widney, Robert Williamson, and all the members of ECE436 and ECE437 at the University of Wisconsin, and ECE467 and ECE468 at Cornell University.

C. RICHARD JOHNSON JR.
WILLIAM A. SETHARES

Telecommunication Breakdown

I

When Is a Digital Radio Like an Onion?

Telecommunication Breakdown: *Concepts of Communication Transmitted via Software-Defined Radio* is structured like an onion. The first chapter presents a sketch of a digital radio; the first layer of the onion. The second chapter peels back the onion to reveal another layer that fills in details and demystifies various pieces of the design. Successive chapters then revisit the same ideas, each layer adding depth and precision. The first functional (though idealized) receiver appears in Chapter 9. Then the idealizing assumptions are stripped away one at a time throughout the remaining chapters, culminating in a sophisticated design in the final chapter. Section 1.3 on page 12 outlines the five layers of the receiver onion and provides an overview of the order in which topics are discussed.

1 A Digital Radio

The fundamental problem of communication is that of reproducing at one point either exactly or approximately a message selected at another point.
—C. Shannon, "A Mathematical Theory of Communication," *Bell System Technical Journal*, Vol. 27, 1948

1.1 A Digital Radio

The fundamental principles of telecommunications have remained much the same since Shannon's time. What has changed, and is continuing to change, is how those principles are deployed in technology. One of the major ongoing changes is the shift from hardware to software—and *Telecommunication Breakdown* reflects this trend by focusing on the design of a digital *software-defined radio* that you will implement in MATLAB.

"Radio" does not literally mean the AM/FM radio in your car, It represents any through-the-air transmission such as television, cell phone, or wireless computer data, though many of the same ideas are also relevant to wired systems such as modems, cable TV, and telephones. "Software defined" means that key elements of the radio are implemented in software. Taking a "software defined" approach mirrors the trend in modern receiver design in which more and more of the system is designed and built in reconfigurable software, rather than in fixed hardware. It also allows the concepts behind the transmission to be introduced, demonstrated, (and hopefully understood) through simulation. For example, when talking about how to translate the frequency of a signal, the procedures are presented mathematically in equations, pictorially in block diagrams, and then concretely as short MATLAB programs.

Our educational philosophy is that it is better to learn by doing: to motivate study with experiments, to reinforce mathematics with simulated examples, to integrate concepts by "playing" with the pieces of the system. Accordingly, each of the later chapters is devoted to understanding one component of the transmission system, and each culminates in a series of tasks that ask you to "build" a particular version of that part of the communication system. In the final chapter, the parts are combined to form a full receiver.

We try to present the essence of each system component in the simplest possible form. We do not intend to show all the most recent innovations (though our presentation and viewpoint are modern), nor do we intend to provide a complete analysis of the various methods. Rather, we ask you to investigate the performance of the subsystems, partly through analysis and partly using the software code that you have created and that we have provided. We do offer insight into all pieces of a complete transmission system. We present the major ideas of communications via a small number of unifying principles such as transforms to teach modulation, and recursive techniques to teach synchronization and equalization. We believe that these basic principles have application far beyond receiver design, and so the time spent mastering them is well worth the effort.

Though far from optimal, the receiver that you will build contains all the elements of a fully functional receiver. It provides a simple way to ask and answer *what if* questions. What if there is noise in the system? What if the modulation frequencies are not exactly as specified? What if there are errors in the received digits? What if the data rate is not high enough? What if there are distortion, reflections, or echoes in the transmission channel? What if the receiver is moving?

The first layer of the *Telecommunication Breakdown* onion begins with a sketch of a digital radio.

1.2 An Illustrative Design

The first design is a brief tour of the outer layer of the onion. If some of the terminology seems obscure or unfamiliar, rest assured that succeeding sections and chapters will revisit the words and refine the ideas. The design is shown in Figures 1.1 through 1.7. While talking about these figures, it will become clear that some ideas are being oversimplified.

Things to worry about later.

Eventually, it will be necessary to come back and examine these more closely. The notes in the margin are reminders to return and think about these areas more deeply later on.

In keeping with Shannon's goal of reproducing at one point a message known at another point, suppose that it is desired to transmit a text message from one place to another. Of course, there is nothing magical about text; however, .mp3 sound files, .jpg photos, snippets of speech, raster scanned television images, or any other kind of information would do, as long as it can be appropriately digitized into ones and zeros.

Can every kind of message be digitized into ones and zeros?

Perhaps the simplest possible scheme would be to transmit a pulse to represent a one and to transmit nothing to represent a zero. With this scheme, however, it is hard to tell the difference between a string of zeroes and no transmission at all. A common remedy is to send a pulse with a positive amplitude to represent a one and a pulse of the same shape, but negative amplitude to represent a zero. In fact, if the receiver could distinguish pulses of different sizes, then it would be possible to send two bits with each symbol, for example, by associating the amplitudes[1] of $+1$, -1, $+3$ and -3 with the four choices 10, 01, 00, and 11. The four symbols $\pm 1, \pm 3$ are called the *alphabet*, and the conversion from the original message (the text) into the symbol alphabet is accomplished by the *coder* in the transmitter diagram Figure 1.1. The first few letters the standard ASCII (binary) representation of these letters and their coding into symbols are as follows:

Some codes are better than others. How can we tell?

letter	binary ASCII code	symbol string	
a	01 10 00 01	$-1,\ 1,\ -3,\ -1$	
b	01 10 00 10	$-1,\ 1,\ -3,\ 1$	
c	01 10 00 11	$-1,\ 1,\ -3,\ 3$	(1.1)
d	01 10 01 00	$-1,\ 1,\ -1,\ -3$	
$\vdots$	$\vdots$	$\vdots$	

---

[1] Many such choices are possible. These particular values were chosen because they are equidistant and so noise would be no more likely to flip a 3 into a 1 than to flip a 1 into a -1.

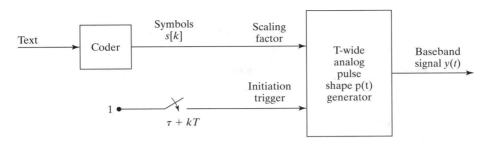

FIGURE 1.1 An idealized baseband transmitter.

In this example, the symbols are clustered into groups of four, and each cluster is called a frame. Coding schemes can be designed to increase the security of a transmission, to minimize the errors, or to maximize the rate at which data are sent. This particular scheme is not optimized in any of these senses, but it is convenient to use in simulation studies.

To be concrete, let

- the *symbol interval* T be the time between successive symbols, and

- the *pulse shape* $p(t)$ be the shape of the pulse that will be transmitted.

For instance, $p(t)$ may be the rectangular pulse

$$p(s) = \begin{bmatrix} 1 & \text{when } 0 \leq s < T, \\ 0 & \text{otherwise} \end{bmatrix} \tag{1.2}$$

which is plotted in Figure 1.2. The transmitter of Figure 1.1 is designed so that every T seconds it produces a copy of $p(\cdot)$ that is scaled by the symbol value $s[\cdot]$. A typical output of the transmitter in Figure 1.1 is illustrated in Figure 1.3 using the rectangular pulse shape. Thus the first pulse begins at some time τ and it is scaled by $s[0]$, producing $s[0]p(t - \tau)$. The second pulse begins at time $\tau + T$ and is scaled by $s[1]$, resulting in $s[1]p(t - \tau - T)$. The third pulse gives $s[2]p(t - \tau - 2T)$, and so on. The complete output of the transmitter is the sum of all these scaled pulses:

$$y(t) = \sum_i s[i]p(t - \tau - iT).$$

Since each pulse ends before the next one begins, successive symbols should not interfere with each other at the receiver. The general method of sending information by scaling a pulse shape with the amplitude of the symbols is called *Pulse Amplitude Modulation* (PAM). When there are four symbols as in (1.1), it is called 4-PAM.

For now, assume that the path between the transmitter and receiver, which is often called the *channel*, is "ideal." This implies that the signal at the receiver is the same as the transmitted signal, though it will inevitably be delayed (slightly) due to the finite speed of the wave, and attenuated by the distance. When the ideal channel has a gain g and a delay δ, the received version of the transmitted signal in Figure 1.3 is shown in Figure 1.4.

There are many ways that a real signal may change as it passes from the transmitter to receiver through a real (nonideal) channel. It may be reflected from mountains or buildings. It may be diffracted as it passes through the atmosphere. The waveform may smear in time so that successive pulses overlap. Other signals may interfere additively

FIGURE 1.2 An isolated rectangular pulse.

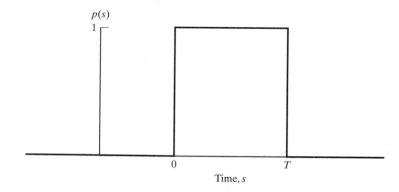

FIGURE 1.3 The transmitted signal consists of a sequence of pulses, one corresponding to each symbol. Each pulse has the same shape as in Figure 1.2, though offset in time (by τ) and scaled in magnitude (by the symbols $s[k]$).

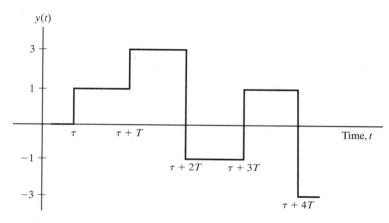

FIGURE 1.4 In the ideal case, the received signal is the same as the transmitted signal of Figure 1.3, though attenuated in magnitude (by g) and delayed in time (by δ).

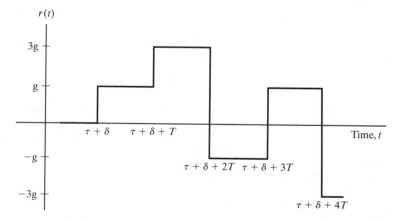

(for instance, a radio station broadcasting at the same frequency in a different city). Noises may enter and change the shape of the waveform.

There are two compelling reasons to consider the telecommunication system in the simplified (idealized) case before worrying about all the things that might go wrong. First, at the heart of any working receiver is a structure that is able to function in the ideal case. The classic approach to receiver design (and also the approach of *Telecommunication*

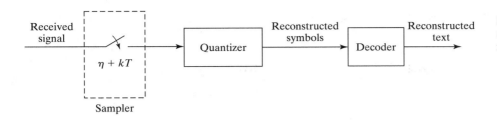

FIGURE 1.5 An idealized baseband receiver.

Breakdown) is to build for the ideal case and later to refine so that the receiver will still work when bad things happen. Second, many of the basic ideas are clearer in the ideal case.

The job of the receiver is to take the received signal (such as that in Figure 1.4) and to recover the original text message. This can be accomplished by an idealized receiver such as that shown in Figure 1.5. The first task this receiver must accomplish is to sample the signal to turn it into computer-friendly digital form. But when should the samples be taken? Comparing Figures 1.3 and 1.4, it is clear that if the received signal were sampled somewhere near the middle of each rectangular pulse segment, then the quantizer could reproduce the sequence of source symbols. This quantizer must either

1. know g so the sampled signal can be scaled by $1/g$ to recover the symbol values, or
2. separate $\pm g$ from $\pm 3g$ and output symbol values ± 1 and ± 3.

Once the symbols have been reconstructed, then the original message can be decoded by reversing the association of letters to symbols used at the transmitter (for example, by reading (1.1) backwards). On the other hand, if the samples were taken at the moment of transition from one symbol to another, then the values might become confused.

To investigate the timing question more fully, let T be the sample interval and τ be the time the first pulse begins. Let δ be the time it takes for the signal to move from the transmitter to the receiver. Thus the $(k+1)$st pulse, which begins at time $\tau + kT$, arrives at the receiver at time $\tau + kT + \delta$. The midpoint of the pulse, which is the best time to sample, occurs at $\tau + kT + \delta + T/2$. As indicated in Figure 1.5, the receiver begins sampling at time η, and then samples regularly at $\eta + kT$ for all integers k. If η were chosen so that

Somehow, the receiver must figure out when to sample.

$$\eta = \tau + \delta + T/2, \qquad (1.3)$$

then all would be well. But there are two problems: the receiver does not know when the transmission began, nor does it know how long it takes for the signal to reach the receiver. Thus both τ and δ are unknown!

Basically, some extra "synchronization" procedure is needed in order to satisfy (1.3). Fortunately, in the ideal case, it is not really necessary to sample exactly at the midpoint; it is necessary only to avoid the edges. Even if the samples are not taken at the center of each rectangular pulse, the transmitted symbol sequence can still be recovered. But if the pulse shape were not a simple rectangle, then the selection of η becomes more critical.

How does the pulse shape interact with timing synchronization?

Just as no two clocks ever tell exactly the same time, no two independent oscillators are ever exactly synchronized. Since the symbol period at the transmitter, call it T_{trans}, is created by a separate oscillator from that creating the symbol period at the receiver, call it T_{rec}, they will inevitably differ. Thus another aspect of timing synchronization that must ultimately be considered is how to automatically adjust T_{rec} so that it aligns with T_{trans}.

What about clock jitter?

Similarly, no clock ticks out each second exactly evenly. Inevitably, there is some jitter, or wobble in the value of T_{trans} and/or T_{rec}. Again, it may be necessary to adjust η to retain sampling near the center of the pulse shape as the clock times wiggle about. The timing adjustment mechanisms are not explicitly indicated in the sampler box in Figure 1.5. For the present idealized transmission system, the receiver sampler period and the symbol period of the transmitter are assumed to be identical (both are called T in Figures 1.1 and 1.5) and the clocks are assumed to be free of jitter.

How to find the start of a frame?

Even under the idealized assumptions above, there is another kind of synchronization that is needed. Imagine joining a broadcast in progress, or one in which the first K symbols have been lost during acquisition. Even if the symbol sequence is perfectly recovered after time K, the receiver would not know which recovered symbol corresponds to the start of each frame. For example, using the letters-to-symbol code of (1.1), each letter of the alphabet is translated into a sequence of four symbols. If the start of the frame is off by even a single symbol, the translation from symbols back into letters will be scrambled. Does this sequence represent a or X?

$$
\begin{array}{c}
\overbrace{-1 \ \ -1, \ \ 1, \ -3, \ -1}^{a} \\
\underbrace{-1, \ -1, \ \ 1, \ -3,}_{X} \ -1
\end{array}
$$

Thus proper decoding requires locating where the frame starts, a step called frame synchronization. Frame synchronization is implicit in Figure 1.5 in the choice of η, which sets the time $t \ (= \eta$ with $k = 0)$ of the first symbol of the first (character) frame of the message of interest.

In the ideal situation, there must be no other signals occupying the same frequency range as the transmission. What bandwidth (what range of frequencies) does the transmitter (1.1) require? Consider transmitting a single T-second wide rectangular pulse. Fourier transform theory shows that any such time-limited pulse cannot be truly band limited, that is, cannot have its frequency content restricted to a finite range. Indeed, the Fourier transform of a rectangular pulse in time is a sinc function in frequency (see Equation (A.20) in Appendix A). The magnitude of this sinc is overbounded by a function that decays as the inverse of frequency (peek ahead to Figure 2.10). Thus, to accommodate this single pulse transmission, all other transmitters must have negligible energy below some factor of $B = 1/T$. For the sake of argument, suppose that a factor of 5 is safe, that is, all other transmitters must have no significant energy within $5B$ Hz. But this is only for a single pulse. What happens when a sequence of T-spaced, T-wide rectangular pulses of various amplitudes is transmitted? Fortunately, as will be established in Section 11.1, the bandwidth requirements remain about the same, at least for most messages.

What is the relation between the pulse shape and the bandwidth?

One fundamental limitation to data transmission is the trade-off between the data rate and the bandwidth. One obvious way to increase the rate at which data are sent is to use shorter pulses, which pack more symbols into a shorter time. This essentially reduces T. The cost is that this would require excluding other transmitters from an even wider range of frequencies since reducing T increases B.

What is the relation between the data rate and the bandwidth?

If the safety factor of $5B$ is excessive, other pulse shapes that would decay faster as a function of frequency could be used. For example, rounding the sharp corners of

Dedicated to Samantha and Franklin

a rectangular pulse reduces its high frequency content. Similarly, if other transmitters operated at high frequencies outside $5B$ Hz, it would be sensible to add a low pass filter at the front end of the receiver. Rejecting frequencies outside the protected $5B$ baseband turf also removes a bit of the higher frequency content of the rectangular pulse. The effect of this in the time domain is that the received version of the rectangle would be wiggly near the edges. In both cases, the timing of the samples becomes more critical as the received pulse deviates further from rectangular.

One shortcoming of the telecommunication system embodied in the transmitter of Figure 1.1 and the receiver of Figure 1.5 is that only one such transmitter at a time can operate in any particular geographical region, since it hogs all the frequencies in the baseband, that is, all frequencies below $5B$ Hz. Fortunately, there is a way to have multiple transmitters operating in the same region simultaneously. The trick is to translate the frequency content so that instead of all transmitters trying to operate in the 0 and $5B$ Hz band, one might use the $5B$ to $10B$ band, another the $10B$ to $15B$ band, etc. Conceivably, this could be accomplished by selecting a different pulse shape (other than the rectangle) that has no low frequency content, but the most common approach is to "modulate" (change frequency) by multiplying the pulse shaped signal by a high frequency sinusoid. Such a "radio frequency" (RF) transmitter is shown in Figure 1.6, though it should be understood that the actual frequencies used may place it in the television band or in the range of frequencies reserved for cell phones, depending on the application.

At the receiver, the signal can be returned to its original frequency (demodulated) by multiplying by another high frequency sinusoid (and then low pass filtering). These frequency translations are described in more detail in Section 2.3, where it is shown that the modulating sinusoid and the demodulating sinusoid must have the same frequencies and the same phases in order to return the signal to its original form. Just as it is impossible to align any two clocks exactly, it is also impossible to generate two independent sinusoids of exactly the same frequency and phase. Hence there will ultimately need to be some kind of "carrier synchronization," a way of aligning these oscillators.

> How can the frequencies and phases of these two sinusoids be aligned?

Adding frequency translation to the transmitter and receiver of Figures 1.1 and 1.5 produces the transmitter in Figure 1.6 and the associated receiver in Figure 1.7. The new block in the transmitter is an analog component that effectively adds the same value (in Hz) to the frequencies of all of the components of the baseband pulse train. As noted, this can be achieved with multiplication by a "carrier" sinusoid with a frequency equal to the desired translation. The new block in the receiver of Figure 1.7 is an analog component that processes the received analog signal prior to sampling in order to subtract the same value (in Hz) from all components of the received signal. The output of this block should be identical to the input to the sampler in Figure 1.5.

> There is no free lunch. How much does the fix cost?

This process of translating the spectrum of the transmitted signal to higher frequencies allows many transmitters to operate simultaneously in the same geographic area. But there is a price. Since the signals are not completely bandlimited to within their assigned $5B$-wide slot, there is some inevitable overlap. Thus the residual energy of one transmitter (the energy outside its designated band) acts as an interference to other transmissions. Solving the problem of multiple transmissions has thus violated one of the assumptions for an ideal transmission. A common theme throughout *Telecommunication Breakdown* is that a solution to one problem often causes another!

In fact, there are many other ways that the transmission channel can deviate from the ideal, and these will be discussed in detail later on (for instance, in Section 4.1 and throughout Chapter 9). Typically, the cluttered electromagnetic spectrum results in a variety of distortions and interferences:

- in-band (within the frequency band allocated to the user of interest)
- out-of-band (frequency components outside the allocated band such as the signals of other transmitters)
- narrowband (spurious sinusoidal-like components)
- broadband (with components at frequencies across the allocated band and beyond)
- fading (when the strength of the received signal fluctuates)
- multipath (when the environment contains many reflective and absorptive objects at different distances, the transmission delay will be different across different paths, smearing the received signal and attenuating some frequencies more than others)

These channel imperfections are all incorporated in the channel model shown in Figure 1.8, which sits in the communication system between Figures 1.6 and 1.7.

Many of these imperfections in the channel can be mitigated by clever use of filtering at the receiver. Narrowband inference can be removed with a notch filter that rejects frequency components in the narrow range of the interferer without removing too much of the broadband signal. Out-of-band interference and broadband noise can be reduced using a bandpass filter that suppresses the signal in the out-of-band frequency range and passes the in-band frequency components without distortion. With regard to Figure 1.7, it

Analog or digital processing?

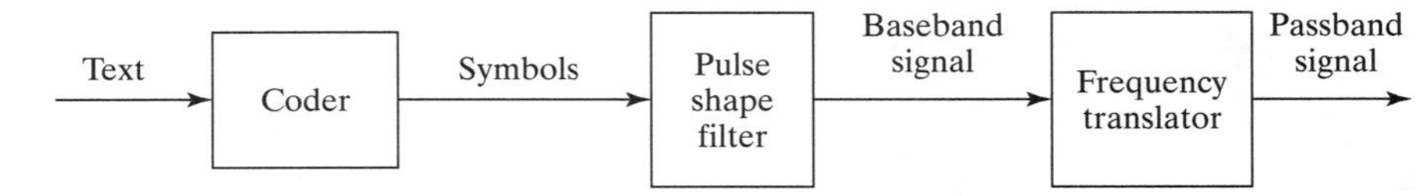

is reasonable to wonder if it is better to perform such filtering before or after the sampler (i.e., by an analog or a digital filter). In modern receivers, the trend is to minimize the amount of analog processing since digital methods are (often) cheaper and (usually)

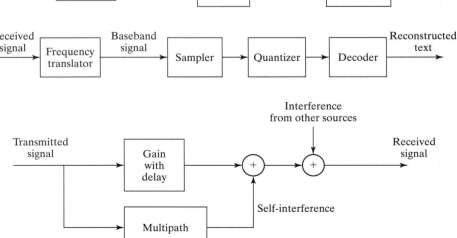

FIGURE 1.6 "Radio frequency" transmitter.

FIGURE 1.7 "Radio frequency" receiver.

FIGURE 1.8 A channel model admitting various kinds of interferences.

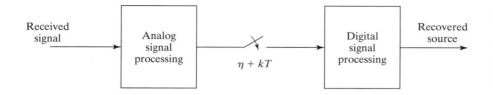

FIGURE 1.9 A generic modern receiver using both analog signal processing (ASP) and digital signal processing (DSP).

more flexible since they can be implemented as reconfigurable software rather than fixed hardware.

Conducting more of the processing digitally requires moving the sampler closer to the antenna. The sampling theorem (discussed in Section 6.1) says that no information is lost as long as the sampling occurs at a rate faster than twice the highest frequency of the signal. Thus, if the signal has been modulated to (say) the band from $20B$ to $25B$ Hz, then the sampler must be able to operate at least as fast as $50B$ samples per second in order to be able to exactly reconstruct the value of the signal at any arbitrary time instant. Assuming this is feasible, the received analog signal can be sampled using a free-running sampler. Interpolation can be used to figure out values of the signal at any desired intermediate instant, such as at time $\eta + kT$ (recall (1.3)) for a particular η that is not an integer multiple of T. Thus, the timing synchronization can be incorporated in the post-sampler digital signal processing box, which is shown generically in Figure 1.9. Observe that Figure 1.7 is one particular version of 1.9.

Use DSP when possible.

How exactly does interpolation work?

However, sometimes it is more cost effective to perform certain tasks in analog circuitry. For example, if the transmitter modulates to a very high frequency, then it may cost too much to sample fast enough. Currently, it is common practice to perform some frequency translation and some out-of-band signal reduction in the analog portion of the receiver. Sometimes the analog portion may translate the received signal all the way back to baseband. Other times, the analog portion translates to some intermediate frequency, and then the digital portion finishes the translation. The advantage of this (seemingly redundant) approach is that the analog part can be made crudely and, hence, cheaply. The digital processing finishes the job, and simultaneously compensates for inaccuracies and flaws in the (inexpensive) analog circuits. Thus, the digital signal processing portion of the receiver may need to correct for signal impairments arising in the analog portion of the receiver as well as from impairments caused by the channel.

Use DSP to compensate for cheap ASP.

The digital signal processing portion of the receiver can perform the following tasks:

- downconvert the sampled signal to baseband
- track any changes in the phase or frequency of the modulating sinusoid
- adjust the symbol timing by interpolation
- compensate for channel imperfections by filtering
- convert modestly inaccurate recovered samples into symbols
- perform frame synchronization via correlation
- decode groups of symbols into message characters

A central task in *Telecommunication Breakdown* is to elaborate on the system structure in Figures 1.6–1.8 to create a working software-defined radio that can perform these tasks. This concludes the illustrative design at the outer, most superficial layer of the onion.

1.3 The Complete Onion

This section provides a whirlwind tour of the complete layered structure of *Telecommunication Breakdown*. Each layer presents the same digital transmission system with the outer layers peeled away to reveal greater depth and detail.

- *The Naive Digital Communications Layer:* As we have just seen, the first layer of the onion introduced the digital transmission of data, and discussed how bits of information may be coded into waveforms, sent across space to the receiver, and then decoded back into bits. Since there is no universal clock, issues of timing become important, and some of the most complex issues in digital receiver design involve the synchronization of the received signal. The system can be viewed as consisting of three parts: the transmitter,

$$
\text{digital message} \rightarrow \text{coding} \rightarrow \text{pulse shaping} \rightarrow \text{frequency translation} \rightarrow
$$

the transmission channel, and the receiver

$$
\text{frequency translation} \rightarrow \text{sampling} \rightarrow \text{decision device} \rightarrow \text{decoding} \rightarrow \text{reconstructed message}
$$

- *The Component Architecture Layer:* The next two chapters provide more depth and detail by outlining a complete telecommunication system. When the transmitted signal is passed through the air using electromagnetic waves, it must take the form of a continuous (analog) waveform. A good way to understand such analog signals is via the Fourier transform, and this is reviewed briefly in Chapter 2. The five basic elements of the receiver will be familiar to many readers, and they are presented in Chapter 3 in a form that will be directly useful when creating MATLAB implementations of the various parts of the communication system. By the end of the second layer, the basic system architecture is fixed; the ordering of the blocks in the system diagram has stabilized.

- *The Idealized System Layer:* The third layer encompasses Chapters 4 through 9. This layer gives a closer look at the idealized receiver—how things work when everything is just right: when the timing is known, when the clocks run at exactly the right speed, when there are no reflections, diffractions, or diffusions of the electromagnetic waves. This layer also integrates ideas from previous systems courses, and introduces a few MATLAB tools that are needed to implement the digital radio. The order in which topics are discussed is precisely the order in which they appear in the receiver:

$$
\underset{Chapter\ 4}{\text{channel}} \rightarrow \underset{Chapter\ 5}{\text{frequency translation}} \rightarrow \underset{Chapter\ 6}{\text{sampling}} \rightarrow \text{frequency translation} \rightarrow
$$

$$
\underbrace{\text{receive filtering} \rightarrow \text{equalization}}_{Chapter\ 7} \rightarrow \underbrace{\text{decision device} \rightarrow \text{decoding}}_{Chapter\ 8} \rightarrow \text{reconstructed message}
$$

Chapter 9 provides a complete (though idealized) software-defined digital radio system.

- *The Adaptive Component Layer:* The fourth layer describes all the practical fixes that are required in order to create a workable radio. One by one the various problems are studied and solutions are proposed, implemented, and tested. These include fixes for additive noise, for timing offset problems, for clock frequency mismatches and jitter, and for multipath reflections. Again, the order in which topics are discussed is the order in which they appear in the receiver:

- *The Integration Layer:* The fifth layer is the final project of Chapter 15 which integrates all the fixes of the fourth layer into the receiver structure of the third layer to create a fully functional digital receiver. The well-fabricated receiver is robust to distortions such as those caused by noise, multipath interference, timing inaccuracies, and clock mismatches.

Please observe that the word "layer" refers to the onion metaphor for the method of presentation (in which each layer of the communication system repeats the essential outline of the last, exposing greater subtlety and complexity), and not to the "layers" of a communication system as might be found in Bertsekas and Gallager's *Data Networks*. In this latter terminology, the whole of *Telecommunication Breakdown* lies within the so-called *physical layer*. Thus we are part of an even larger onion, which is not currently on our plate.

PART

The Component Architecture Layer

The next two chapters provide more depth and detail by outlining a complete telecommunication system. When the transmitted signal is passed through the air using electromagnetic waves, it must take the form of a continuous (analog) waveform. A good way to understand such analog signals is via the Fourier transform, and this is reviewed briefly in Chapter 2. The five basic elements of the receiver will be familiar to many readers, and they are presented in Chapter 3 in a form that will be directly useful when creating MATLAB implementations of the various parts of the communications system. By the end of the second layer, the basic system architecture is fixed; the ordering of the blocks in the system diagram has stabilized.

2 A *Telecommunication System*

The reason digital radio is so reliable is because it employs a smart receiver. Inside each digital radio receiver, there is a tiny computer: a computer capable of sorting through the myriad of reflected and atmospherically distorted transmissions and reconstructing a solid, usable signal for the set to process.
—from http://radioworks.cbc.ca/radio/digital-radio/drri.html (2/2/03)

Telecommunications technologies using electromagnetic transmission surround us: television images flicker, radios chatter, cell phones (and telephones) ring, allowing us to see and hear each other anywhere on the planet. E-mail and the Internet link us via our computers, and a large variety of common devices such as CDs, DVDs, and hard disks augment the traditional pencil and paper storage and transmittal of information. People have always wished to communicate over long distances: to speak with someone in another country, to watch a distant sporting event, to listen to music performed in another place or another time, to send and receive data remotely using a personal computer. In order to implement these desires, a signal (a sound wave, a signal from a TV camera, or a sequence of computer bits) needs to be encoded, stored, transmitted, received, and decoded. Why? Consider the problem of voice or music transmission. Sending sound directly is futile because sound waves dissipate very quickly in air. But if the sound is first transformed into electromagnetic waves, then they can be beamed over great distances very efficiently. Similarly, the TV signal and computer data can be transformed into electromagnetic waves.

2.1 Electromagnetic Transmission of Analog Waveforms

There are some experimental (physical) facts that cause transmission systems to be constructed as they are. First, for efficient wireless broadcasting of electromagnetic energy, an antenna needs to be longer than about 1/10 of a wavelength of the frequency being transmitted. The antenna at the receiver should also be proportionally sized.

The wavelength λ and the frequency f of a sinusoid are inversely proportional. For an electrical signal travelling at the speed of light c ($= 3 \times 10^8$ meters/second), the relationship between wavelength and frequency is

$$\lambda = \frac{c}{f}.$$

For instance, if the frequency of an electromagnetic wave is $f = 10$ KHz, then the length of each wave is

$$\lambda = \frac{3 \times 10^8 \text{ m/s}}{10^4/\text{s}} = 3 \times 10^4 \text{m}.$$

Efficient transmission requires an antenna longer than 0.1λ, which is 3 km! Sinusoids in the speech band would require even larger antennas. Fortunately, there is an alternative to building mammoth antennas. The frequencies in the signal can be translated (shifted, upconverted, or modulated) to a much higher frequency called the *carrier frequency*, where the antenna requirements are easier to meet. For instance,

- AM Radio: $f \approx 600 - 1500$ KHz $\Rightarrow \lambda \approx 500$ m $- 200$ m $\Rightarrow 0.1 \lambda > 20$ m
- VHF (TV): $f \approx 30 - 300$ MHz $\Rightarrow \lambda \approx 10$ m $- 1$ m $\Rightarrow 0.1 \lambda > 0.1$ m
- UHF (TV): $f \approx 0.3 - 3$ GHz $\Rightarrow \lambda \approx 1$ m $- 0.1$ m $\Rightarrow 0.1 \lambda > 0.01$ m
- Cell phones (U.S.): $f \approx 824 - 894$ MHz $\Rightarrow \lambda \approx 0.36 - 0.33$ m $\Rightarrow 0.1 \lambda > 0.03$ m
- PCS: $f \approx 1.8 - 1.9$ GHz $\Rightarrow \lambda \approx 0.167 - 0.158$ m $\Rightarrow 0.1 \lambda > 0.015$ m
- GSM (Europe): $f \approx 890 - 960$ MHz $\Rightarrow \lambda \approx 0.337 - 0.313$ m $\Rightarrow 0.1 \lambda > 0.03$ m
- LEO satellites: $f \approx 1.6$ GHz $\Rightarrow \lambda \approx 0.188$ m $\Rightarrow 0.1 \lambda > 0.0188$ m

Recall that KHz $= 10^3$ Hz; MHz $= 10^6$ Hz; GHz $= 10^9$ Hz.

A second experimental fact is that electromagnetic waves in the atmosphere exhibit different behaviors, depending on the frequency of the waves:

- Below 2 MHz, electromagnetic waves follow the contour of the earth. This is why shortwave (and other) radios can sometimes be heard hundreds or thousands of miles from their source.

- Between 2 and 30 MHz, sky-wave propagation occurs with multiple bounces from refractive atmospheric layers.

- Above 30 MHz, line-of-sight propagation occurs with straight line travel between two terrestrial towers or through the atmosphere to satellites.

- Above 30 MHz, atmospheric scattering also occurs, which can be exploited for long distance terrestrial communication.

Humanmade media in wired systems also exhibit frequency dependent behavior. In the phone system, due to its original goal of carrying voice signals, severe attenuation occurs above 4 KHz.

The notion of frequency is central to the process of long distance communications. Because of its role as a carrier (the AM/UHF/VHF/PCS bands mentioned above) and its role in specifying the bandwidth (the range of frequencies occupied by a given signal), it is important to have tools with which to easily measure the frequency content in a signal. The tool of choice for this job is the Fourier transform (and its discrete counterparts, the DFT and the FFT)[1]. Fourier transforms are useful in assessing energy or power at

---

[1] These are the discrete Fourier transform, which is a computer implementation of the Fourier transform, and the fast Fourier transform, which is a slick, computationally efficient method of calculating the DFT.

particular frequencies. The Fourier transform of a signal $w(t)$ is defined as

$$W(f) = \int_{-\infty}^{\infty} w(t)e^{-j2\pi ft}dt = \mathcal{F}\{w(t)\}, \tag{2.1}$$

where $j = \sqrt{-1}$ and f is given in Hz (i.e., cycles/sec or 1/sec).

Speaking mathematically, $W(f)$ is a function of the frequency f. Thus for each f, $W(f)$ is a complex number and so can be plotted in several ways. For instance, it is possible to plot the real part of $W(f)$ as a function of f and to plot the imaginary part of $W(f)$ as a function of f. Alternatively, it is possible to plot the real part of $W(f)$ versus the imaginary part of $W(f)$. The most common plots of the Fourier transform of a signal are done in two parts: the first graph shows the magnitude $|W(f)|$ versus f (this is called the magnitude spectrum) and second graph shows the phase angle of $W(f)$ versus f (this is called the phase spectrum). Often, just the magnitude is plotted, though this inevitably leaves out information. The relationship between the Fourier transform and the DFT is discussed in considerable detail in Appendix D, and a table of useful properties appears in Appendix A.

2.2 Bandwidth

If, at any particular frequency f_0, the magnitude spectrum is strictly positive ($|W(f_0)| > 0$), then the frequency f_0 is said to be *present* in $w(t)$. The set of all frequencies that are present in the signal is the frequency *content*, and if the frequency content contains only frequencies below some given $f^\dagger$, then the signal is said to be *bandlimited* to $f^\dagger$. Some bandlimited signals are

- Telephone quality speech—maximum frequency $\sim$ 4 KHz and
- Audible music—maximum frequency $\sim$ 20 KHz.

But real world signals are never completely bandlimited, and there is almost always some energy at every frequency. Several alternative definitions of bandwidth are in common use which try to capture the idea that "most of" the energy is contained in a specified frequency region. Usually, these are applied across positive frequencies, with the presumption that the underlying signals are real valued (and hence have symmetric spectra). Here are some of the alternative definitions:

1. *Absolute bandwidth* is the smallest interval $f_2 - f_1$ for which the spectrum is zero outside $f_1 < f < f_2$ along the positive frequency axis.
2. *3-dB (or half-power) bandwidth* is $f_2 - f_1$, where, for frequencies outside $f_1 < f < f_2$, $|H(f)|$ is never greater than $1/\sqrt{2}$ times its maximum value.
3. *Null-to-null (or zero-crossing) bandwidth* is $f_2 - f_1$, where f_2 is first null in $|H(f)|$ above f_0 and, for bandpass systems, f_1 is the first null in the envelope below f_0 where f_0 is frequency of maximum $|H(f)|$. For baseband systems, f_1 is usually zero.
4. *Power bandwidth* is $f_2 - f_1$, where $f_1 < f < f_2$ defines the frequency band in which 99% of the total power resides. Occupied bandwidth is such that 0.5% of power is above f_2 and 0.5% below f_1.

These definitions are illustrated in Figure 2.1.

FIGURE 2.1 Various
ways to define
bandwidth.

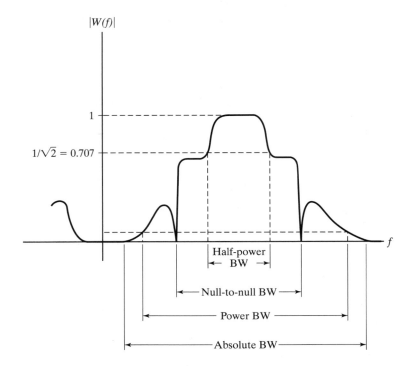

Bandwidth refers to the frequency content of a signal. Since the frequency response of a linear filter is the transform of the impulse response, it can also be used to talk about the bandwidth of a linear system or filter.

2.3 Upconversion at the Transmitter

Suppose that the signal $w(t)$ contains important information that must be transmitted. There are many kinds of operations that can be applied to $w(t)$. *Linear* operations are those for which superposition applies, but linear operations cannot augment the frequency content of a signal—no sine wave can appear at the output of a linear operation if it was not already present in the input.

Thus, the process of modulation (or upconversion), which requires a change of frequencies, must be a nonlinear operation. One useful nonlinearity is multiplication; consider the product of the message waveform $w(t)$ with a cosine wave

$$s(t) = w(t)\,\cos(2\pi f_0 t), \tag{2.2}$$

where f_0 is called the *carrier* frequency. The Fourier transform can now be used to show that this multiplication shifts all frequencies present in the message by exactly f_0 Hz.

Using one of Euler's identities (A.2),

$$\cos(2\pi f_0 t) = \frac{1}{2}\left(e^{j2\pi f_0 t} + e^{-j2\pi f_0 t}\right), \tag{2.3}$$

one can calculate the spectrum (or frequency content) of the signal $s(t)$ from the definition of the Fourier transform given in (2.1). In complete detail, this is

$$S(f) = \mathcal{F}\{s(t)\} = \mathcal{F}\{w(t)\,\cos(2\pi f_0 t)\}$$

$$= \mathcal{F}\left\{ w(t)\left[\frac{1}{2}\left(e^{j2\pi f_0 t} + e^{-j2\pi f_0 t}\right)\right]\right\}$$

$$= \int_{-\infty}^{\infty} w(t)\left[\frac{1}{2}\left(e^{j2\pi f_0 t} + e^{-j2\pi f_0 t}\right)\right]e^{-j2\pi f t}\,dt$$

$$= \frac{1}{2}\int_{-\infty}^{\infty} w(t)\left(e^{-j2\pi(f-f_0)t} + e^{-j2\pi(f+f_0)t}\right)dt$$

$$= \frac{1}{2}\int_{-\infty}^{\infty} w(t)\,e^{-j2\pi(f-f_0)t}\,dt + \frac{1}{2}\int_{-\infty}^{\infty} w(t)\,e^{-j2\pi(f+f_0)t}\,dt$$

$$= \frac{1}{2}W(f-f_0) + \frac{1}{2}W(f+f_0). \tag{2.4}$$

Thus, the spectrum of $s(t)$ consists of two copies of the spectrum of $w(t)$, each shifted in frequency by f_0 (one up and one down) and each half as large. This is sometimes called the *frequency shifting* property of the Fourier transform, and sometimes called the *modulation* property. Figure 2.2 shows how the spectra relate. If $w(t)$ has the magnitude spectrum shown in part (a) (this is shown bandlimited to $f^\dagger$ and centered at zero Hz or *baseband*, though it could be elsewhere), then the magnitude spectrum of $s(t)$ appears as in part (b). This kind of modulation (or *upconversion*, or frequency shift), is ideal for translating speech, music, or other low frequency signals into much higher frequencies (for instance, f_0 might be in the AM or UHF bands) so that it can be transmitted efficiently. It can also be used to convert a high frequency signal back down to baseband when needed, as will be discussed in Section 2.6 and in detail in Chapter 5.

Any sine wave is characterized by three parameters: the amplitude, frequency, and phase. Any of these characteristics can be used as the basis of a modulation scheme: modulating the frequency is familiar from the FM radio, and phase modulation is common in computer modems. The primary example in this book is amplitude modulation as in (2.2), where the message $w(t)$ is multiplied by a sinusoid of fixed frequency and phase. Whatever the modulation scheme used, the idea is the same. A high-frequency sinusoid is used to translate the low frequency message into a form suitable for transmission.

Problems

2.1. Referring to Figure 2.2, find which frequencies are present in $W(f)$ and not in $S(f)$? Which frequencies are present in $S(f)$ and not in $W(f)$?

2.2. Using (2.4), draw analogous pictures for the phase spectrum of $s(t)$ as it relates to the phase spectrum of $w(t)$.

2.3. Suppose that $s(t)$ is modulated again, this time via multiplication with a cosine of frequency f_1. What is the resulting magnitude spectrum? Hint: Let $r(t) = s(t)\cos(2\pi f_1 t)$, and apply (2.4) to find $R(f)$.

FIGURE 2.2 Action
of a modulator: If
the message signal
$w(t)$ has the magni-
tude spectrum shown
in part (a), then the
modulated signal $s(t)$
has the magnitude
spectrum shown in
part (b).

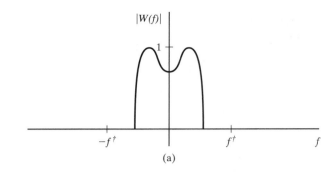

(a)

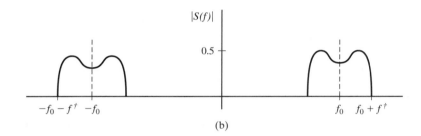

(b)

2.4 Frequency Division Multiplexing

When a signal is modulated, the width (in Hertz) of the replicas is the same as the width
(in Hertz) of the original signal. This is a direct consequence of equation (2.4). For
instance, if the message is bandlimited to $\pm f^*$, and the carrier is f_c, then the modulated
signal has energy in the range from $-f^* - f_c$ to $+f^* - f_c$ and from $-f^* + f_c$ to
$+f^* + f_c$. If $f^* \ll f_c$, then several messages can be transmitted simultaneously by using
different carrier frequencies.

This situation is depicted in Figure 2.3, where three different messages are represented
by the triangular, rectangular, and half-oval spectra, each bandlimited to $\pm f^*$. Each of
these is modulated by a different carrier (f_1, f_2, and f_3), which are chosen so that
they are further apart than the width of the messages. In general, as long as the carrier
frequencies are separated by more than $2f^*$, there will be no overlap in the spectrum of
the combined signal. This process of combining many different signals together is called
multiplexing, and because the frequencies are divided up among the users, the approach
of Figure 2.3 is called frequency division multiplexing (FDM).

Whenever FDM is used, the receiver must separate the signal of interest from all the
other signals present. This can be accomplished with a bandpass filter as in Figure 2.4,
which shows a filter designed to isolate the middle user from the others.

Problem

2.4. Suppose that two carrier frequencies are separated by 1 KHz. Draw the magnitude
spectra if (a) the bandwidth of each message is 200 Hz and (b) the bandwidth
of each message is 2 KHz. Comment on the ability of the bandpass filter at the
receiver to separate the two signals.

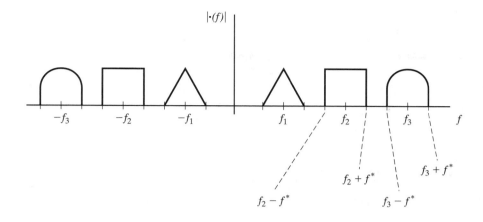

FIGURE 2.3 Three different upconverted signals are assigned different frequency bands. This is called frequency division multiplexing.

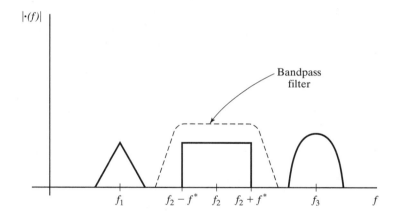

FIGURE 2.4 Separation of a single FDM transmission using a bandpass filter.

Another kind of multiplexing is called time division multiplexing (TDM), in which two (or more) messages use the same carrier frequency but at alternating time instants. More complex multiplexing schemes (such as code division multiplexing) overlap the messages in both time and frequency in such a way that they can be demultiplexed efficiently by appropriate filtering.

2.5 Filters that Remove Frequencies

Each time the signal is modulated, an extra copy (or replica) of the spectrum appears. When multiple modulations are needed (for instance, at the transmitter to convert up to the carrier frequency, and at the receiver to convert back down to the original frequency of the message), copies of the spectrum may proliferate. There must be a way to remove extra copies in order to isolate the original message. This is one of the things that linear filters do very well.

There are several ways of describing the action of a linear filter. In the time domain (the most common method of implementation), the filter is characterized by its impulse response (which is defined to be the output of the filter when the input is an impulse function). By linearity, the output of the filter to any arbitrary input is then the super-position of weighted copies of the impulse response, a procedure known as convolution. Since convolution may be difficult to understand directly in the time domain, the action of a linear filter is often described in the frequency domain.

Perhaps the most important property of the Fourier transform is the duality between convolution and multiplication, which says that

- convolution in time ↔ multiplication in frequency, and
- multiplication in time ↔ convolution in frequency.

This is discussed in detail in Section 4.5. Thus, the convolution of a linear filter can readily be viewed in the frequency (Fourier) domain as a point-by-point multiplication. For instance, an ideal lowpass filter (LPF) passes all frequencies below f_l (which is called the *cutoff* frequency). This is commonly plotted in a curve called the *frequency response* of the filter, which describes the action of the filter.[2] If this filter is applied to a signal $w(t)$, then all energy above f_l is removed from $w(t)$. Figure 2.5 shows this pictorially. If $w(t)$ has the magnitude spectrum shown in part (a), and the frequency response of the lowpass filter with cutoff frequency f_l is as shown in part (b), then the magnitude spectrum of the output appears in part (c).

Problems

2.5. An ideal highpass filter passes all frequencies above some given f_h and removes all frequencies below. Show the result of applying a highpass filter to the signal in Figure 2.5 with $f_h = f_l$.

2.6. An ideal bandpass filter passes all frequencies between an upper limit $\overline{f}$ and a lower limit $\underline{f}$. Show the result of applying a bandpass filter to the signal in Figure 2.5 with $\overline{f} = 2f_l/3$ and $\underline{f} = f_l/3$.

The problem of how to design and implement such filters is considered in detail in Chapter 7.

2.6 Analog Downconversion

Because transmitters typically modulate the message signal with a high frequency carrier, the receiver must somehow remove the carrier from the message that it carries. One way is to multiply the received signal by a cosine wave of the same frequency (and the same phase) as was used at the transmitter. This creates a (scaled) copy of the original signal centered at zero frequency, plus some other high frequency replicas. A lowpass filter can then remove everything but the scaled copy of the original message. This is how the box labelled "frequency translator" in Figure 1.5 is typically implemented.

To see this procedure in detail, suppose that $s(t) = w(t) \cos(2\pi f_0 t)$ arrives at the receiver, which multiplies $s(t)$ by another cosine wave of exactly same frequency and phase to get the demodulated signal

$$d(t) = s(t) \cos(2\pi f_0 t) = w(t) \cos^2(2\pi f_0 t).$$

---

[2] Formally, the frequency response can be calculated as the Fourier transform of the impulse response of the filter.

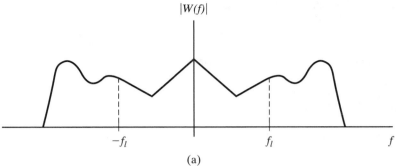

(a)

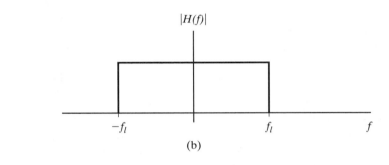

(b)

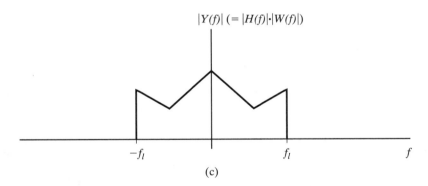

(c)

Using the trigonometric identity (A.4), namely,

$$\cos^2(x) = \frac{1}{2} + \frac{1}{2}\cos(2x),$$

we find that this can be rewritten as

$$d(t) = w(t)\left[\frac{1}{2} + \frac{1}{2}\cos(4\pi f_0 t)\right]$$

$$= \frac{1}{2}w(t) + \frac{1}{2}w(t)\cos(2\pi(2f_0)t).$$

The spectrum of the demodulated signal can be calculated

$$\mathcal{F}\{d(t)\} = \mathcal{F}\left\{\frac{1}{2}w(t) + \frac{1}{2}w(t)\cos(2\pi(2f_0)t)\right\}$$

$$= \frac{1}{2}\mathcal{F}\{w(t)\} + \frac{1}{2}\mathcal{F}\{w(t)\cos(2\pi(2f_0)t)\}$$

by linearity. Now the frequency shifting property (2.4) can be applied to show that

$$\mathcal{F}\{d(t)\} = \frac{1}{2}W(f) + \frac{1}{4}W(f - 2f_0) + \frac{1}{4}W(f + 2f_0). \tag{2.5}$$

Thus, the spectrum of this downconverted received signal has the original baseband component (scaled to 50%) and two matching pieces (each scaled to 25%) centered around twice the carrier frequency f_0 and twice its negative. A lowpass filter can now be used to extract $W(f)$, and hence to recover the original message $w(t)$.

This procedure is shown diagrammatically in Figure 2.6. The spectrum of the original message is shown in (a), and the spectrum of the message modulated by the carrier appears in (b). When downconversion is done as just described, the demodulated signal $d(t)$ has the spectrum shown in (c). Filtering by a lowpass filter (as in part (c)) removes all but a scaled version of the message.

Now consider the FDM transmitted signal spectrum of Figure 2.3. This can be demodulated/downconverted similarly. The frequency-shifting rule (2.4) again ensures that the downconverted spectrum in Figure 2.7 matches (2.5), and the lowpass filter removes all but the desired message from the downconverted signal.

FIGURE 2.6 The message can be recovered by downconversion and lowpass filtering. (a) Shows the original spectrum of the message; (b) shows the message modulated by the carrier f_0; (c) shows the demodulated signal. Filtering with an LPF recovers the original spectrum.

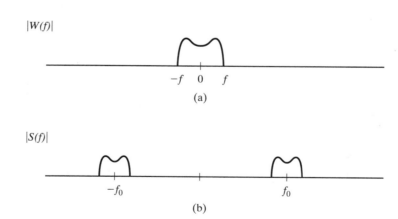

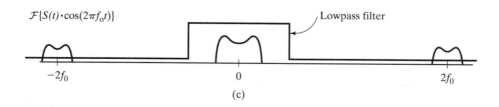

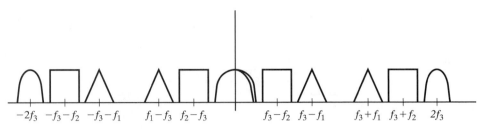

FIGURE 2.7 Downconversion of FDM user to baseband

This is the basic principle of a transmitter and receiver pair. But there are some practical issues that arise. What happens if the oscillator at the receiver is not completely accurate in either frequency or phase? The downconverted received signal becomes $r(t)\cos(2\pi(f_0 + \alpha)t + \beta)$. This can have serious consequences for the demodulated message. What happens if one of the antennas is moving? The Doppler effect suggests that this corresponds to a small nonzero value of α. What happens if the transmitter antenna wobbles due to the wind over a range equivalent to several wavelengths of the transmitted signal? This can alter β. In effect, the baseband component is perturbed from $(1/2)W(f)$, and simply lowpass filtering the downconverted signal results in distortion. Carrier synchronization schemes (which attempt to identify and track the phase and frequency of the carrier) are routinely used in practice to counteract such problems. These are discussed in detail in Chapters 5 and 10.

2.7 Analog Core of Digital Communication System

The signal flow in the AM communication system described in the preceding sections is shown in Figure 2.8. The message is upconverted (for efficient transmission), summed with other FDM users (for efficient use of the electromagnetic spectrum), subjected to possible channel noises (such as thermal noise), bandpass filtered (to extract the desired user), downconverted (requiring carrier synchronization), and lowpass filtered (to recover the actual message).

But no transmission system operates perfectly. Each of the blocks in Figure 2.8 may be noisy, may have components which are inaccurate, and may be subject to fundamental limitations. For instance,

- the bandwidth of a filter may be different from its specification (e.g., the shoulders may not drop off fast enough to avoid passing some of the adjacent signal),
- the frequency of an oscillator may not be exact, and hence the modulation and/or demodulation may not be exact,
- the phase of the carrier is unknown at the receiver, since it depends on the time of travel between the transmitter and the receiver,
- perfect filters are impossible, even in principle,
- no oscillator is perfectly regular, there is always some jitter in frequency.

Even within the frequency range of the message signal, the medium can affect different frequencies in different ways. (These are called *frequency selective effects*.) For example, a signal may arrive at the receiver, and a moment later a copy of the same signal might arrive after having bounced off a mountain or a nearby building. This is called *multipath interference*, and it can be viewed as a sum of weighted and delayed versions

FIGURE 2.8 Analog
AM communication
system

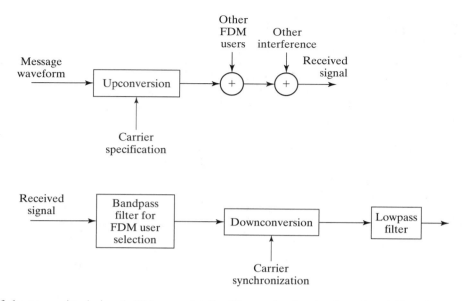

of the transmitted signal. This may be familiar to the (analog broadcast) TV viewer as
"ghosts," misty copies of the original signal that are shifted and superimposed over the
main image. In the simple case of a sinusoid, a delay corresponds to a phase shift, making
it more difficult to reassemble the original message. A special filter called the *equalizer*
is often added to the receiver to help improve the situation. An equalizer is a kind of
"deghosting" circuit,[3] and equalization is addressed in detail in Chapter 13.

2.8 Sampling at the Receiver

Because of the proliferation of inexpensive and capable digital processors, receivers
often contain chips that are essentially special purpose computers. In such receivers,
many of the functions that are traditionally handled by discrete components (such as
analog oscillators and filters) can be handled digitally. Of course, this requires that
the analog received signal be turned into digital information (a series of numbers)
that a computer can process. This analog-to-digital conversion (A/D) is known as
sampling.

Sampling measures the amplitude of the waveform at regular intervals, and then stores
these measurements in memory. Two of the chief design issues in a digital receiver are
the following:

- Where should the signal be sampled?
- How often should the sampling be done?

The answers to these questions are intimately related to each other.

When taking samples of a signal, they must be taken fast enough so that important
information is not lost. Suppose that a signal has no frequency content above f^* Hz. The
widely known Nyquist reconstruction principle (see Section 6.1) says that if sampling

---

[3] We refrain from calling these ghost busters.

occurs at a rate greater than $2f^*$ samples per second, it is possible to reconstruct the original signal from the samples alone. Thus, as long as the samples are taken rapidly enough, no information is lost. On the other hand, when samples are taken too slowly, the signal cannot be reconstructed exactly from the samples, and the resulting distortion is called *aliasing*.

Accordingly, in the receiver, it is necessary to sample at least twice as fast as the highest frequency present in the analog signal being sampled in order to avoid aliasing. Because the receiver contains modulators that change the frequencies of the signals, different parts of the system have different highest frequencies. Hence the answer to the question of how fast to sample is dependent on where the samples will be taken.

The sampling

1. could be done at the input to the receiver at a rate proportional to the carrier frequency,
2. could be done after the downconversion, at a rate proportional to the rate of the symbols, or
3. could be done at some intermediate rate.

Each of these is appropriate in certain situations.

For the first case, consider Figure 2.3, which shows the spectrum of the FDM signal prior to downconversion. Let $f_3 + f^*$ be the frequency of the upper edge of the user spectrum near the carrier at f_3. By the Nyquist principle, the upconverted received signal must be sampled at a rate of at least $2(f_3 + f^*)$ to avoid aliasing. For high frequency carriers, this exceeds the rate of reasonably priced A/D samplers. Thus directly sampling the received signal (and performing all the downconversion digitally) may not feasible, even though it appears desirable for a fully software based receiver.

In the second case, the downconversion (and subsequent lowpass filtering) are done in analog circuitry, and the samples are taken at the output of the lowpass filter. Sampling can take place at a rate twice the highest frequency f^* in the baseband, which is considerably smaller than twice $f_3 + f^*$. Since the downconversion must be done accurately in order to have the shifted spectra of the desired user line up exactly (and overlap correctly), the analog circuitry must be quite accurate. This, too, can be expensive.

In the third case, the downconversion in done in two steps: an analog circuit downconverts to some intermediate frequency, where the signal is sampled. The resulting signal is then digitally downconverted to baseband. The advantage of this (seemingly redundant) method is that the analog downconversion can be performed with minimal precision (and hence inexpensively), while the sampling can be done at a reasonable rate (and hence inexpensively). In Figure 2.9, the frequency f_I of the intermediate downconversion is chosen to be large enough so that the whole FDM band is moved below the upshifted portion. Also, f_I is chosen to be small enough so that the downshifted positive frequency portion lower edge does not reach zero. An analog bandpass filter extracts the whole FDM band at an intermediate frequency (IF), and then it is only necessary to sample at a rate greater than $2(f_3 + f^* - f_I)$.

Downconversion to an intermediate frequency is common since the analog circuitry can be fixed, and the tuning (when the receiver chooses between users) can be done digitally. This is advantageous since tunable analog circuitry is considerably more expensive than tunable digital circuitry.

FIGURE 2.9 FDM
downconversion to an
intermediate frequency

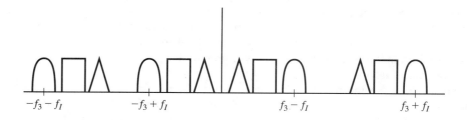

$-f_3 - f_I$ $-f_3 + f_I$ $f_3 - f_I$ $f_3 + f_I$

2.9 Digital Communications Around an Analog Core

The discussion so far in this chapter has concentrated on the classical core of telecommunication systems: the transmission and reception of analog waveforms. In digital systems, as considered in the previous chapter, the original signal consists of a stream of data, and the goal is to send the data from one location to another. The data may be a computer program, ASCII text, pixels of a picture, a digitized MP3 file, or sampled speech from a cell phone. "Data" consist of a sequence of numbers, which can always be converted to a sequence of zeros and ones, called *bits*. How can a sequence of bits be transmitted?

The basic idea is that, since transmission media (such as air, phone lines, the ocean) are analog, the bits are converted into an analog signal. Then this analog signal can be transmitted exactly as before. Thus at the core of every "digital" communication system lies an "analog" system. The output of the transmitter, the transmission medium, and the front end of the receiver are necessarily analog.

Digital methods are not new. Morse code telegraphy (which consists of a sequence of dashes and dots coded into long and short tone bursts) became widespread in the 1850s. The early telephone systems of the 1900s were analog, and then they were digitized in the 1970s.

Digital communications (relative to fully analog) have the following advantages:

- digital circuits are relatively inexpensive,
- data encryption can be used to enhance privacy,
- digital realization supports greater dynamic range,
- signals from voice, video, and data sources can be merged for transmission over a common system,
- noise does not accumulate from repeater to repeater over long distances,
- low error rates are possible, even with substantial noise,
- errors can be corrected via coding.

In addition, digital receivers can easily be reconfigured or upgraded, because they are essentially software driven. For instance, a receiver built for one broadcast standard (say for the American market) could be transformed into a receiver for the European market with little additional hardware.

But there are also some disadvantages of digital communications (relative to fully analog), including the following:

- more bandwidth (is generally) required than with analog,
- synchronization is required.

2.10 Pulse Shaping

In order to transmit a digital data stream, it must be turned into an analog signal. The first step in this conversion is to clump the bits into symbols that lend themselves to translation into analog form. For instance, a mapping from the letters of the English alphabet into bits and then into the 4-PAM symbols ± 1, ± 3 was given explicitly in (1.1). This was converted into an analog waveform using the rectangular pulse shape (1.2), which results in signals that look like Figure 1.3. In general, such signals can be written

$$y(t) = \sum_k s[k]p(t - kT), \tag{2.6}$$

where the $s[k]$ are the values of the symbols, and the function $p(t)$ is the pulse shape. Thus, each member of the 4-PAM data sequence is multiplied by a pulse that is nonzero over the appropriate time window. Adding all the scaled pulses results in an analog waveform that can be upconverted and transmitted. If the channel is perfect (distortionless and noise free), then the transmitted signal will arrive unchanged at the receiver. Is the rectangular pulse shape a good idea?

Unfortunately, though rectangular pulse shapes are easy to understand, they can be a poor choice for a pulse shape because they spread substantial energy into adjacent frequencies. This spreading complicates the packing of users in frequency division multiplexing, and makes it more difficult to avoid having different messages interfere with each other.

To see this, define the rectangular pulse

$$\Pi\left(\frac{t}{T}\right) = \left[\begin{array}{cc} 1 & -T/2 \leq t \leq T/2 \\ 0 & \text{otherwise} \end{array} \right. . \tag{2.7}$$

The shifted pulses (2.7) are sometimes easier to work with than (1.2), and their magnitude spectra are the same by the time shifting property (A.38). The Fourier transform can be calculated directly from the definition (2.1)

$$W(f) = \int_{t=-\infty}^{\infty} \Pi(t/T)e^{-j2\pi ft}dt = \int_{t=-T/2}^{T/2} (1)e^{-j2\pi ft}dt$$

$$= \frac{e^{-j2\pi ft}}{-j2\pi f}\bigg|_{t=-T/2}^{T/2} = \frac{e^{-j\pi fT} - e^{j\pi fT}}{-j2\pi f}$$

$$= T\frac{\sin(\pi fT)}{\pi fT} \equiv T\text{sinc}(fT). \tag{2.8}$$

The sinc function is illustrated in Figure 2.10.

Thus, the Fourier transform of a rectangular pulse in the time domain is a sinc function in the frequency domain. Since the sinc function dies away with an envelope of $1/x$, the frequency content of the rectangular pulse shape is (in principle) infinite. It is not possible to separate messages into different nonoverlapping frequency regions as is required for an FDM implementation as in Figure 2.3.

Alternatives to the rectangular pulse are essential. Consider what is really required of a pulse shape. The pulse is transmitted at time kT and again at time $(k + 1)T$

FIGURE 2.10 The sinc function sinc$(x) \equiv \frac{\sin(\pi x)}{\pi x}$ has zeros at every integer (except zero) and dies away with an envelope of $\frac{1}{\pi x}$.

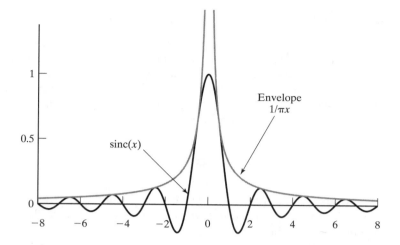

(and again at $(k+2)T \dots$). The received signal is the sum of all these pulses (weighted by the message values). As long as each individual pulse is zero at all integer multiples of T, then the value sampled at those times is just the value of the original pulse (plus many additions of zero). The rectangular pulse of width T seconds satisfies this criterion, as does any other pulse shape that is exactly zero outside a window of width T. But many other pulse shapes also satisfy this condition, without being identically zero outside a window of width T.

In fact, Figure 2.10 shows one such pulse shape—the sinc function itself! It is zero at all integers[4] (except at zero where it is one). Hence, the sinc can be used as a pulse shape. As in (2.6), the shifted pulse shape is multiplied by each member of the data sequence, and then added together. If the channel is perfect (distortionless and noise free), then the transmitted signal will arrive unchanged at the receiver. The original data can be recovered from the received waveform by sampling at exactly the right times. This is one reason why timing synchronization is so important in digital systems. Sampling at the wrong times may garble the data.

To assess the usefulness of the sinc pulse shape, consider its transform. The Fourier transform of the rectangular pulse shape in the time domain is the sinc function in the frequency domain. Analogously, the Fourier transform of the sinc function in the time domain is a rectangular pulse in the frequency domain (see (A.22)). Thus, the spectrum of the sinc is bandlimited, and so it is appropriate for situations requiring bandlimited messages, such as FDM. Unfortunately, the sinc is not entirely practical because it is doubly infinite in time. In any real implementation, it will need to be truncated.

The rectangular and the sinc pulse shapes give two extremes. Practical pulse shapes compromise between a small amount of out-of-band content (in frequency) and an impulse response that falls off rapidly enough to allow reasonable truncation (in the time domain). Commonly used pulse shapes such as the square-root raised cosine shape are described in detail in Chapter 11.

---

[4] In other applications, it may be desirable to have the zero crossings occur at places other than the integers. This can be done by suitably scaling the x.

2.11 Synchronization

Synchronization may occur in several places in the digital receiver:

- Symbol phase synchronization—choosing when (within each interval T) to sample.
- Symbol frequency synchronization—accounting for different clock (oscillator) rates at the transmitter and receiver.
- Carrier phase synchronization—aligning the phase of the carrier at the receiver with the phase of the carrier at the transmitter.
- Carrier frequency synchronization—aligning the frequency of the carrier at the receiver with the frequency of the carrier at the transmitter.
- Frame synchronization—finding the "start" of each message data block.

In digital receivers, it is important to sample the received signal at the appropriate time instants. Moreover, these time instants are not known beforehand; rather, they must be determined from the signal itself. This is the problem of clock recovery. A typical strategy samples several times per pulse and then uses some criterion to pick the best one, to estimate the optimal time, or to interpolate an appropriate value. There must also be a way to deal with the situation when the oscillator defining the symbol clock at the transmitter differs from the oscillator defining the symbol clock at the receiver. Similarly, carrier synchronization is the process of recovering the carrier (in both frequency and phase) from the received signal. This is the same task in a digital receiver as in an analog design (recall that the cosine wave used to demodulate the received signal in (2.5) was aligned in both phase and frequency with the modulating sinusoid at the transmitter), though the details of implementation may differ.

In many applications (such as cell phones), messages come in clusters called packets, and each packet has a header (that is located in some agreed-upon place within each data block) that contains important information. The process of identifying where the header appears in the received signal is called frame synchronization, and is often implemented using a correlation technique.

The point of view adopted in *Telecommunication Breakdown* is that many of these synchronization tasks can be stated quite simply as optimization problems. Accordingly, many of the standard solutions to synchronization tasks can be viewed as solutions to these optimization problems:

- The problem of clock (or timing) recovery can be stated as that of finding a timing offset τ to maximize the energy of the received signal. Solving this optimization problem via a gradient technique leads to a standard algorithm for timing recovery.
- The problem of carrier phase synchronization can be stated as that of finding a phase offset θ to minimize a particular function of the modulated received signal. Solving this optimization problem via a gradient technique leads to the Phase Locked Loop, a standard method of carrier recovery.
- Carrier phase synchronization can also be stated using an alternative performance function that leads directly to the Costas loop, another standard method of carrier recovery.

Our presentation focuses on solving problems using simple recursive (gradient) methods. Once the synchronization problems are correctly stated, techniques for their solution become obvious. With the exception of frame synchronization (which is approached via correlational methods) the problem of designing synchronizers is unified via one simple concept, that of the minimization (or maximization) of an appropriate performance function. Chapters 6, 10, and 12 contain details.

2.12 Equalization

When all is well in the digital receiver, there is no interaction between adjacent data values and all frequencies are treated equally. In most real wireless systems (and many wired systems as well), however, the transmission channel causes multiple copies of the transmitted symbols, each scaled differently, to arrive at the receiver at different times. This *intersymbol interference* can garble the data. The channel may also attenuate different frequencies by different amounts. Thus *frequency selectivity* can render the data indecipherable.

The solution to both of these problems is to build a filter in the receiver that attempts to undo the effects of the channel. This filter, called an *equalizer*, cannot be fixed in advance by the system designer, however, because it must be different to compensate for different channel paths that are encountered when the system is operating. The problem of equalizer design can be stated as a simple optimization problem, that of finding a set of filter parameters to minimize an appropriate function of the error, given only the received data (and perhaps a training sequence). This problem is investigated in detail in Chapter 13, where the same kinds of adaptive techniques used to solve the synchronization problems can also be applied to solve the equalization problem.

2.13 Decisions and Error Measures

In analog systems, the transmitted waveform can attain any value, but in a digital implementation the transmitted message must be one of a small number of values defined by the symbol alphabet. Consequently, the received waveform in an analog system can attain any value, but in a digital implementation the recovered message is meant to be one of a small number of values from the source alphabet. Thus, when a signal is demodulated to a symbol and it is not a member of the alphabet, the difference between the demodulated value (called a *soft* decision) and the nearest element of the alphabet (the *hard* decision) can provide valuable information about the performance of the system.

To be concrete, label the signals at various points as shown in Figure 2.11:

- The binary input message $b(\cdot)$.
- The coded signal $w(\cdot)$ is a discrete-time sequence drawn from a finite alphabet.
- The signal $m(\cdot)$ at the output of the filter and equalizer is continuous valued at discrete times.
- $Q\{m(\cdot)\}$ is a version of $m(\cdot)$ that is quantized to the nearest member of the alphabet.
- The decoded signal $\hat{b}(\cdot)$ is the final (binary) output of the receiver.

If all goes well and the message is transmitted, received, and decoded successfully, then the output should be the same as the input, although there may be some delay δ between the time of transmission and the time when the output is available. When the output differs from the message, then errors have occurred during transmission.

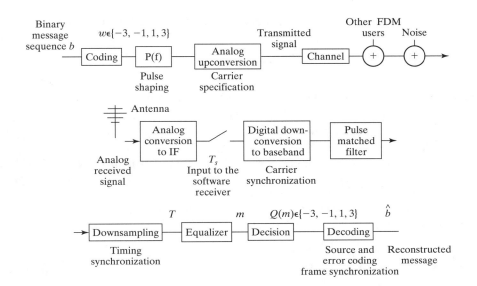

FIGURE 2.11 PAM system diagram.

There are several ways to measure the quality of the system. For instance, the "symbol recovery error"

$$e(kT) = w((k - \delta)T) - m(kT)$$

measures the difference between the message and the soft decision. The average squared error

$$\frac{1}{M} \sum_{k=1}^{M} e^2(kT),$$

gives a measure of the performance of the system. This can be used as in Chapter 13 to adjust the parameters of an equalizer when the source message is known. Alternatively, the difference between the message $w(\cdot)$ and the quantized output of the receiver $Q\{m(\cdot)\}$ can be used to measure the "hard decision error"

$$e(kT) = w((k - \delta)T) - Q\{m(kT)\}.$$

The "decision-directed error" replaces this with

$$e(kT) = Q\{m(kT)\} - m(kT),$$

the error between the soft decisions and the associated hard decisions. This error is used in Section 13.4 as a way to adjust the parameters in an equalizer when the source message is unknown, as a way of adjusting the phase of the carrier in Section 10.5, and as a way of adjusting the symbol timing in Section 12.3.

There are other useful indicators of the performance of digital communication receivers. The indicator

$$c(kT) = \begin{cases} 1 & \text{if } b((k - \delta)T) \neq \hat{b}(kT) \\ 0 & \text{if } b((k - \delta)T) = \hat{b}(kT) \end{cases}$$

counts how many bits have been incorrectly received, and the bit error rate is

$$BER = \frac{1}{M} \sum_{k=1}^{M} c(kT).$$ (2.9)

Similarly, the symbol error rate replaces $c(kT)$ in (2.9) with

$$c(kT) = \begin{cases} 1 & \text{if } w((k-\delta)T)) \neq Q\{m(kT)\} \\ 0 & \text{if } w((k-\delta)T)) = Q\{m(kT)\} \end{cases},$$

which counts the number of alphabet symbols that were transmitted incorrectly. More subjective or context-dependent measures are also possible, such as the percentage of "typical" listeners who can accurately decipher the output of the receiver.

No matter what the exact form of the error measure, the ultimate goal is the accurate and efficient transmission of the message.

2.14 Coding and Decoding

What is information? How much can move across a particular channel in a given amount of time? Claude Shannon proposed a method of measuring information in terms of bits, and a measure of the capacity of the channel in terms of the bit rate—the number of bits transmitted per second (recall the quote at the beginning of the first chapter). This is defined quantitatively by the *channel capacity*, which is dependent on the bandwidth of the channel and on the power of the noise in comparison to the power of the signal. For most receivers, however, the reality is far from the capacity, and this is caused by two factors. First, the data to be transmitted are often redundant, and the redundancy squanders the capacity of the channel. Second, the noise can be unevenly distributed among the symbols. When large noises disrupt the signal, then excessive errors occur.

The problem of redundancy is addressed in Chapter 14 by *source coding*, which strives to represent the data in the most concise manner possible. After demonstrating the redundancy and correlation of English text, Chapter 14 introduces the *Huffman code*, which is a variable-length code that assigns short bit strings to frequent symbols and longer bit strings to infrequent symbols. Like Morse code, this will encode the letter "e" with a short code word, and the letter "z" with a long code word. When correctly applied, the Huffman procedure can be applied to any symbol set (not just the letters of the alphabet), and is "nearly" optimal, that is, it approaches the limits set by Shannon.

The problem of reducing the sensitivity to noise is addressed in Chapter 14 using the idea of *linear block codes*, which cluster a number of symbols together, and then add extra bits. A simple example is the (binary) parity check, which adds an extra bit to each character. If there are an even number of ones then a 1 is added, and if there are an odd number of ones, a 0 is added. The receiver can always detect that a single error has occurred by counting the number of 1's received. If the sum is even, then an error has occurred, while if the sum is odd then no single error can have occurred. More sophisticated versions can not only detect errors, but can also correct them.

Like good equalization and proper synchronization, coding is an essential part of the operation of digital receivers.

2.15 A Telecommunication System

The complete system diagram, including the digital receiver that will be built in this text, is shown in Figure 2.11. This system includes the following:

- A source coding that reduces the redundancy of the message.
- An error coding that allows detection and/or correction of errors that may occur during the transmission.
- A message sequence of T-spaced symbols drawn from a finite alphabet.
- Pulse shaping of the message, designed (in part) to conserve bandwidth.
- Analog upconversion to the carrier frequency (within specified tolerance).
- Channel distortion of transmitted signal.
- Summation with other FDM users, channel noise, and other interferers.
- Analog downconversion to intermediate frequency (including bandpass prefiltering around the desired segment of the FDM passband).
- A/D impulse sampling (preceded by antialiasing filter) at a rate of $\frac{1}{T_s}$ with arbitrary start time. The sampling rate is assumed to be at least as fast as the symbol rate $\frac{1}{T}$.
- Downconversion to baseband (requiring carrier phase and frequency synchronization).
- Lowpass (or pulse-shape-matched) filtering for the suppression of out-of-band users and channel noise.
- Downsampling with timing adjustment to T-spaced symbol estimates.
- Equalization filtering to combat intersymbol interference and narrowband interferers.
- Decision device quantizing soft decision outputs of equalizer to nearest member of the source alphabet (i.e. the hard decision).
- Source and error decoders.

Of course, permutations and variations of this system are possible, but we believe that Figure 2.11 captures the essence of many modern transmission systems. The path taken by *Telecommunication Breakdown* is to break down the telecommunication system into its constituent elements: the modulators and demodulators, the samplers and filters, the coders and decoders. In the various tasks within each chapter, you are asked to build a simulation of the relevant piece of the system. In the early chapters, the parts need to operate only in a pristine, idealized environment, but as we delve deeper into the onion, impairments and noises inevitably intrude. The design evolves to handle the increasingly realistic scenarios.

Throughout this text, we ask you to consider a variety of small questions, some of which are mathematical in nature, most of which are "what if" questions best answered by trial and simulation. We hope that this combination of reflection and activity will be a useful in enlarging your understanding and in training your intuition.

2.16 For Further Reading

There are many books about various aspects of communication systems. Here are some of our favorites. Three basic texts that utilize probability from the outset, and that also pay substantial attention to pragmatic design issues (such as synchronization) are the following:

- J. B. Anderson, *Digital Transmission Engineering*, IEEE Press, 1999.
- J. G. Proakis and M. Salehi, *Communication Systems Engineering*, Prentice Hall, 1994. [This text also has a MATLAB-based companion, *Introduction to Communication Systems Using MATLAB*, Brooks-Cole Pubs., 1999.]
- S. Haykin, *Communication Systems*, 4th edition, John Wiley and Sons, 2001.

Three introductory texts that delay the introduction of probability until the latter chapters are the following:

- L. W. Couch III, *Digital and Analog Communication Systems*, 6th edition, Prentice Hall, 2001.
- B. P. Lathi, *Modern Digital and Analog Communication Systems*, 3rd edition, Oxford University Press, 1998.
- F. G. Stremler, *Introduction to Communication Systems*, 3rd edition, Addison Wesley, 1990.

These references are probably the most compatible with *Telecommunication Breakdown* in terms of the assumed mathematical background.

3 The Five Elements

The Five Elemental Energies of Wood, Fire, Earth, Metal, and Water encompass all the myriad phenomena of nature. It is a paradigm that applies equally to humans.

—The Yellow Emperor's Classic of Internal Medicine

At first glance, block diagrams such as the communication system shown in Figure 2.11 probably appear complex and intimidating. There are so many different blocks and so many unfamiliar names and acronyms! Fortunately, all the blocks can be built from five simple elements:

- *Oscillators*, which create sine and cosine waves,
- *Linear filters*, which augment or diminish particular frequencies or frequency ranges from a signal,
- *Static nonlinearities*, which can change the frequency content of a signal, for instance multipliers, squarers, and quantizers,
- *Samplers*, which change analog (continuous time) signals into discrete-time signals, and
- *Adaptive elements*, which track the desired values of parameters as they slowly change over time.

This section provides a brief overview of these five elements. In doing so, it also reviews some of the key ideas from signals and systems. Later chapters explore how the elements work, how they can be modified to accomplish particular tasks within the communication system, and how they can be combined to create a large variety of blocks such as those that appear in Figure 2.11.

The elements of a communication system have inputs and outputs; the element itself operates on its input signal to create its output signal. The signals that form the inputs and outputs are functions that represent the dependence of some variable of interest (such as a voltage, current, power, air pressure, temperature, etc.) on time.

The action of an element can be described by the manner in which it operates in the "time domain," that is, how the element changes the input waveform moment by moment into the output waveform. Another way of describing the action of an element is by how it operates in the "frequency domain," that is, by how the frequency content of the input relates to the frequency content of the output. Figure 3.1 illustrates these two complementary ways of viewing the elements. Understanding both the time domain and

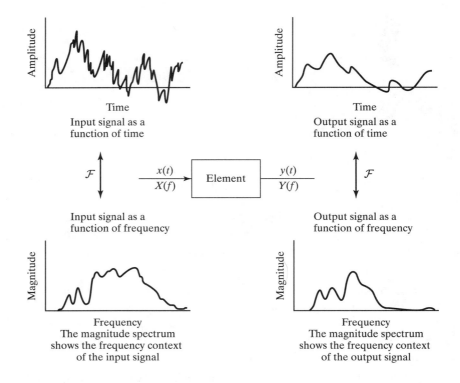

frequency domain behavior is essential. Accordingly, the following sections describe the action of the five elements in both time and frequency.

Readers who have studied signals and systems (often required in electrical engineering degrees), will recognize that the time domain representation of a signal and its frequency domain representation are related by the Fourier transform, which is briefly reviewed in the next section.

3.1 Finding the Spectrum of a Signal

A signal $s(t)$ can often be expressed in analytical form as a function of time t, and the Fourier transform is defined as in (2.1) as the integral of $s(t)e^{-2\pi j f t}$. The resulting transform $S(f)$ is a function of frequency. $S(f)$ is called the spectrum of the signal $s(t)$ and describes the frequencies present in the signal. For example, if the time signal is created as a sum of three sine waves, then the spectrum will have spikes corresponding to each of the constituent sines. If the time signal contains only frequencies between 100 and 200 Hz, then the spectrum will be zero for all frequencies outside of this range. A brief guide to Fourier transforms appears in Appendix D, and a summary of all the transforms and properties that are used throughout *Telecommunication Breakdown* appears in Appendix A.

Often, however, there is no analytical expression for a signal; that is, there is no (known) equation that represents the value of the signal over time. Instead, the signal is defined by measurements of some physical process. For instance, the signal might be the waveform at the input to the receiver, the output of a linear filter, or a sound waveform encoded as an MP3 file. In all these cases, it is not possible to find the spectrum by calculating a Fourier transform.

Rather, the discrete Fourier transform (and its cousin, the more rapidly computable fast Fourier transform, or FFT) can be used to find the spectrum or frequency content of a measured signal. The MATLAB function `plotspec.m`, which plots the spectrum of a signal, is available on the CD. Its help file[1] notes

```
% plotspec(x,Ts) plots the spectrum of the signal x
% Ts = time (in seconds) between adjacent samples in x
```

The function `plotspec.m` is easy to use. For instance, the spectrum of a square wave can be found using the following sequence:

---

specsquare.m: plot the spectrum of a square wave

```
f=10;                      % "frequency" of square wave
time=2;                    % length of time
Ts=1/1000;                 % time interval between samples
t=Ts:Ts:time;             % create a time vector
x=sign(cos(2*pi*f*t));    % square wave = sign of cos wave
plotspec(x,Ts)            % call plotspec to draw spectrum
```

The output of `specsquare.m` is shown[2] in Figure 3.2. The top plot shows `time=2` seconds of a square wave with `f=10` cycles per second. The bottom plot shows a series of spikes that define the frequency content. In this case, the largest spike occurs at ± 10 Hz, followed by smaller spikes at all the odd-integer multiples (i.e., at ± 30, ± 50, ± 70, etc.).

Similarly, the spectrum of a noise signal can be calculated as

---

specnoise.m: plot the spectrum of a noise signal

```
time=1;                    % length of time
Ts=1/10000;                % time interval between samples
x=randn(1,time/Ts);       % Ts points of noise for time seconds
plotspec(x,Ts)            % call plotspec to draw spectrum
```

A typical run of `specnoise.m` is shown in Figure 3.3. The top plot shows the noisy signal as a function of time, while the bottom shows the magnitude spectrum. Because successive values of the noise are generated independently, all frequencies are roughly equal in magnitude. Each run of `specnoise.m` produces plots that are qualitatively similar, though the details will differ.

---

[1] You can view the help file for the MATLAB function xxx by typing `help xxx` at the MATLAB prompt. If you get an error such as `xxx not found`, then this means either that the function does not exist, or that it needs to be moved into the same directory as the MATLAB application. If you don't know what the proper command to do a job is, then use `lookfor`. For instance, to find the command that inverts a matrix, type `lookfor inverse`. You will find the desired command `inv`.

[2] All code listings in *Telecommunication Breakdown* can be found on the CD. We encourage you to open MATLAB and explore the code as you read.

FIGURE 3.2 A square wave and its spectrum, as calculated by using `plotspec.m`.

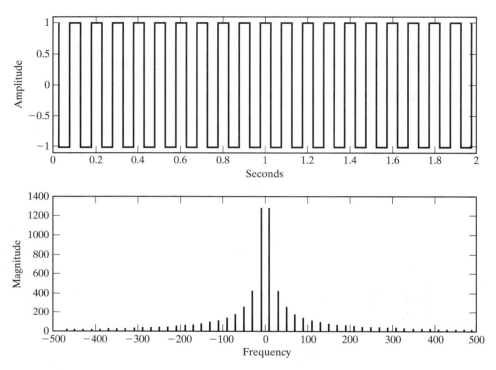

Problems

3.1. Use `specsquare.m` to investigate the relationship between the time behavior of the square wave and its spectrum. The MATLAB command `zoom on` is often helpful for viewing details of the plots.

   (a) Try square waves with different frequencies: `f=20, 40, 100, 300` Hz. How do the time plots change? How do the spectra change?

   (b) Try square waves of different lengths, `time=1, 10, 100` seconds. How does the spectrum change in each case?

   (c) Try different sampling times, `Ts=1/100, 1/10000.` seconds. How does the spectrum change in each case?

3.2. In your *Signals and Systems* course, you probably calculated (analytically) the spectrum of a square wave by using the Fourier series. How does this calculation compare with the discrete data version found by `specsquare.m`?

3.3. Mimic the code in `specsquare.m` to find the spectrum of

   (a) an exponential pulse $s(t) = e^{-t}$,

   (b) a scaled exponential pulse $s(t) = 5e^{-t}$,

   (c) a Gaussian pulse $s(t) = e^{-t^2}$,

   (d) the sinusoids $s(t) = \sin(2\pi f t + \phi)$ for $f = 20, 100, 1000$ and $\phi = 0, \pi/4, \pi/2$.

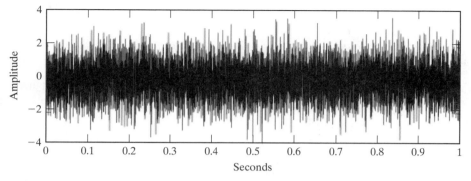

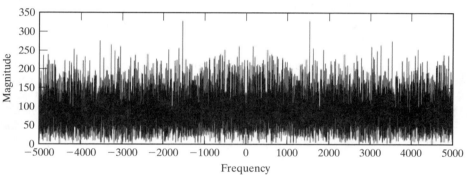

FIGURE 3.3 A noise signal and its spectrum, as calculated using `plotspec.m`.

3.4. MATLAB has several commands that create random numbers:
  (a) Use `rand` to create a signal that is uniformly distributed on $[-1, 1]$. Find the spectrum of the signal by mimicking the code in `specnoise.m`.
  (b) Use `rand` and the `sign` function to create a signal that is $+1$ with probability $1/2$ and -1 with probability $1/2$. Find the spectrum of the signal.
  (c) Use `randn` to create a signal that is normally distributed with mean 0 and variance 3. Find the spectrum of the signal.

While `plotspec.m` can be quite useful, ultimately, it will be necessary to have more flexibility, which, in turn, requires one to understand how the FFT function inside `plotspec.m` works. This will be discussed at length in Chapter 7. The next five sections describe the five elements that are at the heart of communications systems. The elements are described in both the time domain and in the frequency domain.

3.2 The First Element: Oscillators

The Latin word *oscillare* means "to ride in a swing." It is the origin of *oscillate*, which means to move back and forth in steady unvarying rhythm. Thus, a device that creates a signal that moves back and forth in a steady, unvarying rhythm is called an *oscillator*. An electronic oscillator is a device that produces a repetitive electronic signal, usually a sinusoidal wave.

FIGURE 3.4 An oscillator creates a sinusoidal oscillation with a specified frequency f_0 and input ϕ.

A basic oscillator is diagrammed in Figure 3.4. Oscillators are typically designed to operate at a specified frequency f_0, and the input specifies the phase ϕ of the output waveform

$$s(t) = \cos(2\pi f_0 t + \phi).$$

The input may be a fixed number, but it may also be a signal; that is, it may change over time. In this case, the output is no longer a pure sinusoid of frequency f_0. For instance, suppose the phase is a "ramp" or line with slope $2\pi c$; that is, $\phi(t) = 2\pi c t$. Then $s(t) = \cos(2\pi f_0 t + 2\pi c t) = \cos(2\pi (f_0 + c)t)$, and the "actual" frequency of oscillation is $f_0 + c$.

There are many ways to build oscillators from analog components. Generally, there is an amplifier and a feedback circuit that returns a portion of the amplified wave back to the input. When the feedback is aligned properly in phase, sustained oscillations occur.

Digital oscillators are simpler, since they can be directly calculated; no amplifier or feedback are needed. For example, a "digital" sine wave of frequency f Hz and a phase of ϕ radians can be represented mathematically as

$$s(kT_s) = \cos(2\pi f k T_s + \phi), \tag{3.1}$$

where T_s is the time between samples and where k is an integer counter $k = 1, 2, 3, \ldots$. Equation (3.1) can be directly implemented in MATLAB:

---
speccos.m: plot the spectrum of a cosine wave

```
f=10; phi=0;              % specify frequency and phase
time=2;                   % length of time
Ts=1/100;                 % time interval between samples
t=Ts:Ts:time;             % create a time vector
x=cos(2*pi*f*t+phi);      % create cos wave
plotspec(x,Ts)            % draw waveform and spectrum
```

The output of speccos.m is shown in Figure 3.5. As expected, the time plot shows an undulating sinusoidal signal with $f = 10$ repetitions in each second. The actual data points are discrete, with one hundred data points in each second. Do not be fooled by the default method of plotting in which MATLAB "connects the dots" with short line segments for a smoother appearance. The spectrum shows two spikes, one at $f = 10$ Hz and one at $f = -10$ Hz. Why are there *two* spikes? Basic Fourier theory shows that the Fourier transform of a cosine wave is a pair of delta functions at plus and minus the frequency of the cosine wave (see Appendix (A.18)). The two spikes of Figure 3.5 mirror these two delta functions. Alternatively, recall that a cosine wave can be written using Euler's formula as the sum of two complex exponentials, as in (A.2). The spikes of Figure 3.5 represent the magnitudes of these two (complex valued) exponentials.

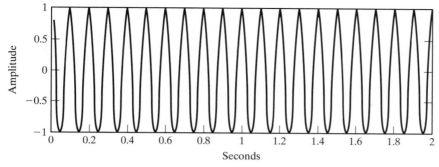

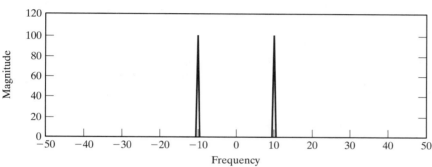

FIGURE 3.5 A sinusoidal oscillator creates a signal that can be viewed in the time domain as in the top plot, or in the frequency domain as in the bottom plot.

Problems

3.5. Mimic the code in `speccos.m` to find the spectrum of a cosine wave
   (a) for different frequencies f=1, 2, 20, 30 Hz,
   (b) for different phases $\phi = 0, 0.1, \pi/8, \pi/2$ radians,
   (c) for different sampling rates Ts=1/10, 1/1000, 1/100000.

3.6. Let $x_1(t)$ be a cosine wave of frequency $f = 10$, $x_2(t)$ be a cosine wave of frequency $f = 18$, and $x_3(t)$ be a cosine wave of frequency $f = 33$. Let $x(t) = x_1(t) + 0.5 * x_2(t) + 2 * x_3(t)$. Find the spectrum of $x(t)$. What property of the Fourier transform does this illustrate?

3.7. Find the spectrum of a cosine wave when
   (a) ϕ is a function of time. Try $\phi(t) = 10\pi t$,
   (b) ϕ is a function of time. Try $\phi(t) = \pi t^2$,
   (c) f is a function of time. Try $f(t) = sin(2\pi t)$,
   (d) f is a function of time. Try $f(t) = t^2$.

3.3 The Second Element: Linear Filters

Linear filters shape the spectrum of a signal. If the signal has too much energy in the low frequencies, a highpass filter can remove them. If the signal has too much high frequency noise, a lowpass filter can reject it. If a signal of interest resides only between f_* and f^*, then a bandpass filter tuned to pass frequencies between f_* and f^* can remove out-of-band interference and noise. More generally, suppose that a signal has frequency

FIGURE 3.6 A "white" signal containing all frequencies is passed through a lowpass filter (LPF) leaving only the low frequencies, a bandpass filter (BPF) leaving only the middle frequencies and a highpass filter (HPF) leaving only the high frequencies.

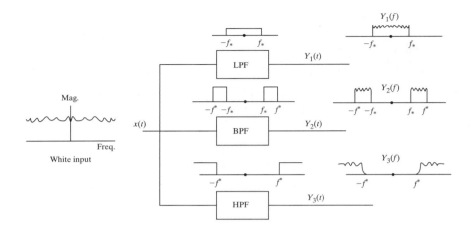

bands in which the magnitude of the spectrum is lower than desired and other bands in which the magnitude is greater than desired. Then a linear filter can compensate by increasing or decreasing the magnitude as needed. This section provides an overview of how to implement simple filters in MATLAB. More thorough treatments of the theory, design, use, and implementation of filters are given in Chapter 7.

While the calculations of a linear filter are usually carried out in the time domain, filters are often specified in the frequency domain. Indeed, the words used to specify filters (such as lowpass, highpass, and bandpass) describe how the filter acts on the frequency content of its input. Figure 3.6, for instance, shows a noisy input entering three different filters. The frequency response of the LPF shows that it allows low frequencies (those below the cutoff frequency f_*) to pass, while removing all frequencies above the cutoff. Similarly, the HPF passes all the high frequencies and rejects those below its cutoff f^*. The action of the BPF is specified by two frequencies. It will remove all frequencies below f_* and remove all frequencies above f^*, leaving only the region between.

Figure 3.6 shows the action of ideal filters. How close are actual implementations? The MATLAB code in `filternoise.m` shows that it is possible to create digital filters that reliably and accurately carry out these tasks.

---

filternoise.m: filter a noisy signal three ways

```
time=3;                                   % length of time
Ts=1/10000;                               % time interval between samples
x=randn(1,time/Ts);                       % generate noise signal
figure(1),plotspec(x,Ts)                  % draw spectrum of input
b=remez(100,[0 0.2 0.21 1],[1 1 0 0]);    % specify the LP filter
ylp=filter(b,1,x);                        % do the filtering
figure(2),plotspec(ylp,Ts)                % plot the output spectrum
b=remez(100,[0 0.24 0.26 0.5 0.51 1],[0 0 1 1 0 0]); % BP filter
ybp=filter(b,1,x);                        % do the filtering
figure(3),plotspec(ybp,Ts)                % plot the output spectrum
b=remez(100,[0 0.74 0.76 1],[0 0 1 1]);   % specify the HP filter
yhp=filter(b,1,x);                        % do the filtering
figure(4),plotspec(yhp,Ts)                % plot the output spectrum
```

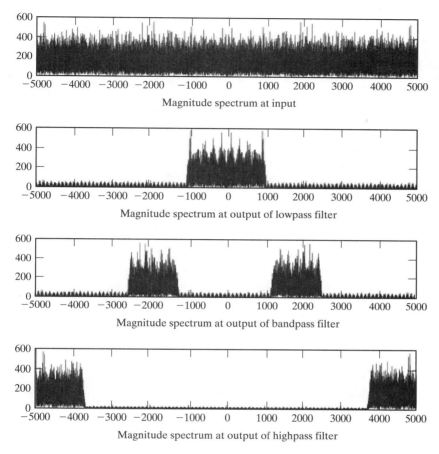

FIGURE 3.7 The spectrum of a "white" signal containing all frequencies is shown in the top figure. This is passed through three filters: a lowpass, a bandpass, and a highpass. The spectra at the outputs of these three filters are shown in the second, third, and bottom plots. The "actual" filters behave much like their idealized counterparts in Figure 3.6.

The output of `filternoise.m` is shown in Figure 3.7. Observe that the spectra at the output of the filters are close approximations to the ideals shown in Figure 3.6. There are some differences, however. While the idealized spectra are completely flat in the passband, the actual ones are rippled. While the idealized spectra completely reject the out-of-band frequencies, the actual ones have small (but nonzero) energy at all frequencies.

Two new MATLAB commands are used in `filternoise.m`. The `remez` command specifies the contour of the filter as a line graph. For instance, typing

```
plot([0 0.24 0.26 0.5 0.51 1],[0 0 1 1 0 0])
```

at the MATLAB prompt draws a box that represents the action of the BPF designed in `filternoise.m` (over the positive frequencies). The frequencies are specified as percentages of $f_{NYQ} = \frac{1}{2T_s}$, which in this case is equal to 5000 Hz. (f_{NYQ} is discussed further in the next section.) Thus the BPF in `filternoise.m` passes frequencies between 0.26x5000 Hz to 0.5x5000 Hz, and rejects all others. The `filter` command uses the output of `remez` to carry out the filtering operation on the vector specified in its third argument. More details about these commands are given in the section on practical filtering in Chapter 7.

FIGURE 3.8 The sampling process is shown in (b) as an evaluation of the signal $x(t)$ at times $\ldots, -2T_s, T_s, 0, T_s, 2T_s, \ldots$. This procedure is schematized in (a) as an element that has the continuous-time signal $x(t)$ as input and the discrete-time signal $x(kT_s)$ as output.

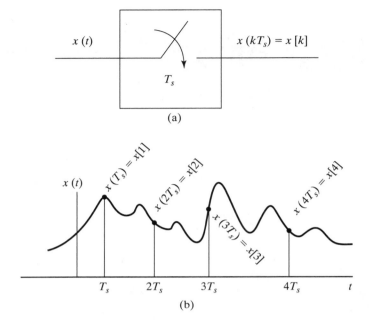

Problems

3.8. Mimic the code in `filternoise.m` to create a filter that
  (a) passes all frequencies above 500 Hz,
  (b) passes all frequencies below 3000 Hz,
  (c) rejects all frequencies between 1500 and 2500 Hz.

3.9. Change the sampling rate to `Ts=1/20000`. Redesign the three filters from Problem 3.8.

3.10. Let $x_1(t)$ be a cosine wave of frequency $f = 800$, $x_2(t)$ be a cosine wave of frequency $f = 2000$, and $x_3(t)$ be a cosine wave of frequency $f = 4500$. Let $x(t) = x_1(t) + 0.5 * x_2(t) + 2 * x_3(t)$. Use $x(t)$ as input to each of the three filters in `filternoise.m`. Plot the spectra, and explain what you see.

3.4 The Third Element: Samplers

Since part of any digital transmission system is analog (transmissions through the air, across a cable, or along a wire, are inherently analog), and part of the system is digital, there must be a way to translate the continuous-time signal into a discrete-time signal and vice versa. The process of sampling an analog signal, sometimes called analog-to-digital conversion, is easy to visualize in the time domain. Figure 3.8 shows how sampling can be viewed as the process of evaluating a continuous-time signal at a sequence of uniformly spaced time intervals, thus transforming the analog signal $x(t)$ into the discrete-time signal $x(kT_s)$.

One of the key ideas in signals and systems is the Fourier series: a signal is periodic in time (it repeats every P seconds), if and only if the spectrum can be written as a sum

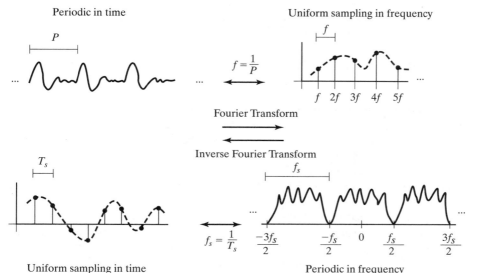

Periodic in time

Uniform sampling in frequency

$f = \frac{1}{P}$

Fourier Transform

Inverse Fourier Transform

$f_s = \frac{1}{T_s}$

Uniform sampling in time

Periodic in frequency

FIGURE 3.9 Fourier's result says that any signal that is periodic in time has a spectrum that consists of a collection of spikes uniformly spaced in frequency. Analogously, any signal whose spectrum is periodic in frequency can be represented in time as a collection of spikes uniformly spaced in time, *and vice versa*.

of complex sinusoids with frequencies at integer multiples of a fundamental frequency f. Moreover, this fundamental frequency can be written in terms of the period as $f = 1/P$. Thus, if a signal repeats 100 times every second ($P = 0.01$ seconds), then its spectrum consists of a sum of sinusoids with frequencies $100, 200, 300, \ldots$ Hz. Conversely, if a spectrum is built from a sum of sinusoids with frequencies $100, 200, 300, \ldots$ Hz, then it must represent a periodic signal in time that has period $P = 0.01$. Said another way, the nonzero portions of the spectrum are uniformly spaced $f = 100$ Hz apart. This uniform spacing can be interpreted as a sampling (in frequency) of an underlying continuous-valued spectrum. This is illustrated in the top portion of Figure 3.9, which shows the time domain representation on the left and the corresponding frequency domain representation on the right.

The basic insight from Fourier series is that any signal which is periodic in time can be reexpressed as a collection of uniformly spaced spikes in frequency; that is,

$$\text{Periodic in Time} \Leftrightarrow \text{Uniform Sampling in Frequency.}$$

The same arguments show the basic result of sampling, which is that

$$\text{Uniform Sampling in Time} \Leftrightarrow \text{Periodic in Frequency.}$$

Thus, whenever a signal is uniformly sampled in time (say, with sampling interval T_s seconds), the spectrum will be periodic; that is, it will repeat every $f_s = 1/T_s$ Hz.

Two conventions are often observed when drawing periodic spectra that arise from sampling. First, the spectrum is usually drawn centered at 0 Hz. Thus, if the period of repetition is f_s, this is drawn from $-f_s/2$ to $f_s/2$, rather than from 0 to f_s. This makes sense because the spectra of individual real valued sinusoidal components contain two spikes symmetrically located around 0 Hz (as we saw in Section 3.2). Accordingly, the highest frequency that can be represented unambiguously is $f_s/2$, and this frequency is often called the *Nyquist* frequency f_{NYQ}.

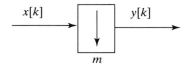

FIGURE 3.10 The discrete signal $x[k]$ is downsampled by a factor of m by removing all but one of every m samples.

The second convention is to draw only one period of the spectrum. After all, the others are identical copies that contain no new information. This is evident in the bottom right diagram of Figure 3.9 where the spectrum between $-3f_s/2$ and $-f_s/2$ is the same as the spectrum between $f_s/2$ and $3f_s/2$. In fact, we have been observing this convention throughout sections 3.2 and 3.3, since all of the figures of spectra (Figures 3.2, 3.3, 3.5, and 3.7) show just one period of the complete spectrum.

Perhaps you noticed that plotspec.m changes the frequency axis when the sampling interval Ts is changed. (If not, go back and redo Problem 3.1(c).) By the second convention, plotspec.m shows exactly one period of the complete spectrum. By the first convention, the plots are labelled from $-f_{NYQ}$ to f_{NYQ}.

What happens when the frequency of the signal is too high for the sampling rate? The representation becomes ambiguous. This is called *aliasing*, and is investigated by simulation in the problems below. Aliasing and other sampling-related issues (such as reconstructing an analog signal from its samples) are covered in more depth in Chapter 6.

Closely related to the digital sampling of an analog signal is the (digital) downsampling of a digital signal, which changes the rate at which the signal is represented. The simplest case downsamples by a factor of m, removing all but one out of every m samples. This is written

$$y[k] = x[mk + i],$$

where i is an integer between 0 and $m - 1$, and is commonly drawn in block form as in Figure 3.10. If the spectrum of $x[k]$ is bandlimited to $1/m$ of the Nyquist rate, then downsampling by m loses no information. Otherwise, aliasing occurs. Like analog-to-digital sampling, downsampling is a time varying operation.

Problems

3.11. Mimicking the code in speccos.m with the sampling interval Ts=1/100, find the spectrum of a cosine wave $\cos(2\pi f t)$ when f=30, 40, 49, 50, 51, 60 Hz. Which of these show aliasing?

3.12. Create a cosine wave with frequency 50 Hz. Plot the spectrum when this wave is sampled at Ts=1/50, 1/90, 1/100, 1/110, and 1/200. Which of these show aliasing?

3.13. Mimic the code in speccos.m with sampling interval Ts=1/100 to find the spectrum of a square wave with fundamental f=10, 20, 30, 33, 43 Hz. Can you predict where the spikes will occur in each case? Which of the square waves show aliasing?

3.5 The Fourth Element: Static Nonlinearities

Linear functions such as filters cannot add new frequencies to a signal (though they can remove unwanted frequencies). Even simple nonlinearities such as squaring and quantizing can and will add new frequencies. Some nonlinearities can be useful in the communication system in a variety of ways.

Perhaps the simplest nonlinearity is the square, which takes its input at each time instant and multiplies it by itself. Suppose the input is a sinusoid at frequency f, that is, $x(t) = \cos(2\pi f t)$. Then the output is the sinusoid squared, which can be rewritten using the cosine-cosine product (A.4) as

$$y(t) = x^2(t) = \cos^2(2\pi f t) = \frac{1}{2} + \frac{1}{2}\cos(2\pi(2f)t).$$

The spectrum of $y(t)$ has a spike at 0 Hz due to the constant, and a spike at $\pm 2f$ Hz from the double frequency term. Unfortunately, the action of a squaring element is not always as simple as this example might suggest. The following exercises encourage you to explore the kinds of changes that occur in the spectra when using a variety of simple nonlinear elements.

Problems

3.14. Mimic the code in `speccos.m` with `Ts=1/1000` to find the spectrum of the output $y(t)$ of a squaring block when the input is
   (a) $x(t) = \cos(2\pi f t)$ for $f = 100$ Hz,
   (b) $x(t) = \cos(2\pi f_1 t) + \cos(2\pi f_2 t)$ for $f_1 = 100$ and $f_2 = 150$ Hz,
   (c) a filtered noise sequence with nonzero spectrum between $f_1 = 100$ and $f_2 = 300$ Hz. Hint: generate the input by modifying `filternoise.m`.

   Can you explain the large DC (zero frequency) component?

3.15. Try different values of f_1 and f_2 in Problem 3.14. Can you predict what frequencies will occur in the output. When is aliasing an issue?

3.16. Repeat Problem 3.15 when the input is a sum of three sinusoids.

3.17. Suppose that the output of a nonlinear block is $y(t) = g(x(t))$, where

$$g(t) = \begin{cases} 1 & x(t) > 0 \\ -1 & x(t) \le 0 \end{cases}$$

is a quantizer that outputs positive one when the input is positive and outputs minus one when the input is negative. Find the spectrum of the output when the input is
   (a) $x(t) = \cos(2\pi f t)$ for $f = 100$ Hz,
   (b) $x(t) = \cos(2\pi f_1 t) + \cos(2\pi f_2 t)$ for $f_1 = 100$ and $f_2 = 150$ Hz.

3.18. Suppose that the output of a nonlinear block with input $x(t)$ is $y(t) = x^2(t)$. Find the spectrum of the output when the input is
   (a) $x(t) = \cos(2\pi f t + \phi)$ for $f = 100$ Hz and $\phi = 0.5$,
   (b) $x(t) = \cos(2\pi f_1 t) + \cos(2\pi f_2 t)$ for $f_1 = 100$ and $f_2 = 150$ Hz.

(c) $x(t) = \cos(2\pi f_1 t + \phi) + n(t)$ where $f_1 = 100$, $\phi = 0.5$, and where $n(t)$ is a white noise.

3.19. The MATLAB function `quantalph.m` (available on the CD) quantizes a signal to the nearest element of a desired set. Its help file reads

```
% function  y=quantalph(x,alphabet)
% quantize the input signal x to the alphabet
% using nearest neighbor method
% input x - vector to be quantized
% alphabet - vector of discrete values that y can take on
% sorted in ascending order
% output y - quantized vector
```

Let x be a random vector `x=randn(1,n)` of length n. Quantize x to the nearest $[-3, -1, 1, 3]$.
  (a)   What percentage of the outputs are 1's? 3's?
  (b)   Plot the magnitude spectrum of x and the magnitude spectrum of the output.
  (c)   Now let `x=3*randn(1,n)` and answer the same questions.

One of the most useful nonlinearities is multiplication by a cosine wave. As shown in Chapter 2, such modulation blocks can be used to change the frequency of a signal. The following MATLAB code implements a simple modulation nonlinearity.

---
modulate.m: change the frequency of the input

```
time=.5; Ts=1/10000;              % total time and sampling interval
t=Ts:Ts:time;                     % define a "time" vector
fc=1000; cmod=cos(2*pi*fc*t);     % create cos of freq fc
fi=100; x=cos(2*pi*fi*t);         % input is cos of freq fi
y=cmod.*x;                        % multiply input by cmod
figure(1), plotspec(cmod,Ts)      % find spectra and plot
figure(2), plotspec(x,Ts)
figure(3), plotspec(y,Ts)
```

The first four lines of the code create the modulating sinusoid (i.e., an oscillator). The next line specifies the input (in this case another cosine wave). The MATLAB syntax `.*` calculates a point-by-point multiplication of the two vectors `cmod` and `x`. The output of `modulate.m` is shown in Figure 3.11. The spectrum of the input contains spikes representing the input sinusoid at ± 100 Hz and the spectrum of the modulating sinusoid contains spikes at ± 1000 Hz. As expected from the modulation property of the transform, the output contains sinusoids at $\pm 1000 \pm 100$ Hz, which appear in the spectrum as the two pairs of spikes at ± 900 and ± 1100 Hz. Of course, this modulation can be applied to any signal, not just to an input sinusoid. In all cases, the output will contain two copies of the input, one shifted up in frequency and the other shifted down in frequency.

Problem

3.20. Mimic the code in `modulate.m` to find the spectrum of the output $y(t)$ of a modulator block (with modulation frequency $f_c = 1000$ Hz) when

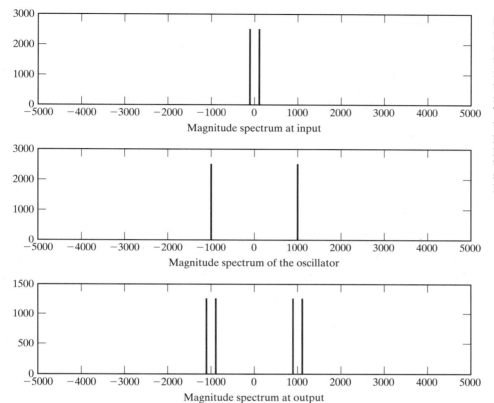

FIGURE 3.11 The spectrum of the input sinusoid is shown in the top figure. The middle figure shows the spectrum of the modulating wave. The bottom shows the spectrum of the point-by-point multiplication (in time) of the two, which is the same as their convolution (in frequency).

(a)   the input is $x(t) = \cos(2\pi f_1 t) + \cos(2\pi f_2 t)$ for $f_1 = 100$ and $f_2 = 150$ Hz,
(b)   the input is a square wave with fundamental $f = 150$ Hz,
(c)   the input is a noise signal with all energy below 300 Hz,
(d)   the input is a noise signal bandlimited to between 2000 and 2300 Hz,
(e)   the input is a noise signal with all energy below 1500 Hz.

3.6 The Fifth Element: Adaptation

Adaptation is a primitive form of learning. The adaptive elements of a communication system find approximate values of unknown parameters. A common strategy is to guess a value, to assess how good the guess is, and to then refine the estimate. Over time, the guesses (hopefully) converge to a useful estimate of the unknown value.

Figure 3.12 shows an adaptive element containing two parts. The adaptive subsystem parameterized by a changes the input into the output. The quality assessment mechanism monitors the output (and other relevant signals) and tries to determine whether a should be increased or decreased. The arrow through the system indicates that the a value is then adjusted accordingly.

Adaptive elements occur in a number of places in the communication system, including the following:

FIGURE 3.12 The adaptive element is a subsystem that transforms the input into the output (parameterized by a) and a quality assessment mechanism that evaluates how to alter a, in this case, whether to increase or decrease a.

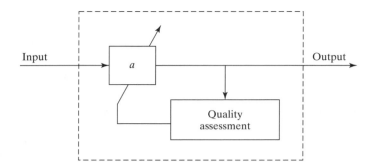

- In an automatic gain control, the "adaptive subsystem" is multiplication by a constant a. The quality assessment mechanism gauges whether the power at the output of the AGC is too large or too small, and adjusts a accordingly.

- In a phase-locked loop, the "adaptive subsystem" contains a sinusoid with an unknown phase shift a. The quality assessment mechanism adjusts a to maximize a filtered version of the product of the sinusoid and its input.

- In a timing recovery setting, the "adaptive subsystem" is a fractional delay given by a. The quality assessment mechanism monitors the power of the output, and adjusts a to maximize this power.

- In an equalizer, the "adaptive subsystem" is a linear filter parameterized by a set of a's. The quality assessment mechanism monitors the deviation of the output of the system from a target set and adapts the a's accordingly.

Chapter 6 provides an introduction to adaptive elements in communication systems, and a detailed discussion of their implementation is postponed until then.

3.7 Summary

The bewildering array of blocks and acronyms in a typical communication system diagram really consists of just a handful of simple elements: oscillators, linear filters, static nonlinearities, samplers, and adaptive elements. For the most part, these are ideas that the reader will have encountered to some degree in previous studies, but they have been summarized here in order to present them in the same form and using the same notation as in later chapters. In addition, this chapter has emphasized the "how-to" aspects by providing a series of MATLAB exercises, which will be useful when creating simulations of the various parts of a receiver.

3.8 For Further Reading

The intellectual background of the material presented here is often called *Signals and Systems*. One of the most accessible books is

- J. H. McClellan, R. W. Schafer, and M. A. Yoder, *DSP First: A Multimedia Approach*, Prentice Hall, 1998.

Other books provide greater depth and detail about the theory and uses of Fourier transforms. We recommend these as both background and supplementary reading:

- A. V. Oppenheim, A. S. Willsky, and S.H. Nawab, *Signals and Systems,* Second Edition, Prentice Hall, 1997.

- F. J. Taylor, *Signals and Systems*, McGraw-Hill, Inc., NY, 1994.

- S. Haykin and B. Van Veen, *Signals and Systems*, Wiley, 2002.

There are also many wonderful new books about digital signal processing, and these provide both depth and detail about basic issues such as sampling and filter design. Some of the best are the following:

- A. V. Oppenheim, R. W. Schafer, and J. R. Buck, *Discrete-Time Signal Processing*, Prentice Hall, 1999.

- B. Porat, *A Course in Digital Signal Processing*, Wiley, 1997.

- S. Mitra, *Digital Signal Processing: A Computer Based Approach*, McGraw-Hill, 1998.

Finally, since MATLAB is fundamental to our presentation, it is worth mentioning some books that describe the uses (and abuses) of the MATLAB language. Some are:

- V. Stonick and K. Bradley, *Labs for Signals and Systems Using MATLAB*, PWS Publishing, 1996.

- D. Hanselman and B. Littlefield, *Understanding MATLAB 6*, Prentice Hall, 2001.

- C. S. Burrus, J. H. McClellan, A. V. Oppenheim, T. W. Parks, R. W. Schafer, H. W. Schessler, *Computer-Based Exercises for Signal Processing Using MATLAB*, Prentice Hall, 1994.

The Idealized System Layer

The next layer encompasses Chapters 4 through 9. This gives a closer look at the idealized receiver—how things work when everything is just right: when the timing is known, when the clocks run at exactly the right speed, when there are no reflections, diffractions, or diffusions of the electromagnetic waves. This layer also introduces a few MATLAB tools that are needed to implement the digital radio. The order in which topics are discussed is precisely the order in which they appear in the receiver:

$$\text{channel} \rightarrow \begin{array}{c}\text{frequency}\\\text{translation}\end{array} \rightarrow \text{sampling} \rightarrow \begin{array}{c}\text{frequency}\\\text{translation}\end{array} \rightarrow$$

$$\textit{Chapter } 4 \qquad\qquad \textit{Chapter } 5 \qquad\qquad \textit{Chapter } 6$$

$$\underbrace{\begin{array}{c}\text{receive}\\\text{filtering}\end{array} \rightarrow \text{equalization}}_{\textit{Chapter } 7} \rightarrow \underbrace{\begin{array}{c}\text{decision}\\\text{device}\end{array} \rightarrow \text{decoding}}_{\textit{Chapter } 8} \rightarrow \begin{array}{c}\text{reconstructed}\\\text{message}\end{array}$$

Channel—impairments and linear systems   Chapter 4
Frequency translation—amplitude modulation and IF   Chapter 5
Sampling and automatic gain control   Chapter 6
Receive filtering—digital filtering   Chapter 7
Symbols to bits to signals   Chapter 8

Chapter 9 provides a complete (though idealized) software-defined digital radio system.

4 *Modelling Corruption*

From there to here, from here to there, funny things are everywhere.
—Dr. Seuss, *One Fish, Two Fish, Red Fish, Blue Fish,* 1960

If every signal that went from here to there arrived at its intended receiver unchanged, the life of a communications engineer would be easy. Unfortunately, the path between here and there can be degraded in several ways, including multipath interference, changing (fading) channel gains, interference from other users, broadband noise, and narrowband interference.

This chapter begins by describing these problems, which are diagrammed in Figure 4.1. More important than locating the sources of the problems is fixing them. The received signal can be processed using linear filters to help reduce the interferences and to undo, to some extent, the effects of the degradations. The central question is how to specify filters that can successfully mitigate these problems, and answering this requires a fairly detailed understanding of filtering. Thus, a discussion of linear filters occupies the bulk of this chapter, which also provides a background for other uses of filters throughout the receiver, such as the lowpass filters used in the demodulators of Chapter 5, the pulse shaping and matched filters of Chapter 11, and the equalizing filters of Chapter 13.

4.1 When Bad Things Happen to Good Signals

The path from the transmitter to the receiver is not simple, as Figure 4.1 suggests. Before the signal reaches the receiver, it is subject to a series of possible "funny things," events that may corrupt the signal and degrade the functioning of the receiver. This section discusses five kinds of corruption that are used throughout the chapter to motivate and explain the various purposes that linear filters may serve in the receiver.

4.1.1 Other Users

Many different users must be able to broadcast at the same time. This requires that there be a way for a receiver to separate the desired transmission from all the others (for instance, to tune to a particular radio or TV station among a large number that may be broadcasting simultaneously in the same geographical region). One standard method is to allocate different frequency bands to each user. This was called frequency division multiplexing (FDM) in Chapter 2, and was shown diagrammatically in Figure 2.3 on page 23. The signals from the different users can be separated using a bandpass filter, as

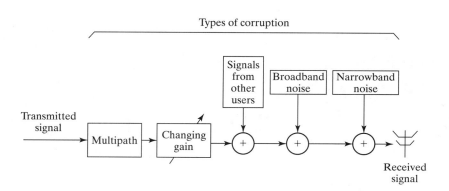

in Figure 2.4 on page 23. Of course, practical filters do not completely remove out-of-band signals, nor do they pass in-band signals completely without distortions. Recall the three filters in Figure 3.7 on page 47.

4.1.2 Broadband Noise

When the signal arrives at the receiver, it is small and must be amplified. While it is possible to build high-gain amplifiers, the noises and interferences will also be amplified along with the signal. In particular, any noise *in the amplifier itself* will be increased. This is often called "thermal noise" and is usually modelled as white (independent) broadband noise. Thermal noise is inherent in any electronic components and is caused by small random motions of electrons, like the Brownian motion of small particles suspended in water.

Such broadband noise is another reason that a bandpass filter is applied at the front end of the receiver. By applying a suitable filter, the total power in the noise (compared to the total power in the signal) can often be improved. Figure 4.2 shows the spectrum of the signal as a pair of triangles centered at the carrier frequency $\pm f_c$ with bandwidth $2B$. The total power in the signal is the area under the triangles. The spectrum of the noise is the flat region, and its power is the shaded area. After applying the bandpass filter, the power in the signal remains (more or less) unchanged, while the power in the noise is greatly reduced. Thus, the signal-to-noise ratio (SNR) improves.

4.1.3 Narrowband Noise

Noises are not always white; that is, the spectrum may not always be flat. Stray sine waves (and other signals with narrow spectra) may also impinge on the receiver. These may be caused by errant transmitters that accidently broadcast in the frequency range of the signal, or they may be harmonics of a lower frequency wave as it experiences nonlinear distortion. If these narrowband disturbances occur out of band, they will automatically be attenuated by the bandpass filter just as if they were a component of the wideband noise. However, if they occur in the frequency region of the signal, they decrease the SNR in proportion to their power. Judicious use of a "notch" filter (one designed to remove just the offending frequency) can be an effective tool.

Figure 4.3 shows the spectrum of the signal as the pair of triangles, along with three narrowband interferers represented by the three pairs of spikes. After the bandpass filter, the two pairs of out-of-band spikes are removed, but the in-band pair remains.

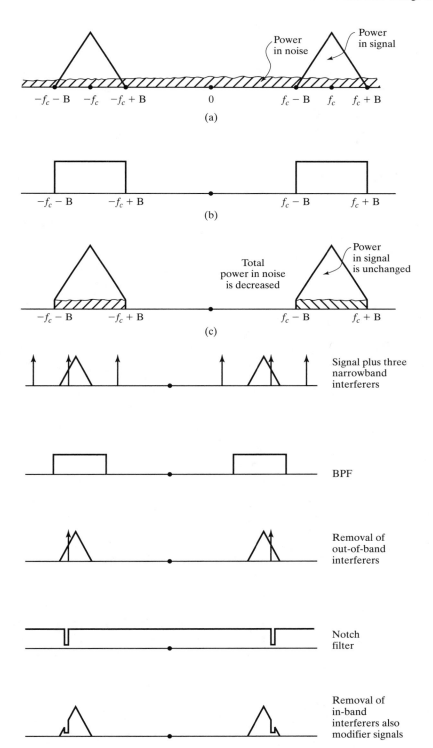

FIGURE 4.2 The signal-to-noise ratio is depicted graphically as the ratio of the power of the signal (the area under the triangles) to the power in the noise (the shaded area). After the bandpass filter, the power in the noise decreases, and so the SNR increases.

FIGURE 4.3 Three narrowband interferers are shown in the top figure (the three pairs of spikes). The BPF cannot remove the in-band interferer, though a narrow notch filter can, at the expense of changing the signal in the region where the narrow band noise occurred.

Applying a narrow notch filter tuned to the frequency of the interferer allows its removal, although this cannot be done without also affecting the signal somewhat.

4.1.4 Multipath Interference

In some situations, an electromagnetic wave can propagate directly from one place to another. For instance, when a radio signal from a spacecraft is transmitted back to Earth, the vacuum of space guarantees that the wave will arrive more or less intact (though greatly attenuated by distance). Often, however, the wave reflects, refracts, or diffracts, and the signal arriving is quite different from the one that was sent.

These distortions can be thought of as a combination of scaled and delayed reflections of the transmitted signal, which occur when there are different propagation paths from the transmitter to the receiver. Between two transmission towers, for instance, the paths may include one along the line-of-sight, reflections from the atmosphere, reflections from nearby hills, and bounces from a field or lake between the towers. For indoor digital TV reception, there are many (local) time-varying reflectors, including people in the receiving room, nearby vehicles, and the buildings of an urban environment. Figure 4.4, for instance, shows multiple reflections that arrive after bouncing off a cloud, after bouncing off a mountain, and others that are scattered by multiple bounces from nearby buildings.

The strength of the reflections depends on the physical properties of the reflecting surface, while the delay of the reflections is primarily determined by the length of the transmission path. Let $s(t)$ be the transmitted signal. If N delays are represented by

FIGURE 4.4 The received signal may be a combination of several copies of the original transmitted signal, each with a different attenuation and delay.

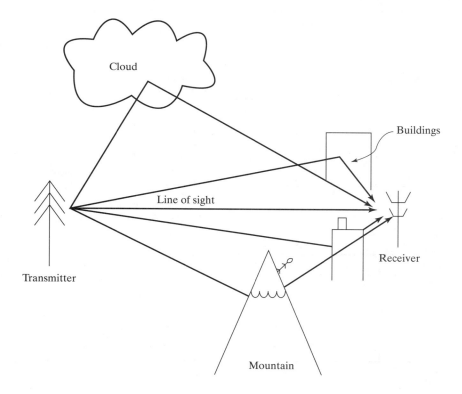

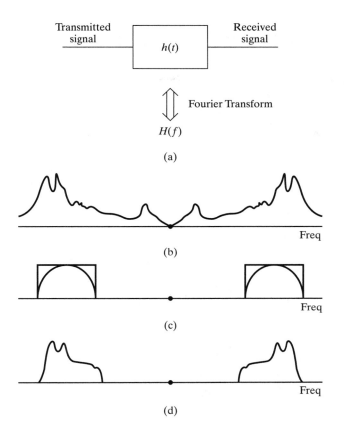

FIGURE 4.5 (a) The channel model (4.1) as a filter. (b) The frequency response of the filter. (c) The BPF and spectrum of the signal. The product of (b) and (c) gives (d), the distorted spectrum at the receiver.

Δ_1, Δ_2, ..., Δ_N, and the strengths of the reflections are h_1, h_2, ..., h_N, then the received signal $r(t)$ is

$$r(t) = h_1 s(t - \Delta_1) + h_2 s(t - \Delta_2) + \ldots + h_N s(t - \Delta_N). \qquad (4.1)$$

As will become clear in Section 4.4, this model of the channel has the form of a linear filter (since the expression on the right hand side is a convolution of the transmitted signal and the h_i's). This is shown in part (a) of Figure 4.5. Since this channel model is a linear filter, it can also be viewed in the frequency domain, and part (b) shows its frequency response. When this is combined with the BPF and the spectrum of the signal (shown in (c)), the result is the distorted spectrum shown in (d).

What can be done?

If the kinds of distortions introduced by the channel are known (or can somehow be determined), then the bandpass filter at the receiver can be modified in order to undo the effects of the channel. This can be seen most clearly in the frequency domain, as in Figure 4.6. Observe that the BPF is shaped (part (d)) to approximately invert the debilitating effects of the channel (part (a)) in the frequency band of the signal and to remove all the out-of-band frequencies. The resulting received signal spectrum (part (e)) is again a close copy of the transmitted signal spectrum, in stark contrast to the received signal spectrum in Figure 4.5 where no shaping was attempted.

FIGURE 4.6 (a) The frequency response of the channel. (b) The spectrum of the signal. (c) The product of (a) and (b) which is the spectrum of the received signal. (d) A BPF filter that has been shaped to undo the effect of the channel. (e) The product of (c) and (d), which combine to give a clean representation of the original spectrum of the signal.

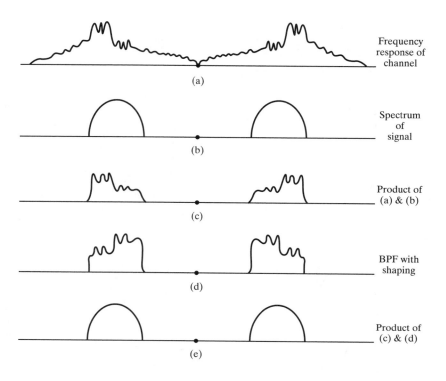

Thus, filtering in the receiver can be used to reshape the received signal within the frequency band of the transmission as well as to remove unwanted out-of-band frequencies.

4.1.5 Fading

Another kind of corruption that a signal may encounter on its journey from the transmitter to the receiver is called "fading," where the frequency response of the channel changes slowly over time. This may be caused because the transmission path changes. For instance, a reflection from a cloud might disappear when the cloud dissipates, an additional reflection might appear when a truck moves into a narrow city street, or in a mobile device such as a cell phone the operator might turn a corner and cause a large change in the local geometry of reflections. Fading may also occur when the transmitter and/or the receiver are moving. The Doppler effect shifts the frequencies slightly, causing interferences that may slowly change.

Such time-varying problems cannot be fixed by a single fixed filter; rather, the filter must somehow compensate differently at different times. This is an ideal application for the adaptive elements of Section 3.6, though results from the study of linear filters will be crucial in understanding how the time variations in the frequency response can be represented as time-varying coefficients in the filter that represents the channel.

4.2 Linear Systems: Linear Filters

Linear systems appear in many places in communication systems. The transmission channel is often modeled as a linear system as in (4.1). The bandpass filters used in the front end to remove other users (and to remove noises) are linear. Lowpass filters are crucial to the operation of the demodulators of Chapter 5. The equalizers of Chapter 13 are linear filters that are designed during the operation of the receiver on the basis of certain characteristics of the received signal.

Linear systems can be described in any one of three equivalent ways:

- The *impulse response* $h(t)$ is a function of time that defines the output of a linear system when the input is an impulse (or δ) function. When the input to the linear system is more complicated than a single impulse, the output can be calculated from the impulse response via the *convolution* operator.

- The *frequency response $H(f)$* is a function of frequency that defines how the spectrum of the input is changed into the spectrum of the output. The frequency response and the impulse response are intimately related: $H(f)$ is the Fourier transform of $h(t)$. Sometimes $H(f)$ is called the *transfer function*.

- A linear *difference or differential equation* (such as (4.1)) shows explicitly how the linear system can be implemented and can be useful in assessing stability and performance.

This chapter describes the three representations of linear systems and shows how they interrelate. The discussion begins by exploring the δ-function, and then showing how it is used to define the impulse response. The convolution property of the Fourier transform then shows that the transform of the impulse response describes how the system behaves in terms of the input and output spectra, and so it is called the frequency response. The final step is to show how the action of the linear system can be redescribed in the time domain as a difference (or as a differential) equation. This is postponed to Chapter 7, and is also discussed in some detail in Appendix F.

4.3 The Delta "Function"

One way to see how a system behaves is to kick it and see how it responds. Some systems react sluggishly, barely moving away from their resting state, while others respond quickly and vigorously. Defining exactly what is meant mathematically by a "kick" is trickier than it seems because the kick must occur over a very short amount of time, yet must be energetic in order to have any effect. This section defines the impulse (or delta) function $\delta(t)$, which is a useful "kick" for the study of linear systems.

The criterion that the impulse be energetic is translated to the mathematical statement that its integral over all time must be nonzero, and it is typically scaled to unity, that is,

$$\int_{-\infty}^{\infty} \delta(t)dt = 1. \tag{4.2}$$

The criterion that it occur over a very short time span is translated to the statement that, for every positive ϵ,

$$\delta(t) = \begin{cases} 0 & t < -\epsilon \\ 0 & t > \epsilon \end{cases} . \tag{4.3}$$

Thus, the impulse $\delta(t)$ is explicitly defined to be equal to zero for all $t \neq 0$. On the other hand, $\delta(t)$ is implicitly defined when $t = 0$ by the requirement that its integral be unity. Together, these guarantee that $\delta(t)$ is no ordinary function.[1]

The most important consequence of the definitions (4.2) and (4.3) is the *sifting property*

$$\int_{-\infty}^{\infty} w(t)\delta(t - t_0)dt = w(t)|_{t=t_0} = w(t_0), \tag{4.4}$$

which says that the delta function picks out the value of the function $w(t)$ from under the integral at exactly the time when the argument of the δ function is zero, that is, when $t = t_0$. To see this, observe that $\delta(t - t_0)$ is zero whenever $t \neq t_0$, and hence $w(t)\delta(t - t_0)$ is zero whenever $t \neq t_0$. Thus,

$$\int_{-\infty}^{\infty} w(t)\delta(t - t_0)dt = \int_{-\infty}^{\infty} w(t_0)\delta(t - t_0)dt$$

$$= w(t_0) \int_{-\infty}^{\infty} \delta(t - t_0)dt = w(t_0) \cdot 1 = w(t_0).$$

Sometimes it is helpful to think of the impulse as a limit. For instance, define the rectangular pulse of width $1/n$ and height n by

$$\delta_n(t) = \begin{cases} 0, & t < -1/2n \\ n, & -1/2n \leq t \leq 1/2n \\ 0, & t > 1/2n \end{cases}.$$

Then $\delta(t) = \lim_{n \to \infty} \delta_n(t)$ fulfills both criteria (4.2) and (4.3). Informally, it is not unreasonable to think of $\delta(t)$ as being zero everywhere except at $t = 0$, where it is infinite. While it is not really possible to "plot" the delta function $\delta(t - t_0)$, it can be represented in graphical form as zero everywhere except for an up-pointing arrow at t_0. When the δ function is scaled by a constant, the value of the constant is often placed in parenthesis near the arrowhead. Sometimes, when the constant is negative, the arrow is drawn pointing down. For instance, Figure 4.7 shows a graphical representation of the function $w(t) = \delta(t + 10) - 2\delta(t + 1) + 3\delta(t - 5)$.

What is the spectrum (Fourier transform) of $\delta(t)$? This can be calculated directly from the definition by replacing $w(t)$ in (2.1) with $\delta(t)$:

$$\mathcal{F}\{\delta(t)\} = \int_{-\infty}^{\infty} \delta(t)e^{-j2\pi ft}dt. \tag{4.5}$$

Apply the sifting property (4.4) with $w(t) = e^{-j2\pi ft}$ and $t_0 = 0$. Thus $\mathcal{F}\{\delta(t)\} = e^{-j2\pi ft}|_{t=0} = 1$.

Alternatively, suppose that δ is a function of frequency, that is, a spike at zero frequency. The corresponding time domain function can be calculated analogously using the definition of the inverse Fourier transform, that is, by substituting $\delta(f)$ for $W(f)$ in (A.16) and integrating:

$$\mathcal{F}^{-1}\{\delta(f)\} = \int_{-\infty}^{\infty} \delta(f)e^{j2\pi ft}df = e^{-j2\pi ft}|_{f=0} = 1.$$

Thus a spike at frequency zero is a "DC signal" (a constant) in time.

---

[1] The impulse is called a *distribution* and is the subject of considerable mathematical investigation.

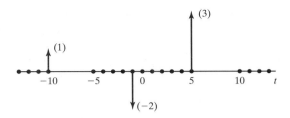

FIGURE 4.7 The function $w(t) = \delta(t+10) - 2\delta(t+1) + 3\delta(t-5)$ consisting of three weighted δ functions is represented graphically as three weighted arrows at $t = -10, -1, 5$, weighted by the appropriate constants.

The discrete time counterpart of $\delta(t)$ is the (discrete) delta function

$$\delta[k] = \begin{cases} 1 & k = 0 \\ 0 & k \neq 0 \end{cases} .$$

While there are a few subtleties (i.e., differences) between $\delta(t)$ and $\delta[k]$, for the most part they act analogously. For example, the program `specdelta.m` calculates the spectrum of the (discrete) delta function.

---

specdelta.m plot the spectrum of a delta function

```
time=2;                 % length of time
Ts=1/100;               % time interval between samples
t=Ts:Ts:time;           % create time vector
x=zeros(size(t));       % create signal of all zeros
x(1)=1;                 % delta function
plotspec(x,Ts)          % draw waveform and spectrum
```

The output of `specdelta.m` is shown in Figure 4.8. As expected from (4.5), the magnitude spectrum of the delta function is equal to 1 at all frequencies.

Problems

4.1. Calculate the Fourier transform of $\delta(t - t_0)$ from the definition. Now calculate it using the time shift property (A.38). Are they the same? Hint: They had better be.

4.2. Use the definition of the IFT (A.16) to show that

$$\delta(f - f_0) \Leftrightarrow e^{j2\pi f_0 t}.$$

4.3. Mimic the code in `specdelta.m` to find the spectrum of the discrete delta function when:
   (a) The delta does not occur at the start of x. Try `x[10]=1`, `x[100]=1`, and `x[110]=1`. How do the spectra differ? Can you use the time shift property (A.38) to explain what you see?
   (b) The delta changes magnitude x. Try `x[1]=10`, `x[10]=3`, and `x[110]= 0.1`. How do the spectra differ? Can you use the linearity property (A.31) to explain what you see?

FIGURE 4.8 A (discrete) delta function at time 0 has a magnitude spectrum equal to 1 for all frequencies.

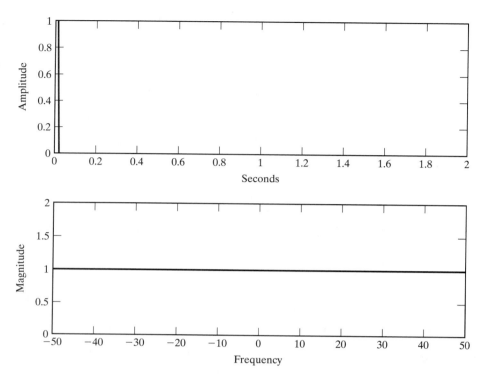

FIGURE 4.8 A (discrete) delta function at time 0 has a magnitude spectrum equal to 1 for all frequencies.

4.4. Mimic the code in `specdelta.m` to find the spectrum of a signal containing two delta functions when:

(a) The deltas are located at the start and the end (i.e., `x(1)=1; x(end)=1`).

(b) The deltas are located symmetrically from the start and end, for instance, `x(90)=1; x(end-90)=1`.

(c) The deltas are located arbitrarily, for instance, `x(33)=1; x(120)=1`.

4.5. Mimic the code in `specdelta.m` to find the spectrum of a train of equally spaced pulses. For instance, `x(1:20:end)=1` spaces the pulses 20 samples apart, and `x(1:25:end)=1` places the pulses 25 samples apart.

(a) Can you predict how far apart the resulting pulses in the spectrum will be?

(b) Show that

$$\sum_{k=-\infty}^{\infty} \delta(t - kT_s) \quad \Leftrightarrow \quad \frac{1}{T_s} \sum_{n=-\infty}^{\infty} \delta(f - nf_s), \qquad (4.6)$$

where $f_s = 1/T_s$. Hint: Let $w(t) = 1$ in (A.27) and (A.28).

(c) Now can you predict how far apart the pulses in the spectrum are? Your answer should be in terms of how far apart the pulses are in the time signal.

In Section 3.2, the spectrum of a sinusoid was shown to consist of two symmetrical spikes in the frequency domain, (recall Figure 3.5 on page 45). The next example shows why this is true by explicitly taking the Fourier transform.

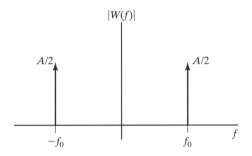

$|W(f)|$

$A/2$

$A/2$

$-f_0$

f_0

f

FIGURE 4.9 The magnitude spectrum of a sinusoid with frequency f_0 and amplitude A contains two δ function spikes, one at $f = f_0$ and the other at $f = -f_0$.

Example 4.1 Spectrum of a Sinusoid

Let $w(t) = A\sin(2\pi f_0 t)$, and use Euler's identity (A.3) to rewrite $w(t)$ as

$$w(t) = \frac{A}{2j}\left[e^{j2\pi f_0 t} - e^{-j2\pi f_0 t}\right].$$

Applying the linearity property (A.31) and the result of Problem 4.2 gives

$$\mathcal{F}\{w(t)\} = \frac{A}{2j}\left[\mathcal{F}\{e^{j2\pi f_0 t}\} - \mathcal{F}\{e^{-j2\pi f_0 t}\}\right]$$

$$= j\frac{A}{2}\left[-\delta(f - f_0) + \delta(f + f_0)\right]. \tag{4.7}$$

Thus, the magnitude spectrum of a sine wave is a pair of δ functions with opposite signs, located symmetrically about zero frequency, as shown in Figure 4.9. This magnitude spectrum is at the heart of one important interpretation of the Fourier transform: it shows the frequency content of any signal by displaying which frequencies are present (and which frequencies are absent) from the waveform. For example, Figure 4.10(a) shows the magnitude spectrum $W(f)$ of a real-valued signal $w(t)$. This can be interpreted as saying that $w(t)$ contains (or is made up of) "all the frequencies" up to B Hz, and that it contains no sinusoids with higher frequency. Similarly, the modulated signal $s(t)$ in Figure 4.10(b) contains all positive frequencies between $f_c - B$ and $f_c + B$, and no others.

Note that the Fourier transform in (4.7) is purely imaginary, as it must be because $w(t)$ is odd (see A.37). The phase spectrum is a flat line at $-90°$ because of the factor j.

Problems

4.6. What is the magnitude spectrum of $\sin(2\pi f_0 t + \theta)$? Hint: Use the frequency shift property (A.34). Show that the spectrum of $\cos(2\pi f_0 t)$ is $\frac{1}{2}(\delta(f - f_0) + \delta(f + f_0))$. Compare this analytical result to the numerical results from Problem 3.5.

FIGURE 4.10 The magnitude spectrum of a message signal $w(t)$ is shown in (a). When $w(t)$ is modulated by a cosine at frequency f_c, the spectrum of the resulting signal $s(t) = w(t)\cos(2\pi f_c t + \phi)$ is shown in (b).

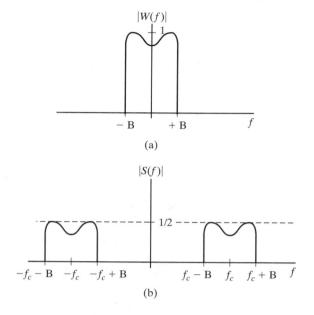

(a)

(b)

4.7. Let $w_i(t) = a_i \sin(2\pi f_i t)$ for $i = 1, 2, 3$. Without doing any calculations, write down the spectrum of $v(t) = w_1(t) + w_2(t) + w_3(t)$. Hint: Use linearity. Graph the magnitude spectrum of $v(t)$ in the same manner as in Figure 4.9. Verify your results with a simulation mimicking that in Problem 3.6.

4.8. Let $W(f) = \sin(2\pi f t_0)$. What is the corresponding time function?

4.4 Convolution in Time: It's What Linear Systems Do

Suppose that a system has impulse response $h(t)$, and that the input consists of a sum of three impulses occurring at times t_0, t_1, and t_2, with amplitudes a_0, a_1, and a_2 (for example, the signal $w(t)$ of Figure 4.7). By linearity of the Fourier transform, property (A.31), the output is a superposition of the outputs due to each of the input pulses. The output due to the first impulse is $a_0 h(t - t_0)$, which is the impulse response scaled by the size of the input and shifted to begin when the first input pulse arrives. Similarly, the outputs to the second and third input impulses are $a_1 h(t - t_1)$ and $a_2 h(t - t_2)$, respectively, and the complete output is the sum $a_0 h(t - t_0) + a_1 h(t - t_1) + a_2 h(t - t_2)$.

Now suppose that the input is a continuous function $x(t)$. At any time instant λ, the input can be thought of as consisting of an impulse scaled by the amplitude $x(\lambda)$, and the corresponding output will be $x(\lambda)h(t - \lambda)$, which is the impulse response scaled by the size of the input and shifted to begin at time λ. The complete output is then given by summing over all λ. Since there is a continuum of possible values of λ, this "sum" is actually an integral, and the output is

$$y(t) = \int_{-\infty}^{\infty} x(\lambda)h(t - \lambda)d\lambda \equiv x(t) * h(t). \tag{4.8}$$

This integral defines the convolution operator $*$ and provides a way of finding the output $y(t)$ of any linear system, given its impulse response $h(t)$ and the input $x(t)$.

MATLAB has several functions that simplify the numerical evaluation of convolutions. The most obvious of these is `conv`, which is used in `convolex.m` to calculate the convolution of an input x (consisting of two delta functions at times $t = 1$ and $t = 3$) and a system with impulse response h that is an exponential pulse. The convolution gives the output of the system.

---

convolex.m: example of numerical convolution

```
Ts=1/100; time=10;              % sampling interval and total time
t=0:Ts:time;                    % create time vector
h=exp(-t);                      % define impulse response
x=zeros(size(t));               % input is sum of two delta functions...
x(1/Ts)=3; x(3/Ts)=2;           % ...at times t=1 and t=3
y=conv(h,x);                    % do convolution
subplot(3,1,1), plot(t,x)       % and plot
subplot(3,1,2), plot(t,h)
subplot(3,1,3), plot(t,y(1:length(t)))
```

Figure 4.11 shows the input to the system in the top plot, the impulse response in the middle plot, and the output of the system in the bottom plot. Nothing happens before time $t = 1$, and the output is zero. When the first spike occurs, the system responds by jumping to 3 and then decaying slowly at a rate dictated by the shape of $h(t)$. The decay continues smoothly until time $t = 3$, when the second spike enters. At this point,

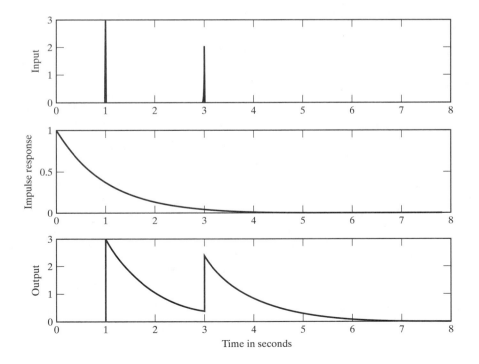

FIGURE 4.11 The convolution of the input (the top plot) with the impulse response of the system (the middle plot) gives the output in the bottom plot.

the output jumps up by 2, and is the sum of the response to the second spike, plus the remainder of the response to the first spike. Since there are no more inputs, the output slowly dies away.

Problems

4.9. Suppose that the impulse response $h(t)$ of a linear system is the exponential pulse

$$h(t) = \begin{cases} e^{-t} & t \geq 0 \\ 0 & t < 0 \end{cases}. \tag{4.9}$$

Suppose that the input to the system is $3\delta(t-1) + 2\delta(t-3)$. Use the definition of convolution (4.8) to show that the output is $3h(t-1) + 2h(t-3)$, where

$$h(t - t_0) = \begin{bmatrix} e^{-t+t_0} & t \geq t_0 \\ 0 & t < t_0 \end{bmatrix}.$$

How does your answer compare to Figure 4.11?

4.10. Suppose that a system has an impulse response that is an exponential pulse. Mimic the code in `convolex.m` to find its output when the input is a white noise (recall `specnoise.m` on page 41).

4.11. Mimic the code in `convolex.m` to find the output of a system when the input is an exponential pulse and the impulse response is a sum of two delta functions at times $t = 1$ and $t = 3$.

The next two problems show that linear filters commute with differentiation, and with each other.

4.12. Use the definition to show that convolution is commutative (i.e., that $w_1(t) * w_2(t) = w_2(t) * w_1(t)$). Hint: Apply the change of variables $\tau = t - \lambda$ in (4.8).

4.13. Suppose a filter has impulse response $h(t)$. When the input is $x(t)$, the output is $y(t)$. If the input is $x_d(t) = \frac{\partial x(t)}{\partial t}$, the output is $y_d(t)$. Show that $y_d(t)$ is the derivative of $y(t)$. Hint: Use (4.8) and the result of Problem 4.12.

4.14. Let $w(t) = \Pi\left(\frac{t}{T}\right)$ be the rectangular pulse of (2.7). What is $w(t) * w(t)$? Hint: A pulse shaped like a triangle.

4.15. Redo Problem 4.14 numerically by suitably modifying `convolex.m`. Let $T = 1.5$ seconds.

4.16. Suppose that a system has an impulse response that is a sinc function (as defined in (2.8)), and that the input to the system is a white noise (as in `specnoise.m` on page 41).
  (a) Mimic `convolex.m` to numerically find the output.
  (b) Plot the spectrum of the input and the spectrum of the output (using `plot-spec.m`). What kind of filter would you call this?

4.5 Convolution ⇔ Multiplication

While the convolution operator (4.8) describes mathematically how a linear system acts on a given input, time domain approaches are often not particularly revealing about the general behavior of the system. Who would guess, for instance in Problems 4.10 and 4.16, that convolution with exponentials and sinc functions would act like lowpass filters? By working in the frequency domain, however, the convolution operator is transformed into a simpler point-by-point multiplication, and the generic behavior of the system becomes clearer.

The first step is to understand the relationship between convolution in time, and multiplication in frequency. Suppose that the two time signals $w_1(t)$ and $w_2(t)$ have Fourier transforms $W_1(f)$ and $W_2(f)$. Then,

$$\mathcal{F}\{w_1(t) * w_2(t)\} = W_1(f)W_2(f). \tag{4.10}$$

To justify this property, begin with the definition of the Fourier transform (2.1) and apply the definition of convolution (4.8) to obtain

$$\mathcal{F}\{w_1(t) * w_2(t)\} = \int_{t=-\infty}^{\infty} w_1(t) * w_2(t) e^{-j2\pi ft} dt$$

$$= \int_{t=-\infty}^{\infty} \left[\int_{\lambda=-\infty}^{\infty} w_1(\lambda) w_2(t-\lambda) d\lambda \right] e^{-j2\pi ft} dt.$$

Reversing the order of integration and using the time shift property (A.38) produces

$$\mathcal{F}\{w_1(t) * w_2(t)\} = \int_{\lambda=-\infty}^{\infty} w_1(\lambda) \left[\int_{t=-\infty}^{\infty} w_2(t-\lambda) e^{-j2\pi ft} dt \right] d\lambda$$

$$= \int_{\lambda=-\infty}^{\infty} w_1(\lambda) \left[W_2(f) e^{-j2\pi f\lambda} \right] d\lambda$$

$$= W_2(f) \int_{\lambda=-\infty}^{\infty} w_1(\lambda) e^{-j2\pi f\lambda} d\lambda = W_1(f)W_2(f).$$

Thus, convolution in the time domain is the same as multiplication in the frequency domain. See (A.40).

The companion to the convolution property is the multiplication property, which says that multiplication in the time domain is equivalent to convolution in the frequency domain (see (A.41)); that is,

$$\mathcal{F}\{w_1(t)w_2(t)\} = W_1(f) * W_2(f) = \int_{-\infty}^{\infty} W_1(\lambda) W_2(f-\lambda) d\lambda. \tag{4.11}$$

The usefulness of these convolution properties is apparent when applying them to linear systems. Suppose that $H(f)$ is the Fourier transform of the impulse response $h(t)$. Suppose that $X(f)$ is the Fourier transform of the input $x(t)$ that is applied to the system. Then (4.8) and (4.10) show that the Fourier transform of the output is exactly equal to the product of the transforms of the input and the impulse response, that is,

$$Y(f) = \mathcal{F}\{y(t)\} = \mathcal{F}\{x(t) * h(t)\} = \mathcal{F}\{h(t)\}\mathcal{F}\{x(t)\} = H(f)X(f).$$

This can be rearranged to solve for

$$H(f) = \frac{Y(f)}{X(f)}, \qquad (4.12)$$

which is called the *frequency response* of the system because it shows, for each frequency f, how the system responds. For instance, suppose that $H(f_1) = 3$ at some frequency f_1. Then whenever a sinusoid of frequency f_1 is input into the system, it will be amplified by a factor of 3. Alternatively, suppose that $H(f_2) = 0$ at some frequency f_2. Then whenever a sinusoid of frequency f_2 is input into the system, it is removed from the output (because it has been multiplied by a factor of 0).

The frequency response shows how the system treats inputs containing various frequencies. In fact, this property was already used repeatedly in Chapter 1 when drawing curves that describe the behavior of lowpass and bandpass filters. For example, the filters of Figures 2.5, 2.4, and 2.6 are used to remove unwanted frequencies from the communications system. In each of these cases, the plot of the frequency response describes concretely and concisely how the system (or filter) affects the input, and how the frequency content of the output relates to that of the input. Sometimes, the frequency response $H(f)$ is called the *transfer function* of the system, since it "transfers" the input $x(t)$ (with transform $X(f)$) into the output $y(t)$ (with transform $Y(f)$).

Thus, the impulse response describes how a system behaves directly in time, while the frequency response describes how it behaves in frequency. The two descriptions are intimately related because the frequency response is the Fourier transform of the impulse response. This will be used repeatedly in Section 7.2 to design filters for the manipulation (augmentation or removal) of specified frequencies.

Example 4.2

In Problem 4.16, a system was defined to have an impulse response that is a sinc function. The Fourier transform of a sinc function in time is a rect function in frequency (A.22). Hence, the frequency response of the system is a rectangle that passes all frequencies below $f_c = 1/T$ and removes all frequencies above (i.e., the system is a lowpass filter).

MATLAB can help to visualize the relationship between the impulse response and the frequency response. For instance, the system in `convolex.m` is defined via its impulse response, which is a decaying exponential. Figure 4.11 shows its output when the input is a simple sum of delta functions, and Problem 4.10 explores the output when the input is a white noise. In `freqresp.m`, the behavior of this system is explained by looking at its frequency response.

---

freqresp.m: numerical example of impulse and frequency response

```
Ts=1/100; time=10;        % sampling interval and total time
t=0:Ts:time;              % create time vector
h=exp(-t);                % define impulse response
plotspec(h,Ts)            % find and plot frequency response
```

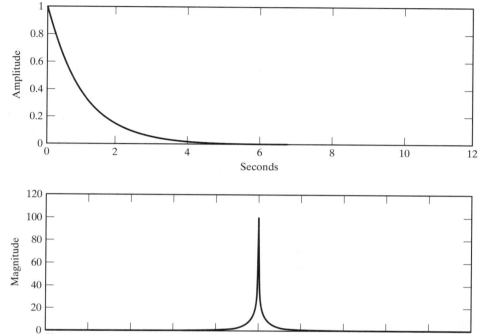

FIGURE 4.12 The action of a system in time is defined by its impulse response (in the top plot). The action of the system in frequency is defined by its frequency response (in the bottom plot), a kind of lowpass filter.

The output of `freqresp.m` is shown in Figure 4.12. The frequency response of the system (which is just the magnitude spectrum of the impulse response) is found using `plotspec.m`. In this case, the frequency response amplifies low frequencies and attenuates other frequencies more as the frequency increases. This explains, for instance, why the output of the convolution in Problem 4.10 contained (primarily) lower frequencies, as evidenced by the slower undulations in time.

Problems

4.17. Suppose a system has an impulse response that is a sinc function. Using `freqresp.m`, find the frequency response of the system. What kind of filter does this represent? Hint: center the sinc in time; for instance, use `h=sinc(10*(t-time/2))`.

4.18. Suppose a system has an impulse response that is a sin function. Using `freqresp.m`, find the frequency response of the system. What kind of filter does this represent? Can you predict the relationship between the frequency of the sine wave and the location of the peaks in the spectrum? Hint: try `h=sin(25*t)`.

4.19. Create a simulation (analogous to `convolex.m`) that inputs white noise into a system with impulse response that is a sinc function (as in Problem 4.17). Calculate the spectra of the input and output using `plotspec.m`. Verify that the system behaves as suggested by the frequency response in Problem 4.17.

4.20. Create a simulation (analogous to `convolex.m`) that inputs white noise into a system with impulse response that is a sin function (as in Problem 4.18). Calculate the spectra of the input and output using `plotspec.m`. Verify that the system behaves as suggested by the frequency response in Problem 4.18.

So far, Section 4.5 has emphasized the idea of finding the frequency response of a system as a way to understand its behavior. Reversing things suggests another use. Suppose it was necessary to build a filter with some special characteristic in the frequency domain (for instance, in order to accomplish one of the goals of bandpass filtering in Section 4.1). It is easy to specify the filter in the frequency domain. Its impulse response can then be found by taking the inverse Fourier transform, and the filter can be implemented using convolution. Thus, the relationship between impulse response and frequency response can be used both to study and to design systems.

In general, this method of designing filters is not optimal (in the sense that other design methods can lead to more efficient designs), but it does show clearly what the filter is doing, and why. Whatever the design procedure, the representation of the filter in the time domain and its representation in the frequency domain are related by nothing more than a Fourier transform.

4.6 Improving SNR

Section 4.1 described several kinds of corruption that a signal may encounter as it travels from the transmitter to the receiver. This section shows how linear filters can help. Perhaps the simplest way a linear bandpass filter can be used is to remove broadband noise from a signal. (Recall Section 4.1.2 and especially Figure 4.2.)

A common way to quantify noise is the signal-to-noise ratio (SNR) which is the ratio of the power of the signal to the power of the noise at a given point in the system. If the SNR at one point is larger than the SNR at another point, then the performance is better because there is more signal in comparison to the amount of noise. For example, consider the SNR at the input and the output of a BPF as shown in Figure 4.13. The signal at the input ($r(t)$ in part (a)) is composed of the message signal $x(t)$ and the noise signal $n(t)$, and the SNR at the input is therefore

$$\text{SNR}_{\text{input}} = \frac{\text{power in } x(t)}{\text{power in } n(t)}.$$

Similarly, the output $y(t)$ is composed of a filtered version of the message ($y_x(t)$ in part (b)) and a filtered version of the noise ($y_n(t)$ in part (b)). The SNR at the output can therefore be calculated as

$$\text{SNR}_{\text{output}} = \frac{\text{power in } y_x(t)}{\text{power in } y_n(t)}.$$

Observe that the SNR at the output cannot be calculated directly from $y(t)$ (since the two components are scrambled together). But, since the filter is linear,

$$y(t) = BPF\{x(t) + n(t)\} = BPF\{x(t)\} + BPF\{n(t)\} = y_x + y_n,$$

which effectively shows the equivalence of parts (a) and (b) of Figure 4.13.

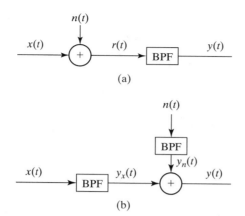

FIGURE 4.13 Two equivalent ways to draw the same system. In part (a) it is easy to calculate the SNR at the input, while the alternative form (b) allows easy calculation of the SNR at the output of the BPF.

The MATLAB program `improvesnr.m` explores this scenario concretely. The signal `x` is a bandlimited signal, containing only frequencies between 3000 and 4000 Hz. This is corrupted by a broadband noise `n` (perhaps caused by an internally generated thermal noise) to form the received signal. The SNR of this input `snrinp` is calculated as the ratio of the power of the signal `x` to the power of the noise `n`. The output of the BPF at the receiver is `y`, which is calculated as a BPF version of `x+n`. The BPF is created using the `remez` command just like the bandpass filter in `filternoise.m` on page 46. To calculate the SNR of `y`, however, the code also implements the system in the alternative form of part (b) of Figure 4.13. Thus, `yx` and `yn` represent the signal `x` filtered through the BPF and the noise `n` passed through the same BPF. The SNR at the output is then the ratio of the power in `yx` to the power in `yn`, which is calculated using the function `pow.m`, available on the CD.

---

improvesnr.m: using a linear filter to improve SNR

```
time=3; Ts=1/20000;              % length of time and sampling interval
b=remez(100,[0 0.29 0.3 0.4 0.41 1],[0 0 1 1 0 0]); % BP filter
n=0.25*randn(1,time/Ts);         % generate white noise signal
x=filter(b,1,2*randn(1,time/Ts));  % bandlimited signal between 3K and 4K
y=filter(b,1,x+n);               % (a) filter the received signal+noise
yx=filter(b,1,x); yn=filter(b,1,n); % (b) filter signal and noise separately
z=yx+yn;                         % add them
diffzy=max(abs(z-y))             % and make sure y and z are equal
snrinp=pow(x)/pow(n)             % SNR at input
snrout=pow(yx)/pow(yn)           % SNR at output
```

Since the data generated in `improvesnr.m` is random, the numbers are slightly different each time the program is run. Using the default values, the SNR at the input is about 7.8, while the SNR at the output is about 61. This is certainly a noticeable improvement. The variable `diffzy` shows the largest difference between the two ways of calculating the output (that is, between parts (a) and (b) of Figure 4.13). This is on the order of 10^{-15}, which is effectively the numerical resolution of MATLAB calculations, indicating that the two are (effectively) the same.

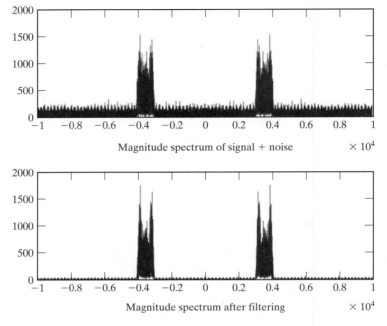

FIGURE 4.14 The spectrum of the input to the BPF is shown in the top plot. The spectrum of the output is shown in the bottom. The overall improvement in SNR is clear.

Figure 4.14 plots the spectra of the input and the output of a typical run of `improvesnr.m`. Observe the large noise floor in the top plot, and how this is reduced by passage through the BPF. Observe also that the signal is still changed by the noise in the pass band between 3000 and 4000 Hz, since the BPF has no effect there.

The program `improvesnr.m` can be thought of as a simulation of the effect of having a BPF at the receiver for the purposes of improving the SNR when the signal is corrupted by broadband noise, as was described in Section 4.1.2. The following problems ask you to mimic the code in `improvesnr.m` to simulate the benefit of applying filters to the other problems presented in Section 4.1.

Problems

4.21. Suppose that the noise in `improvesnr.m` is replaced with narrowband noise (as discussed in Section 4.1.3). Investigate the improvements in SNR

    (a) when the narrowband interference occurs outside the 3000 to 4000 Hz pass band.

    (b) when the narrowband interference occurs inside the 3000 to 4000 Hz pass band.

4.22. Suppose that the noise in `improvesnr.m` is replaced with "other users" who occupy different frequency bands (as discussed in Section 4.1.1). Are there improvements in the SNR?

The other two problems posed in Section 4.1 were multipath interference and fading. These require more sophisticated processing because the design of the filters depends on the operating circumstances of the system. These situations will be discussed in detail in Chapters 6 and 13.

4.7 For Further Reading

An early description of the linearity of communication channels can be found in

- P. A. Bello, "Characterization of Randomly Time-Variant Linear Channels," *IEEE Transactions on Communication Systems*, December 1963.

5 Analog (De)modulation

Beam me up, Scotty.
—attributed to James T. Kirk, Starship *Enterprise*, "Star Trek"

Several parts of a communication system modulate the signal and change the underlying frequency band in which the signal lies. These frequency changes must be reversible; after processing, the receiver must be able to reconstruct (a close approximation to) the transmitted signal.

The input message $w(kT)$ in Figure 5.1 is a discrete-time sequence drawn from a finite alphabet. The ultimate output $m(kT)$ produced by the decision device (or quantizer) is also discrete-time and is drawn from the same alphabet. If all goes well and the message is transmitted, received, and decoded successfully, then the output should be the same as the input, although there may be some delay δ between the time of transmission and the time when the output is available. Though the system is digital in terms of the message communicated and the performance assessment, the middle of the system is inherently analog from the (pulse-shaping) filter of the transmitter to the sampler at the receiver.

At the transmitter in Figure 5.1, the digital message has already been turned into an analog signal by the pulse shaping (which was discussed briefly in Section 2.10 and is considered in detail in Chapter 11). For efficient transmission, the analog version of the message must be shifted in frequency, and this process of changing frequencies is called modulation or upconversion. At the receiver, the frequency must be shifted back down, and this is called demodulation or downconversion. Sometimes the demodulation is done in one step (all analog) and sometimes the demodulation proceeds in two steps; an analog downconversion to the intermediate frequency and then a digital downconversion to the baseband. This two step procedure is shown in Figure 5.1.

There are many ways that signals can be modulated. Perhaps the simplest is *amplitude modulation*, which is discussed in two forms (large and small carrier) in the next two sections. This is generalized to the simultaneous transmission of two signals using *quadrature modulation* in Section 5.3, and it is shown that quadrature modulation uses bandwidth more efficiently than amplitude modulation. This gain in efficiency can also be obtained using *single sideband* and *vestigial sideband* methods, which are discussed in the document titled *Other Modulations*, available on the CD-ROM. Demodulation can also be accomplished using sampling as discussed in Section 6.2, and amplitude modulation can also be accomplished with a simple squaring and filtering operation as in Problem 5.8.

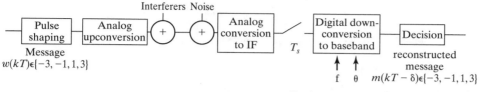

FIGURE 5.1 Digital communication system.

Throughout, the chapter contains a series of exercises that prepare readers to create their own modulation and demodulation routines in MATLAB. These lie at the heart of the software receiver that will be assembled in Chapters 9 and 15.

5.1 Amplitude Modulation with Large Carrier

Perhaps the simplest form of (analog) transmission system modulates the message signal by a high frequency carrier in a two-step procedure: multiply the message by the carrier, then add the product to the carrier. At the receiver, the message can be demodulated by extracting the envelope of the received signal.

Consider the transmitted/modulated signal

$$v(t) = A_c w(t)\cos(2\pi f_c t) + A_c\cos(2\pi f_c t)$$

diagrammed in Figure 5.2. The process of multiplying the signal in time by a (co)sinusoid is called *mixing*. This can be rewritten in the frequency domain by mimicking the development from (2.2) to (2.4) on page 21. Using the frequency shift property of the Fourier transform (A.33) and the transform of the cosine (A.18), the Fourier transform of $v(t)$ is

$$V(f) = \frac{1}{2}A_c W(f + f_c) + \frac{1}{2}A_c W(f - f_c) + \frac{1}{2}A_c\delta(f - f_c) + \frac{1}{2}A_c\delta(f + f_c). \quad (5.1)$$

The spectra of $|W(f)|$ and $|V(f)|$ are sketched in Figure 5.3. The vertical arrows in the bottom figure represent the transform of the cosine carrier at frequency f_c (i.e., a pair of delta functions at $\pm f_c$). The scaling by $\frac{A_c}{2}$ is indicated next to the arrowheads.

If $w(t) \geq -1$, then the envelope of $v(t)$ is the same as $w(t)$ and an *envelope detector* can be used as a demodulator (envelopes are discussed in detail in Appendix C). An example is given in the following MATLAB program. The "message" signal is a sinusoid with a drift in the DC offset, and the carrier wave is at a much higher frequency.

AMlarge.m: large carrier AM demodulated with "envelope"

```
time=.33; Ts=1/10000;                    % sampling interval and time
t=0:Ts:time; lent=length(t);             % define a "time" vector
fc=1000; c=cos(2*pi*fc*t);               % define carrier at freq fc
fm=20; w=10/lent*[1:lent]+cos(2*pi*fm*t); % create "message" > -1
v=c.*w+c;                                % modulate with large carrier
fbe=[0 0.05 0.1 1]; damps=[1 1 0 0]; fl=100; % low pass filter design
b=remez(fl,fbe,damps);                   % impulse response of LPF
envv=(pi/2)*filter(b,1,abs(v));          % find envelope
```

FIGURE 5.2 Large
carrier amplitude
modulation.

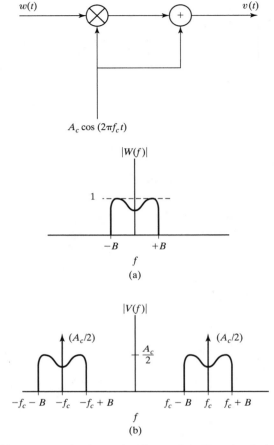

FIGURE 5.3 Spectrum
of large carrier ampli-
tude modulation.

The output of this program is shown in Figure 5.4. The slowly increasing sinusoidal "message" $w(t)$ is modulated by the carrier $c(t)$ at $f_c = 1000$ Hz. The heart of the modulation is the point-by-point multiplication of the message and the carrier in the fifth line. This product $v(t)$ is shown in Figure 5.4(c). The enveloping operation is accomplished by applying a lowpass filter to $2v(t)e^{j2\pi f_c t}$ (as discussed in Appendix C), which recovers the original message signal, though it is offset by 1 and delayed by the linear filter.

Problems

5.1. Using `AMlarge.m`, plot the spectrum of the message $w(t)$, the spectrum of the carrier $c(t)$, and the spectrum of the received signal $v(t)$. What is the spectrum of the envelope? How close are your results to the theoretical predictions in (5.1)?

5.2. One of the advantages of transmissions using AM with large carrier is that there is no need to know the (exact) phase or frequency of the transmitted signal. Verify this using `AMlarge.m`.

    (a) Change the phase of the transmitted signal; for instance, let `c=cos(2*pi* fc*t+phase)` with `phase=0.1, 0.5, pi/3, pi/2, pi`, and verify that the recovered envelope remains unchanged.

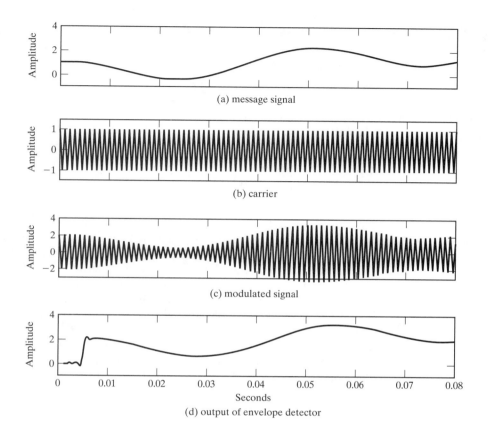

FIGURE 5.4 An undulation message (top) is modulated by a carrier (b). The composite signal is shown in (c), and the output of an envelope detector is shown in (d).

(a) message signal

(b) carrier

(c) modulated signal

(d) output of envelope detector

(b) Change the frequency of the transmitted signal; for instance, let `c=cos(2*pi*(fc+g)*t)` with `g=10, -10, 100, -100,` and verify that the recovered envelope remains unchanged. Can `g` be too large?

5.3. Create your own message signal $w(t)$, and rerun `AMlarge.m`. Repeat Problem 5.1 with this new message. What differences do you see?

5.4. In `AMlarge.m`, verify that the original message `w` and the recovered envelope `envv` are offset by 1, except at the end points where the filter does not have enough data. Hint: the delay induced by the linear filter is approximately `fl/2`.

The principle advantage of transmission systems that use AM with a large carrier is that exact synchronization is not needed; the phase and frequency of the transmitter need not be known at the receiver, as was demonstrated in Problem 5.2. This means that the receiver can be simpler than when synchronization circuitry is required. The main disadvantage is that adding the carrier into the signal increases the power needed for transmission but does not increase the amount of useful information transmitted. Here is a clear engineering tradeoff; the value of the wasted signal strength must be balanced against the cost of the receiver.

5.2 Amplitude Modulation with Suppressed Carrier

It is also possible to use AM without adding the carrier. Consider the transmitted/modulated signal

$$v(t) = A_c w(t)\cos(2\pi f_c t)$$

diagrammed in Figure 5.5(a), in which the message $w(t)$ is mixed with the cosine carrier. Direct application of the frequency shift property of Fourier transforms (A.33) shows that the spectrum of the received signal is

$$V(f) = \frac{1}{2}A_c W(f + f_c) + \frac{1}{2}A_c W(f - f_c).$$

As with AM with large carrier, the upconverted signal $v(t)$ for AM with suppressed carrier has twice the bandwidth of the original message signal. If the original message occupies the frequencies between $\pm B$ Hz, then the modulated message has support between $f_c - B$ and $f_c + B$, a bandwidth of $2B$. See Figure 4.10.

As illustrated in (2.5) on page 26, the received signal can be demodulated by mixing with a cosine that has the same frequency and phase as the modulating cosine, and the original message can then be recovered by low pass filtering. But, as a practical matter, the frequency and phase of the modulating cosine (located at the transmitter) can never be known exactly at the receiver.

Suppose that the frequency of the modulator is f_c but that the frequency at the receiver is $f_c + \gamma$, for some small γ. Similarly, suppose that the phase of the modulator is 0 but that the phase at the receiver is ϕ. Figure 5.5(b) shows this downconverter, which can be described by

$$x(t) = v(t)\,\cos(2\pi(f_c + \gamma)t + \phi) \tag{5.2}$$

and

$$m(t) = \text{LPF}\{x(t)\},$$

where LPF represents a lowpass filtering of the demodulated signal $x(t)$ in an attempt to recover the message. Thus, the downconversion described in (5.3) acknowledges that the

FIGURE 5.5 AM suppressed carrier communication system: (a) the transmitter/modulator, (b) the receiver/demodulator.

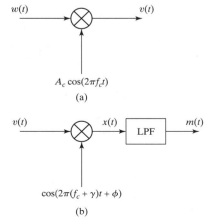

receiver's local oscillator may not have the same frequency or phase as the transmitter's local oscillator. In practice, accurate *a priori* information is available for carrier frequency, but (relative) phase could be anything, since it depends on the distance between the transmitter and receiver as well as when the transmission begins. Because the frequencies are high, the wavelengths are small and even small motions can change the phase significantly.

The remainder of this section investigates what happens when the frequency and phase are not known exactly, that is, when either γ or ϕ (or both) are nonzero. Using the frequency shift property of Fourier transforms (5.3) on $x(t)$ in (5.2) produces

$$X(f) = \frac{A_c}{4} \left[e^{j\phi}\{W(f + f_c - (f_c + \gamma)) + W(f - f_c - (f_c + \gamma))\} \right.$$
$$+ e^{-j\phi}\{W(f + f_c + (f_c + \gamma)) + W(f - f_c + (f_c + \gamma))\} \Big]$$
$$= \frac{A_c}{4} \left[e^{j\phi} W(f - \gamma) + e^{j\phi} W(f - 2f_c - \gamma) \right.$$
$$+ e^{-j\phi} W(f + 2f_c + \gamma) + e^{-j\phi} W(f + \gamma) \Big]. \tag{5.3}$$

If there is no frequency offset (i.e., if $\gamma = 0$), then

$$X(f) = \frac{A_c}{4} \left[(e^{j\phi} + e^{-j\phi}) W(f) + e^{j\phi} W(f - 2f_c) + e^{-j\phi} W(f + 2f_c) \right].$$

Because $\cos(x) = (1/2)(e^{jx} + e^{-jx})$ from (A.2), this can be rewritten

$$X(f) = \frac{A_c}{2} W(f)\cos(\phi) + \frac{A_c}{4} \left[e^{j\phi} W(f - 2f_c) + e^{-j\phi} W(f + 2f_c) \right].$$

Thus, a perfect lowpass filtering of $x(t)$ with cutoff below $2f_c$ removes the high frequency portions of the signal near $\pm 2f_c$ to produce

$$m(t) = \frac{A_c}{2} w(t) \cos(\phi). \tag{5.4}$$

The factor $\cos(\phi)$ attenuates the received signal (except for the special case when $\phi = 0 \pm 2\pi k$ for integers k). If ϕ were sufficiently close to $0 \pm 2\pi k$ for some integer k, then this would be tolerable. But there is no way to know the relative phase, and hence $\cos(\phi)$ can assume any possible value within $[-1, 1]$. The worst case occurs as ϕ approaches $\pm \pi/2$, when the message is attenuated to zero! A scheme for carrier phase synchronization, which automatically tries to align the phase of the cosine at the receiver with the phase at the transmitter is vital. This is discussed in detail in Chapter 10.

To continue the investigation, suppose that the carrier phase offset is zero, (i.e., $\phi = 0$), but that the frequency offset γ is not. Then the spectrum of $x(t)$ from (5.3) is

$$X(f) = \frac{A_c}{4} \left[W(f - \gamma) + W(f - 2f_c - \gamma) + W(f + 2f_c + \gamma) + W(f + \gamma) \right],$$

and the lowpass filtering of $x(t)$ produces

$$M(f) = \frac{A_c}{4} \left[W(f - \gamma) + W(f + \gamma) \right].$$

FIGURE 5.6 When
there is a carrier fre-
quency offset in the
receiver oscillator, the
two images of $W(\cdot)$
do not align properly.
Their sum is *not* equal
to $\frac{A_c}{2} W(f)$.

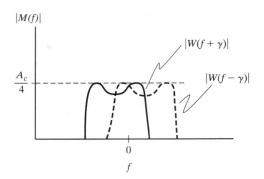

This is shown in Figure 5.6. Recognizing this spectrum as a frequency shifted version of $w(t)$, it can be translated back into the time domain using (A.33) to give

$$m(t) = \frac{A_c}{2} w(t) \, \cos(2\pi\gamma t). \tag{5.5}$$

Instead of recovering the message $w(t)$, the frequency offset causes the receiver to recover a low frequency amplitude modulated version of it. This is bad with even a small carrier frequency offset. While $\cos(\phi)$ in (5.4) is a fixed scaling, $\cos(2\pi\gamma t)$ in (5.5) is a time-varying scaling that will alternately recover $m(t)$ (when $\cos(2\pi\gamma t) \approx 1$) and make recovery impossible (when $\cos(2\pi\gamma t) \approx 0$). Transmitters are typically expected to maintain suitable accuracy to a nominal carrier frequency setting known to the receiver. Ways of automatically tracking (inevitable) small frequency deviations are discussed at length in Chapter 10.

The following code AM.m generates a message $w(t)$ and modulates it with a carrier at frequency f_c. The demodulation is done with a cosine of frequency $f_c + \gamma$ and a phase offset of ϕ. When $\gamma = 0$ and $\phi = 0$, the output (a lowpass version of the demodulated signal) is nearly identical to the original message, except for the inevitable delay caused by the linear filter. Figure 5.7 shows four plots: the message $w(t)$ on top, followed by the upconverted signal $v(t) = w(t)\cos(2\pi f_c t)$, followed in turn by the downconverted signal $x(t)$. The lowpass filtered version is shown in the bottom plot; observe that it is nearly identical to the original message, albeit with a slight delay.

AM.m suppressed carrier with (possible) freq and phase offset

```
time=.3; Ts=1/10000;                              % sampling interval and time base
t=Ts:Ts:time; lent=length(t);                     % define a "time" vector
fc=1000; c=cos(2*pi*fc*t);                         % define the carrier at freq fc
fm=20; w=5/lent*(1:lent)+cos(2*pi*fm*t);           % create "message"
v=c.*w;                                            % modulate with carrier
gamma=0; phi=0;                                    % freq & phase offset
c2=cos(2*pi*(fc+gamma)*t+phi);                     % create cosine for demod
x=v.*c2;                                           % demod received signal
fbe=[0 0.1 0.2 1]; damps=[1 1 0 0]; fl=100;        % low pass filter design
b=remez(fl,fbe,damps);                             % impulse response of LPF
m=2*filter(b,1,x);                                 % LPF the demodulated signal
```

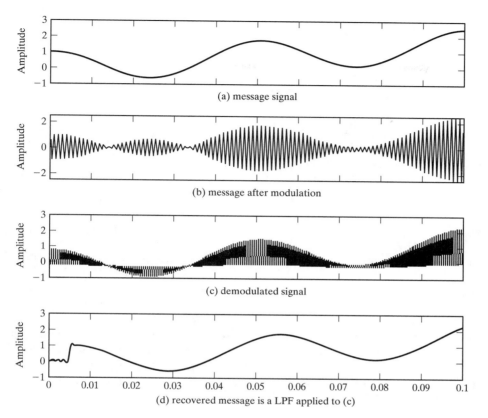

FIGURE 5.7 The message signal in the top frame is modulated to produce the signal in the second plot. Demodulation gives the signal in the third plot, and the LPF recovers the original message (with delay) in the bottom plot.

Problems

5.5. Using AM.m as a starting point, plot the spectra of $w(t)$, $v(t)$, $x(t)$, and $m(t)$.

5.6. Try different phase offsets $\phi = [-\pi, -\pi/2, -\pi/3, -\pi/6, 0, \pi/6, \pi/3, \pi/2, \pi]$. How well does the recovered message $m(t)$ match the actual message $w(t)$? For each case, what is the spectrum of $m(t)$?

5.7. Try different frequency offsets $\gamma = [.01, 0.1, 1.0, 10]$. How well does the recovered message $m(t)$ match the actual message $w(t)$? For each case, what is the spectrum of $m(t)$? Hint: look over more than just the first $1/10$ second to see the effect.

5.8. Consider the system shown in Figure 5.8. Show that the output of the system is $A_0 w(t) \cos(2\pi f_0 t)$, as indicated.

5.9. Create a MATLAB routine to implement the square-law mixing modulator of Figure 5.8.
   (a) Create a signal $w(t)$ that has bandwidth 100 Hz
   (b) Modulate the signal to 1000 Hz.
   (c) Demodulate using the AM demodulator from AM.m (to recover the original $w(t)$).

FIGURE 5.8 Square-law mixing transmitter.

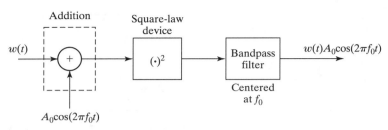

$A_0\cos(2\pi f_0 t)$

5.10. Use the square-law modulator from Problem 5.9 to answer the following questions:

   (a) How sensitive is the system to errors in the frequency of the cosine wave?

   (b) How sensitive is the system to an unknown phase offset in the cosine wave?

5.3 Quadrature Modulation

In AM transmission where the baseband signal and its modulated passband version are real valued, the spectrum of the modulated signal has twice the bandwidth of the baseband signal. As pictured in Figure 4.10 on page 70, the spectrum of the baseband signal is nonzero only for frequencies between $-B$ and B. After modulation, the spectrum is nonzero in the interval $[-f_c - B, -f_c + B]$ and in the interval $[f_c - B, f_c + B]$. Thus the total width of frequencies occupied by the modulated signal is twice that occupied by the baseband signal. This represents a kind of inefficiency or redundancy in the transmission. Quadrature modulation provides one way of removing this redundancy by sending two messages in the frequency ranges between $[-f_c - B, -f_c + B]$ and $[f_c - B, f_c + B]$, thus utilizing the spectrum more efficiently.

To see how this can work, suppose that there are two message streams $m_1(t)$ and $m_2(t)$. Modulate one message with a cosine, and the other with (the negative of) a sine to form

$$v(t) = A_c[m_1(t)\cos(2\pi f_c t) - m_2(t)\sin(2\pi f_c t)].$$

The signal $v(t)$ is then transmitted. A receiver structure that can recover the two messages is shown in Figure 5.9. The signal $s_1(t)$ at the output of the receiver is intended to recover the first message $m_1(t)$. It is often called the "in-phase" signal. Similarly, the signal $s_2(t)$ at the output of the receiver is intended to recover the (negative of the) second message $m_2(t)$. It is often called the "quadrature" signal. These are also sometimes modeled as the "real" and the "imaginary" parts of a single "complex valued" signal.[1]

To examine the recovered signals $s_1(t)$ and $s_2(t)$ in Figure 5.9, first evaluate the signals before the lowpass filtering. Using the trigonometric identities (A.4) and (A.8), $x_1(t)$ becomes

$$x_1(t) = v(t)\cos(2\pi f_c t)$$
$$= A_c m_1(t)\cos^2(2\pi f_c t) - A_c m_2(t)\sin(2\pi f_c t)\cos(2\pi f_c t)$$
$$= \frac{A_c m_1(t)}{2}(1 + \cos(4\pi f_c t)) - \frac{A_c m_2(t)}{2}(\sin(4\pi f_c t)).$$

---

[1] This complex representation is explored more fully in Appendix C.

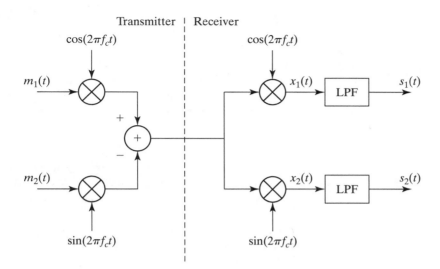

FIGURE 5.9 Quadrature modulation transmitter and receiver.

Lowpass filtering $x_1(t)$ produces

$$s_1(t) = \frac{A_c m_1(t)}{2}.$$

Similarly, $x_2(t)$ can be rewritten using (A.5) and (A.8)

$$x_2(t) = v(t) \, \sin(2\pi f_c t)$$
$$= A_c m_1(t) \, \cos(2\pi f_c t) \, \sin(2\pi f_c t) - A_c m_2(t) \, \sin^2(2\pi f_c t)$$
$$= \frac{A_c m_1(t)}{2} \, \sin(4\pi f_c t) - \frac{A_c m_2(t)}{2}(1 - \cos(4\pi f_c t)),$$

and lowpass filtering $x_2(t)$ produces

$$s_2(t) = \frac{-A_c m_2(t)}{2}.$$

Thus, in the ideal situation in which the phases and frequencies of the modulation and the demodulation are identical, both messages can be recovered. But if the frequencies and/or phases are not exact, then problems analogous to those encountered with AM will occur in the quadrature modulation. For instance, if the phase of (say) the demodulator $x_1(t)$ is not correct, then there will be some distortion or attenuation in $s_1(t)$. However, problems in the demodulation of $s_1(t)$ may also cause problems in the demodulation of $s_2(t)$. This is called cross-interference between the two messages.

Problems

5.11. Use AM.m as a starting point to create a quadrature modulation system that implements the block diagram of Figure 5.9.

(a) Examine the effect of a phase offset in the demodulating sinusoids of the receiver, so that $x_1(t) = v(t)\cos(2\pi f_c t + \phi)$ and $x_2(t) = v(t)\sin(2\pi f_c t + \phi)$ for a variety of ϕ. Refer to Problem 5.6.

(b) Examine the effect of a frequency offset in the demodulating sinusoids of the receiver, so that $x_1(t) = v(t)\cos(2\pi(f_c + \gamma)t)$ and $x_2(t) = v(t)\sin(2\pi(f_c + \gamma)t)$ for a variety of γ. Refer to Problem 5.7.

(c) Confirm that a $\pm 1°$ phase error in the receiver oscillator corresponds to more than 1% cross-interference.

Thus the inefficiency of real-valued double-sided AM transmission can be reduced using complex valved quadrature modulation, which recaptures the lost bandwidth by sending two messages simultaneously. For simplicity and clarity, *Telecommunication Breakdown* focuses on the real PAM case. The complex QAM case is discussed at length an the CD-ROM. There are other ways of recapturing the lost bandwidth: both single side band and vestigial sideband (discussed in the document *Other Modulations* on the CD-ROM) send a single message, but use only half the bandwidth.

5.4 Injection to Intermediate Frequency

All the modulators and demodulators discussed in the previous sections downconvert to baseband in a single step, that is, the spectrum of the received signal is shifted by mixing with a cosine of frequency f_c that matches the transmission frequency f_c. As suggested in Section 2.8, it is also possible to downconvert to some desired intermediate frequency (IF) f_I (as depicted in Figure 2.9), and to then later downconvert to baseband by mixing with a cosine of the intermediate frequency f_I. There are several advantages to such a two-step procedure:

- all frequency bands can be downconverted to the same IF, which allows use of standardized amplifiers, modulators and filters on the IF signals, and

- sampling can be done at the Nyquist rate of the IF rather than the Nyquist rate of the transmission.

The downconversion to an intermediate frequency (followed by bandpass filtering to extract the passband around the IF) can be accomplished in two ways: by a local oscillator modulating from above the carrier frequency (called high-side injection) or from below (low-side injection). To see this, consider the double sideband modulation (from Section 5.2) that creates the transmitted signal

$$v(t) = 2w(t)\cos(2\pi f_c t)$$

from the message signal $w(t)$ and the downconversion to IF via

$$x(t) = 2[v(t) + n(t)]\cos(2\pi f_I t),$$

where $n(t)$ represents interference such as noise and spurious signals from other users. By the frequency shifting property (A.33),

$$V(f) = W(f + f_c) + W(f - f_c), \tag{5.6}$$

and the spectrum of the IF signal is

$$X(f) = V(f + f_I) + V(f - f_I) + N(f + f_I) + N(f - f_I)$$
$$= W(f + f_c - f_I) + W(f - f_c - f_I) + W(f + f_c + f_I)$$
$$+ W(f - f_c + f_I) + N(f + f_I) + N(f - f_I). \qquad (5.7)$$

Example 5.1

Consider a message spectrum $W(f)$ that has a bandwidth of 200 kHz, an upconversion carrier frequency $f_c = 850$ kHz, and an objective to downconvert to an intermediate frequency of 455 kHz. For low-side injection (with $f_I < f_c$), the goal is to center $W(f - f_c + f_I)$ in (5.7) at 455 kHz, such that $455 - f_c + f_I = 0$. Hence, $f_I = f_c - 455 = 395$. For high-side injection (with $f_I > f_c$), the goal is to center $W(f + f_c - f_I)$ at 455 kHz, such that $455 + f_c - f_I = 0$ or $f_I = f_c + 455 = 1305$. For illustrative purposes, suppose that the interferers represented by $N(f)$ are a pair of delta functions at $\pm$ 105 and 1780 kHz. Figure 5.10 sketches $|V(f)|$ and $|X(f)|$ for both high-side and low-side injection. In this example, both methods end up with unwanted narrowband interferences in the passband.

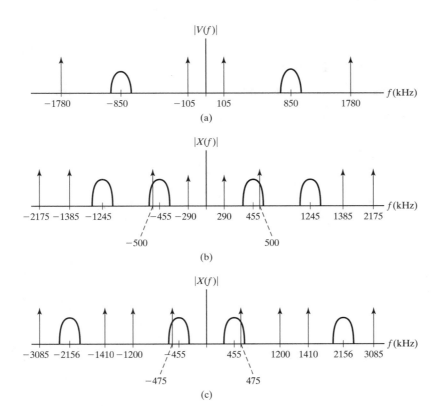

FIGURE 5.10 Example of high-side and low-side injection to IF (a) transmitted spectrum, (b) low-side injected spectrum, and (c) high-side injected spectrum.

The following observations can be regarding high-side and low-side injection:

- Low-side injection results in symmetry in the translated message spectrum about $\pm f_c$ on each of the positive and negative half-axes.

- High-side injection separates the undesired images further from the lower frequency portion (which will ultimately be retained to reconstruct the message). This eases the requirements on the bandpass filter.

- Both high-side and low-side injection can place frequency interferers in undesirable places. This highlights the need for adequate out-of-band rejection by a bandpass filter before downconversion to IF.

Problems

5.12. A transmitter operates as a standard AM with suppressed carrier transmitter (as in AM.m). Create a demodulation routine that operates in two steps: by mixing with a cosine of frequency $3f_c/4$ and subsequently mixing with a cosine of frequency $f_c/4$. Where must pass/reject filters be placed in order to ensure reconstruction of the message? Let $f_c = 2000$.

5.13. Using your MATLAB code from Problem 5.12, investigate the effect of a sinusoidal interference:
  (a) at frequency $\frac{f_c}{6}$,
  (b) at frequency $\frac{f_c}{3}$,
  (c) at frequency $3f_c$.

5.5 For Further Reading

A friendly and readable introduction to analog transmission systems can be found in

- P. J. Nahin, *On the Science of Radio*, AIP Press, 1996.

6 Sampling with Automatic Gain Control

The James Brown canon represents a vast catalogue of recordings—
the mother lode of beats—a righteously funky legacy of grooves for
us to soak in, sample, and quote.

—John Ballon in "MustHear Review,"
http://www.musthear.com/reviews/funkypeople.html

As foreshadowed in Section 2.8, transmission systems cannot be fully digital because the medium through which the signal propagates is analog. Hence, whether the signal begins as analog (such as voice or music) or whether it begins as digital (such as mpeg, jpeg or wav files), it will be converted to a high frequency analog signal when it is transmitted. In a digital receiver, the received signal must be transformed into a discrete-time signal in order to allow subsequent digital processing.

This chapter begins by considering the sampling process in both the time domain and in the frequency domain. Then Section 6.3 discusses how MATLAB can be used to simulate the sampling process. This is not completely obvious because analog signals cannot be represented exactly in the computer. Two simple tricks are suggested. The first expresses the analog signal in functional form and takes "samples" of the function by evaluating it at the desired times. The second *oversamples* the analog signal so that it is represented at a high data rate; the "sampling" can then be done on the oversampled signal.

Sampling is important because it translates the signal from analog to digital. It is equally important to be able to translate from digital back into analog, and the celebrated *Nyquist sampling theorem* shows that this is possible for any bandlimited signal, assuming the sampling rate is fast enough. When the goal of this translation is to rebuild a copy of the transmitted signal, this is called *reconstruction*. When the goal is to determine the value of the signal at some particular point, it is called *interpolation*. Techniques (and MATLAB code) for both reconstruction and interpolation appear in Section 6.4.

Figure 6.1 shows the received signal passing through a BPF (which removes out-of-band interference and isolates the desired frequency range) followed by a fixed demodulation to the intermediate frequency (IF) where sampling takes place. The automatic gain control (AGC) accounts for changes in the strength of the received signal. When the received signal is powerful, the gain a is small; when the signal strength is low, the gain a is high. The goal is to guarantee that the analog to digital converter does not saturate (the signal does not routinely surpass the highest level that can be represented), and that it does not lose dynamic range (the digitized signal does not always remain in a small

FIGURE 6.1 The front end of the receiver. After filtering and demodulation, the signal is sampled. An automatic gain control (AGC) is needed to utilize the full dynamic range of the sampler.

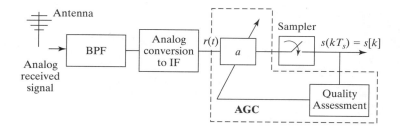

number of the possible levels). The key in the AGC is that the gain must automatically adjust to account for the signal strength, which may vary slowly over time.

The AGC provides the simplest example of a system element that must adapt to changes in its environment (recall the "fifth element" of Chapter 3). How can such elements be designed? *Telecommunication Breakdown* suggests a general method based on gradient-directed optimization. First, a "goal" and an associated "objective function" are chosen. Since it is desired to maintain the output of the AGC at a roughly constant power, the associated objective function is defined to be the average deviation of the power from that constant; the goal is to minimize the objective function. The gain parameter is then adjusted according to a "steepest descent" method that moves the estimate "downhill" towards the optimal value that minimizes the objective. In this case the adaptive gain parameter is increased (when the average power is too small) or decreased (when the average power is too large), thus maintaining a steady power. While it would undoubtedly be possible to design a successful AGC without recourse to such a general optimization method, the framework developed in Sections 6.5 through 6.7 will also be useful in designing other adaptive elements such as the phase tracking loops of Chapter 10, the clock recovery algorithms of Chapter 12, and the equalization schemes of Chapter 13.

6.1 Sampling and Aliasing

Sampling can be modelled as a point-by-point multiplication in the time domain by a pulse train (a sequence of impulses). (Recall Figure 3.8 on page 48.) While this is intuitively plausible, it is not terribly insightful. The effects of sampling become apparent when viewed in the frequency domain. When the sampling is done correctly, no information is lost. However, if the sampling is done too slowly, aliasing artifacts are inevitable. This section shows the "how" and "why" of sampling.

Suppose an analog waveform $w(t)$ is to be sampled every T_s seconds to yield a discrete-time sequence $w[k] = w(kT_s) = w(t)|_{t=kT_s}$ for all integers k.[1] This is called *point sampling* because it picks off the value of the function $w(t)$ at the points kT_s. One way to model point sampling is to create a continuous valued function that consists of a train of pulses that are scaled by the values $w(kT_s)$. The *impulse sampling* function is

$$w_s(t) = w(t) \sum_{k=-\infty}^{\infty} \delta(t - kT_s)$$

---

[1] Observe the notation $w(kT_s)$ means $w(t)$ evaluated at the time $t = kT_s$. This is also notated $w[k]$ (with the square brackets), where the sampling rate T_s is implicit.

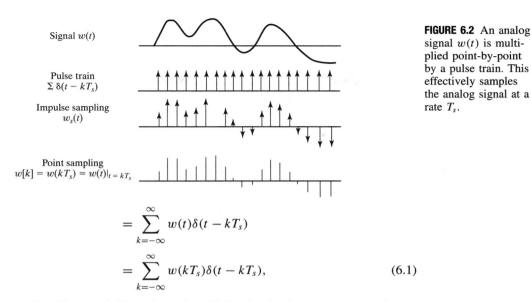

Signal $w(t)$

Pulse train
$\Sigma\, \delta(t - kT_s)$

Impulse sampling
$w_s(t)$

Point sampling
$w[k] = w(kT_s) = w(t)|_{t\,=\,kT_s}$

FIGURE 6.2 An analog signal $w(t)$ is multiplied point-by-point by a pulse train. This effectively samples the analog signal at a rate T_s.

$$= \sum_{k=-\infty}^{\infty} w(t)\delta(t - kT_s)$$

$$= \sum_{k=-\infty}^{\infty} w(kT_s)\delta(t - kT_s), \tag{6.1}$$

and it is illustrated in Figure 6.2. The effect of multiplication by the pulse train is clear in the time domain. But the relationship between $w_s(t)$ and $w(t)$ is clearer in the frequency domain, which can be understood by writing $W_s(f)$ as a function of $W(f)$.

The transform $W_s(f)$ is given in (A.27) and (A.28). With $f_s = 1/T_s$, this is

$$W_s(f) = f_s \sum_{n=-\infty}^{\infty} W(f - nf_s). \tag{6.2}$$

Thus, the spectrum of the sampled signal $w_s(t)$ differs from the spectrum of the original $w(t)$ in two ways:

- Amplitude scaling—each term in the spectrum $W_s(f)$ is multiplied by the factor f_s.
- Replicas—for each n, $W_s(f)$ contains a copy of $W(f)$ shifted to $f - nf_s$.

Sampling creates an infinite sequence of replicas, each separated by f_s Hz. Said another way, sampling in time is the same as periodic in frequency, where the period is defined by the sampling rate. Readers familiar with Fourier series will recognize this as the dual of the property that periodic in time is the equivalent of sampling in frequency. Indeed, Equation (6.2) shows why the relationships in Figure 3.9 on page 49 hold.

Figure 6.3 shows these replicas in two possible cases. In (a), $f_s \geq 2B$, where B is the bandwidth of $w(t)$, and the replicas do not overlap. Hence, it is possible to extract the one replica centered at zero by using a lowpass filter. Assuming that the filtering is without error, $W(f)$ is recovered from the sampled version $W_s(f)$. Since the transform is invertible, this means that $w(t)$ can be recovered from $w_s(t)$. Therefore, no loss of information occurs in the sampling process.[2] Such a result is known as

---

[2] Be clear about this: The analog signal $w(t)$ is sampled to give $w_s(t)$, which is nonzero only at the sampling instants kT_s. If $w_s(t)$ is then input into a perfect analog lowpass filter, its output is the same as the original $w(t)$. Such filtering cannot be done with any digital filter operating at the sampling rate f_s. In terms of Figure 6.3, the digital filter can remove and reshape the frequencies between the bumps, but can never remove the periodic bumps.

FIGURE 6.3 The spectrum of a sampled signal is periodic with period equal to f_s. In case (a), the original spectrum $W(f)$ is bandlimited to less than $\frac{f_s}{2}$ and there is no overlapping of the replicas. When $W(f)$ is not bandlimited to less than $\frac{f_s}{2}$, as in (b), the overlap of the replicas is called aliasing.

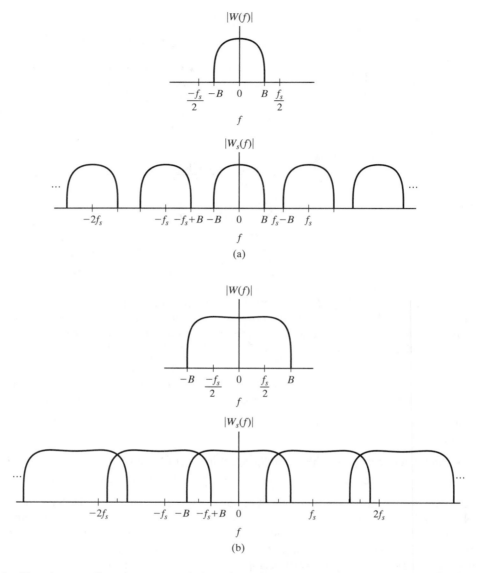

the Nyquist sampling theorem, and the minimum allowable sampling rate is called the *Nyquist rate*.

> *Nyquist Sampling Theorem:* If the signal $w(t)$ is bandlimited to B, $(W(f) = 0$ for all $|f| > B)$ and if the sampling rate is faster than $f_s = 2B$, then $w(t)$ can be reconstructed exactly from its samples $w(kT_s)$.

On the other hand, in part (b) of Figure 6.3, the replicas overlap because the repetitions are narrower than the width of the spectrum $W(f)$. In this case, it is impossible to recover the original spectrum perfectly from the sampled spectrum, and hence it is impossible to exactly recover the original waveform from the sampled version. The overlapping of the replicas and the resulting distortions in the reconstructed waveform are called *aliasing*.

Bandwidth can also be thought of as limiting the rate at which data can flow over a channel. When a channel is constrained to a bandwidth $2B$, then the output of the channel is a signal with bandwidth no greater than $2B$. Accordingly, the output can contain no frequencies above f_s, and symbols can be transmitted no faster than one every T_s seconds, where $\frac{1}{T_s} = f_s$.

Problems

6.1. Human hearing extends up to about 20 KHz. What is the minimum sampling rate needed to fully capture a musical performance? Compare this to the CD sampling rate of 44.1 KHz. Some animal sounds, such as the singing of dolphins and the chirping of bats, occur at frequencies up to about 50 KHZ. What does this imply about CD recordings of dolphin or bat sounds?

6.2. U.S. high-definition (digital) television (HDTV) is transmitted in the same frequency bands as conventional television (for instance, Channel 2 is at 54 MHz), and each channel has a bandwidth of about 6 MHz. What is the minimum sampling rate needed to fully capture the HDTV signal once it has been modulated to baseband?

6.2 Downconversion via Sampling

The processes of modulation and demodulation, which shift the frequencies of a signal, can be accomplished by mixing with a cosine wave that has a frequency equal to the amount of the desired shift, as was demonstrated repeatedly throughout Chapter 5. But this is not the only way. Since sampling creates a collection of replicas of the spectrum of a waveform, it changes the frequencies of the signal.

When the message signal is analog and bandlimited to $\pm B$, this can be used for demodulation. Suppose that the signal is transmitted with a carrier at frequency f_c. Direct sampling of this signal creates a collection of replicas, one near DC. This procedure is shown in Figure 6.4 for $f_s = \frac{f_c}{2}$, though beware: when f_s and f_c are not simply related, the replica may not land exactly at DC.

This demodulation-by-sampling is diagrammed in Figure 6.5 (with $f_s = \frac{f_c}{n}$, where n is a small positive integer), and can be thought of as an alternative to mixing with a cosine (that must be synchronized in frequency and phase with the transmitter oscillator). The magnitude spectrum $|W(f)|$ of a message $w(t)$ is shown in Figure 6.4(a), and the spectrum after upconversion is shown in part (b); this is the transmitted signal $s(t)$. At the receiver, $s(t)$ is sampled, which can be modelled as a multiplication with a train of delta functions in time

$$y(t) = s(t) \sum_{n=-\infty}^{\infty} \delta(t - nT_s),$$

FIGURE 6.4 Spectra in a sampling downconverter. The (bandlimited analog) signal $W(f)$ shown in (a) is upconverted to the transmitted signal in (b). Directly sampling this (at a rate equal to $f_s = \frac{f_c}{2}$) results in the spectrum shown in the bottom plot.

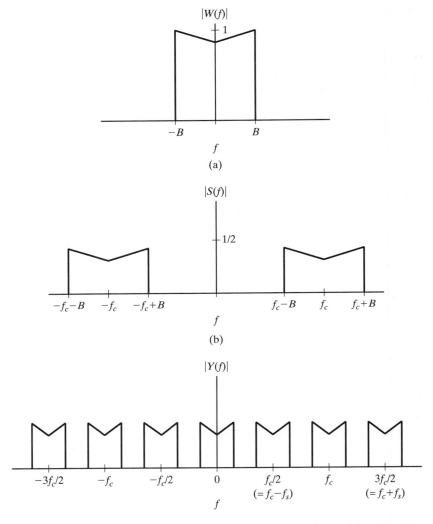

where T_s is the sample period. Using (6.2), this can be transformed into the frequency domain as

$$Y(f) = \frac{1}{T_s} \sum_{n=-\infty}^{\infty} S(f - nf_s),$$

where $f_s = 1/T_s$. The magnitude spectrum of $Y(f)$ is illustrated in Figure 6.4(c) for the particular choice $f_s = f_c/2$ (and $T_s = 2/f_c$) with $B < \frac{f_c}{4} = \frac{f_s}{2}$.

There are three ways that the sampling can proceed:

1. sample faster than the Nyquist rate of the IF frequency,
2. sample slower than the Nyquist rate of the IF frequency, and then downconvert the replica closest to DC, and
3. sample so that one of the replicas is directly centered at DC.

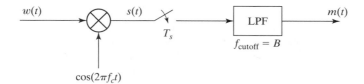

FIGURE 6.5 System diagram of sampling as downconversion.

The first is a direct imitation of the analog situation where no aliasing will occur. This may be expensive because of the high sample rates required to achieve Nyquist sampling. The third is the situation depicted in Figures 6.4 and 6.5, which permit downconversion to baseband without an additional oscillator. This may be sensitive to small deviations in frequency (for instance, when f_s is not exactly $f_c/2$). The middle method downconverts part of the way by sampling and part of the way by mixing with a cosine. The middle method is used in the $\mathcal{M}^6$ receiver project in Chapter 15.

Problems

6.3. Create a simulation of a sampling-based modulator that takes a signal with bandwidth 100 Hz and transforms it into the "same" signal centered at 5000 Hz. Be careful; there are two "sampling rates" in this problem. One reflects the assumed sampling rate for the modulation and the other represents the sampling rate that is used in MATLAB to represent a "continuous time" signal. You may wish to reuse code from `sine100Hzsamp.m`. What choices have you made for these two sampling rates?

6.4. Implement the procedure diagrammed in Figure 6.5. Comment on the choice of sampling rates. How have you specified the LPF?

6.5. Using your code from Problem 6.4, examine the effect of "incorrect" sampling rates by demodulating with $f_s + \gamma$ instead of f_s. This is analogous to the problem that occurs in cosine mixing demodulation when the frequency is not accurate. Is there an analogy to the phase problem that occurs, for instance, with nonzero ϕ in (5.4)?

6.3 Exploring Sampling in MATLAB

It is not possible to capture all of the complexities of analog-to-digital conversion inside a computer program, because all signals within a (digital) computer are already "sampled." Nonetheless, most of the key ideas can be illustrated by using two tricks to simulate the sampling process:

- Evaluate a function at appropriate values (or times).
- Represent a data waveform by a large number of samples and then reduce the number of samples.

The first is useful when the signal can be described by a known function, while the second is necessary whenever the procedure is data driven, that is, when no functional form is available. This section explores both approaches via a series of MATLAB experiments.

Consider representing a sine wave of frequency $f = 100$ Hz. The sampling theorem asserts that the sampling rate must be greater than the Nyquist rate of 200 samples per second. But in order to visualize the wave clearly, it is often useful to sample considerably faster. The following MATLAB code calculates and plots the first $1/10$ second of a 100 Hz sine wave with a sampling rate of $f_s = \frac{1}{T_s} = 10000$ samples per second.

---

sine100hz.m: generate 100 Hz sine wave with sampling rate fs=1/Ts

```
f=100;                      % frequency of wave
time=0.1;                   % total time in seconds
Ts=1/10000;                 % sampling interval
t=Ts:Ts:time;              % define a "time" vector
w=sin(2*pi*f*t);           % define the sine wave
plot(t,w)                   % plot the sine vs. time
xlabel('seconds')           % label the x axis
ylabel('amplitude')         % label the y axis
```

Running `sine100hz.m` plots the first 10 periods of the sine wave. Each period lasts 0.01 seconds, and each period contains 100 points, as can be verified by looking at w(1:100). Changing the variables `time` or `Ts` displays different numbers of cycles of the same sine wave, while changing `f` plots sine waves with different underlying frequencies.

Problems

6.6. What must the sampling rate be so that each period of the wave is represented by 20 samples? Check your answer using the program above.

6.7. Let `Ts=1/500`. How does the plot of the sine wave appear? Let `Ts=1/100`, and answer the same question. How large can `Ts` be if the plot is to retain the appearance of a sine wave? Compare your answer to the theoretical limit. Why are they different?

When the sampling is rapid compared to the underlying frequency of the signal (for instance, the program `sine100hz.m` creates 100 samples in each period), then the plot appears and acts much like an analog signal, even though it is still, in reality, a discrete time sequence. Such a sequence is called *oversampled* relative to the signal period. The following program simulates the process of sampling the 100 Hz oversampled sine wave. This is downsampling, as shown in Figure 3.10 on page 50.

---

sine100hzsamp.m: simulated sampling of the 100 Hz sine wave

```
f=100; time=0.05; Ts=1/10000; t=Ts:Ts:time;   % freq and time vectors
w=sin(2*pi*f*t);                                % create sine wave w(t)
ss=10;                                          % take 1 in ss samples
wk=w(1:ss:end);                                 % the "sampled" sequence
ws=zeros(size(w)); ws(1:ss:end)=wk;            % sampled waveform ws(t)
plot(t,w)                                       % plot the waveform
hold on, plot(t,ws,'r'), hold off              % plot "sampled" wave
xlabel('seconds'), ylabel('amplitude')          % label the axes
```

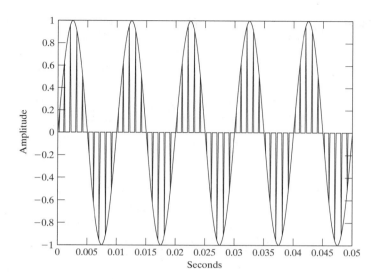

FIGURE 6.6 Removing all but one of each N points from an over-sampled waveform simulates the sampling process.

Running `sine100hzsamp.m` results in the plot shown in Figure 6.6, where the "continuous" sine wave w is downsampled by a factor of `ss=10`; that is, all but one of each `ss` samples is removed. Thus, the waveform w represents the analog signal that is to be sampled at the *effective sampling interval* `ss*Ts`. The spiky signal ws corresponds to the sampled signal $w_s(t)$, while the sequence wk contains just the amplitude values at the tips of the spikes.

Problems

6.8. Modify `sine100hzsamp.m` to create an oversampled sinc wave, and then sample this with `ss=10`. Repeat this exercise with `ss=30`, `ss=100`, and `ss=200`. Comment on what is happening. Hint: In each case, what is the effective sampling interval?

6.9. Plot the spectrum of the 100 Hz sine wave when it is created with different downsampling rates `ss=10`, `ss=11`, `ss=30`, and `ss=200`. Explain what you see.

6.4 Interpolation and Reconstruction

The previous sections explored how to convert analog signals into digital signals. The central result is that if the sampling is done faster than the Nyquist rate, then no information is lost. In other words, the complete analog signal $w(t)$ can be recovered from its discrete samples $w[k]$. When the goal is to find the complete waveform, this is called *reconstruction*; when the goal is to find values of the waveform at particular points between the sampling instants, it is called *interpolation*. This section explores bandlimited interpolation and reconstruction in theory and practice.

The samples $w(kT_s)$ form a sequence of numbers that represent an underlying continuous valued function $w(t)$ at the time instants $t = kT_s$. The sampling interval T_s is presumed to have been chosen so that the sampling rate $f_s > 2B$ where B is the highest frequency present in $w(t)$. The Nyquist sampling theorem presented in Section 6.1 states

that the values of $w(t)$ can be recovered exactly at any time τ. The formula (which is justified subsequently) for recovering $w(\tau)$ from the samples $w(kT_s)$ is

$$w(\tau) = \int_{t=-\infty}^{\infty} w_s(t) \, \text{sinc}(\tau - t)dt,$$

where $w_s(t)$ (defined in (6.1)) is zero everywhere except at the sampling instants $t = kT_s$. Substituting (6.1) into $w(\tau)$ shows that this integral is identical to the sum

$$w(\tau) = \sum_{k=-\infty}^{\infty} w(kT_s) \, \text{sinc}(\tau - kT_s). \tag{6.3}$$

In principle, if the sum is taken over all time, the value of $w(\tau)$ is exact. As a practical matter, the sum must be taken over a suitable (finite) time window.

To see why interpolation works, note that the formula (6.3) is a convolution (in time) of the signal $w(kT_s)$ and the sinc function. Since convolution in time is the same as multiplication in frequency by (A.40), the transform of $w(\tau)$ is equal to the product of $\mathcal{F}\{w_s(kT_s)\}$ and the transform of the sinc. By (A.22), the transform of the sinc function in time is a rect function in frequency. This rect function is a lowpass filter, since it passes all frequencies below $\frac{f_s}{2}$ and removes all frequencies above. Since the process of sampling a continuous time signal generates replicas of the spectrum at integer multiples of f_s by (6.2), the lowpass filter removes all but one of these replicas. In effect, the sampled data is passed through an analog lowpass filter to create a continuous-time function, and the value of this function at time τ is the required interpolated value. When $\tau = nT_s$, then $\text{sinc}(\tau - nT_s) = 1$, and $\text{sinc}(\tau - nT_s) = 0$ for all kT_s with $k \neq n$. When τ is between sampling instants, the sinc is nonzero at all kT_s, and (6.3) combines them to recover $w(\tau)$.

To see how (6.3) works, the following code generates a sine wave w of frequency 20 Hz with a sampling rate of 100 Hz. This is a modestly sampled sine wave, having only five samples per period, and its graph is jumpy and discontinuous. Because the sampling rate is greater than the Nyquist rate, it is possible in principle to recover the underlying smooth sine wave from which the samples are drawn. Running `sininterp.m` shows that it is also possible in practice. The plot in Figure 6.7 shows the original wave (which appears choppy because it is sampled only five times per period), and the reconstructed or smoothed waveform (which looks just like a sine wave). The variable `intfac` specifies how many extra interpolated points are calculated, and need not be an integer. Larger numbers result in smoother curves but also require more computation.

---

sininterp.m: demonstrate interpolation/reconstruction using sin wave

```
f=20; Ts=1/100; time=20;            % freq, sampling interval, and time
t=Ts:Ts:time;                       % time vector
w=sin(2*pi*f*t);                    % w(t) = a sine wave of f Hertz
over=100;                           % # of data points to use in smoothing
intfac=10;                          % how many interpolated points
tnow=10.0/Ts:1/intfac:10.5/Ts;      % smooth/interpolate from 10 to 10.5 sec
wsmooth=zeros(size(tnow));          % save smoothed data here
for i=1:length(tnow)
  wsmooth(i)=interpsinc(w,tnow(i),over);
end                                 % and loop for next point
```

(handwritten margin note:) sinc $f=f_s$ then $B = \frac{f_s}{2}$

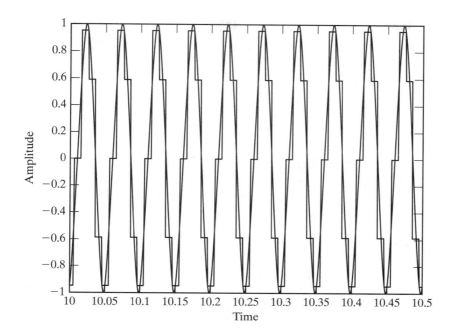

FIGURE 6.7 A convincing sine wave can be reconstructed from its samples using sinc interpolation. The choppy wave represents the samples, and the smooth wave shows the reconstruction.

In implementing (6.3), some approximations are used. First, the sum cannot be calculated over an infinite time horizon, and the variable `over` replaces the sum $\sum_{k=-\infty}^{\infty}$ with $\sum_{k=-over}^{over}$. Each pass through the `for` loop calculates one point of the smoothed curve `wsmooth` using the MATLAB function `interpsinc.m`, which is shown below. The value of the sinc is calculated at each time using the function `SRRC.m` with the appropriate offset `tau`, and then the convolution is performed by the `conv` command. This code is slow and unoptimized. A clever programmer will see that there is no need to calculate the sinc for every point, and efficient implementations use sophisticated look-up tables to avoid the calculation of transcendental functions completely.

```
function y=interpsinc(x, t, l, beta)
% interpolate to find a single point using the direct method
%        x = sampled data
%        t = place at which value desired
%        l = one sided length of data to interpolate
%        beta = rolloff factor for SRRC function
%             = 0 is a sinc
if nargin==3, beta=0; end;              % if unspecified, beta is 0
tnow=round(t);                          % create indices tnow=integer part
tau=t-round(t);                         % plus tau=fractional part
s_tau=SRRC(l,beta,1,tau);              % interpolating sinc at offset tau
x_tau=conv(x(tnow-l:tnow+l),s_tau);    % interpolate the signal
y=x_tau(2*l+1);                         % the new sample
```

While the indexing needed in `interpsinc.m` is a bit tricky, the basic idea is not: the sinc interpolation of (6.3) is just a linear filter with impulse response $h(t) = \text{sinc}(t)$. (Remember, convolutions are the hallmark of linear filters.) Thus, it is a lowpass filter, since the frequency response is a rect function. The delay τ is proportional to the phase of the frequency response.

Problems

6.10. In `sininterp.m`, what happens when the sampling rate is too low? How large can the sampling interval Ts be? How high can the frequency f be?

6.11. In `sininterp.m`, what happens when the window is reduced? Make `over` smaller and find out. What happens when too few points are interpolated? Make `intfac` smaller and find out.

6.12. Create a more interesting (more complex) wave $w(t)$. Answer the above questions for this $w(t)$.

6.13. Let $w(t)$ be a sum of five sinusoids for t between -10 and 10 seconds. Let $w(kT)$ represent samples of $w(t)$ with $T = .01$. Use `interpsinc.m` to interpolate the values $w(0.011)$, $w(0.013)$, and $w(0.015)$. Compare the interpolated values to the actual values. Explain any discrepancies.

Observe that $\text{sinc}(t)$ dies away (slowly) in time at a rate proportional to $\frac{1}{\pi t}$. This is one of the reasons that so many terms are used in the convolution (i.e., why the variable `over` is large). A simple way to reduce the number of terms is to use a function that dies away more quickly than the sinc; a common choice is the *square-root raised cosine* (SRRC) function, which plays an important role in pulse shaping in Chapter 11. The functional form of the SRRC is given in Equation (11.8). The SRRC can easily be incorporated into the interpolation code by replacing the code `interpsinc(w,tnow(i),over)` with `interpsinc(w,tnow(i),over,beta)`.

Problem

6.14. With `beta=0`, the SRRC is exactly the sinc. Redo the above exercises trying various values of `beta` between 0 and 1.

The function `srrc.m` is available on the CD. Its help file is

```
% s=srrc(syms, beta, P, t_off);
% Generate a Square-Root Raised Cosine Pulse
%        'syms' is 1/2 the length of srrc pulse in symbol durations
%        'beta' is the rolloff factor: beta=0 gives the sinc function
%        'P' is the oversampling factor
%        t_off is the phase (or timing) offset
```

MATLAB also has a built-in function called `resample`, which has the following help file:

```
%RESAMPLE  Change the sampling rate of a signal.
%   Y = RESAMPLE(X,P,Q) resamples the sequence in vector X at P/Q times
```

```
%    the original sample rate using a polyphase implementation.  Y is P/Q
%    times the length of X (or the ceiling of this if P/Q is not an integer).
%    P and Q must be positive integers.
```

This technique is different from that used in (6.3). It is more efficient numerically at reconstructing entire waveforms, but only it works when the desired resampling rate is rationally related to the original. The method of (6.3) is far more efficient when isolated (not necessarily evenly spaced) interpolating points are required, which is crucial for synchronization tasks in Chapter 12.

6.5 Iteration and Optimization

An important practical part of the sampling procedure is that the dynamic range of the signal at the input to the sampler must remain within bounds. This can be accomplished using an automatic gain control, which is depicted in Figure 6.1 as multiplication by a scalar a, along with a "quality assessment" block that adjusts a in response to the power at the output of the sampler. This section discusses the background needed to understand how the quality assessment works. The essential idea is to state the goal of the assessment mechanism as an optimization problem.

Many problems in communications (and throughout engineering) can be framed in terms of an optimization problem. Solving such problems requires three basic steps:

1. Setting a goal—choosing a "performance" or "objective" function.
2. Choosing a method of achieving the goal—minimizing or maximizing the objective function.
3. Testing to make sure the method works as anticipated.

"Setting the goal" usually consists of finding a function that can be minimized (or maximized), and for which locating the minimum (or maximum) value provides useful information about the problem at hand. Moreover, the function must be chosen carefully so that it (and its derivative) can be calculated based on quantities that are known, or which can be derived from signals that are easily obtainable. Sometimes the goal is obvious, and sometimes it is not.

There are many ways of carrying out the minimization or maximization procedure. Some of these are direct. For instance, if the problem is to find the point at which a polynomial function achieves its minimum value, this can be solved directly by finding the derivative and setting it equal to zero. Often, however, such direct solutions are impossible, and even when they are possible, recursive (or adaptive) approaches often have better properties when the signals are noisy. This chapter focuses on a recursive method called *steepest descent*, which is the basis of many adaptive elements used in communications systems (and of all the elements used in *Telecommunication Breakdown*).

The final step in implementing any solution is to check that the method behaves as desired, despite any simplifying assumptions that may have been made in its derivation. This may involve a detailed analysis of the resulting methodology, or it may involve simulations. Thorough testing would involve both analysis and simulation in a variety of settings that mimic, as closely as possible, the situations in which the method will be used.

Imagine being lost on a mountainside on a foggy night. Your goal is to get to the village which lies at the bottom of a valley below. Though you cannot see far, you can reach out and feel the nearby ground. If you repeatedly step in the direction that heads downhill most steeply, you eventually reach a depression in which all directions lead up. If the contour of the land is smooth, and without any local depressions that can trap you, then you will eventually arrive at the village. The optimization procedure called "steepest descent" implements this scenario mathematically where the mountainside is defined by the "performance" function and the optimal answer lies in the valley at the minimum value. Many standard communications algorithms (adaptive elements) can be viewed in this way.

6.6 An Example of Optimization: Polynomial Minimization

This first example is too simple to be of practical use, but it does show many of the ideas starkly. Suppose that the goal is to find the value at which the polynomial

$$J(x) = x^2 - 4x + 4 \tag{6.4}$$

achieves its minimum value. Thus step (1) is set. As any calculus book will suggest, the direct way to find the minimum is to take the derivative, set it equal to zero, and solve for x. Thus, $\frac{dJ(x)}{dx} = 2x - 4 = 0$ is solved when $x = 2$, which is indeed the value of x where the parabola $J(x)$ reaches bottom. Sometimes (one might truthfully say "often"), however, such direct approaches are impossible. Maybe the derivative is just too complicated to solve (which can happen when the functions involved in $J(x)$ are extremely nonlinear). Or maybe the derivative of $J(x)$ cannot be calculated precisely from the available data, and instead must be estimated from a noisy data stream.

One alternative to the direct solution technique is an adaptive method called "steepest descent" (when the goal is to minimize), and called "hill climbing" (when the goal is to maximize). Steepest descent begins with an initial guess of the location of the minimum, evaluates which direction from this estimate is most steeply "downhill," and then makes a new estimate along the downhill direction. Similarly, hill climbing begins with an initial guess of the location of the maximum, evaluates which direction climbs the most rapidly, and then makes a new estimate along the uphill direction. With luck, the new estimates are better than the old. The process repeats, hopefully getting closer to the optimal location at each step. The key ingredient in this procedure is to recognize that the uphill direction is defined by the gradient evaluated at the current location, while the downhill direction is the negative of this gradient.

To apply steepest descent to the minimization of the polynomial $J(x)$ in (6.4), suppose that a current estimate of x is available at time k, which is denoted $x[k]$. A new estimate of x at time $k + 1$ can be made using

$$x[k+1] = x[k] - \mu \left. \frac{dJ(x)}{dx} \right|_{x=x[k]}, \tag{6.5}$$

where μ is a small positive number called the stepsize, and where the gradient (derivative) of $J(x)$ is evaluated at the current point $x[k]$. This is then repeated again and again

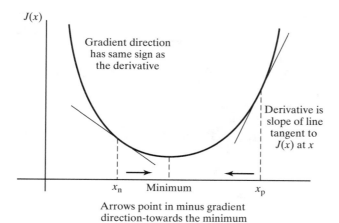

FIGURE 6.8 Steepest descent finds the minimum of a function by always pointing in the direction that leads downhill.

as k increments. This procedure is shown in Figure 6.8. When the current estimate $x[k]$ is to the right of the minimum, the negative of the gradient points left. When the current estimate is to the left of the minimum, the negative gradient points to the right. In either case, as long as the stepsize is suitably small, the new estimate $x[k+1]$ is closer to the minimum than the old estimate $x[k]$; that is, $J(x[k+1])$ is less than $J(x[k])$.

To make this explicit, the iteration defined by (6.5) is

$$x[k+1] = x[k] - \mu(2x[k] - 4),$$

or, rearranging,

$$x[k+1] = (1 - 2\mu)x[k] + 4\mu. \tag{6.6}$$

In principle, if (6.6) is iterated over and over, the sequence $x[k]$ should approach the minimum value $x = 2$. Does this actually happen?

There are two ways to answer this question. It is straightforward to simulate the process. Here is some MATLAB code that takes an initial estimate of x called $x(1)$ and iterates Equation (6.6) for $N=500$ steps.

---

polyconverge.m: find the minimum of $J(x) = x^2 - 4x + 4$ via steepest descent

---

```
N=500;                          % number of iterations
mu=.01;                         % algorithm stepsize
x=zeros(size(1,N));             % initialize x to zero
x(1)=3;                         % starting point x(1)
for k=1:N-1
  x(k+1)=(1-2*mu)*x(k)+4*mu;    % update equation
end
```

Figure 6.9 shows 50 different $x(1)$ starting values superimposed; all converge smoothly to the minimum at $x = 2$.

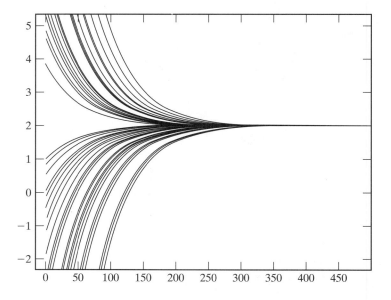

Problem

6.15. Explore the behavior of steepest descent by running `polyconverge.m` with different parameters.

(a) Try `mu=-.01, 0, .0001, .02, .03, .05, 1.0, 10.0`. Can `mu` be too large or too small?

(b) Try `N=5, 40, 100, 5000`. Can `N` be too large or too small?

(c) Try a variety of values of `x(1)`. Can `x(1)` be too large or too small?

As an alternative to simulation, observe that the process (6.6) is itself a linear time invariant system, of the general form

$$x[k + 1] = ax[k] + b, \qquad (6.7)$$

which is stable as long as $|a| < 1$. For a constant input, the final value theorem of z-Transforms (see (A.55)) can be used to show that the asymptotic (convergent) output value is $\lim_{k \to \infty} x_k = \frac{b}{1-a}$. To see this without reference to arcane theory, observe that if x_k is to converge, then it must converge to some value, say x^*. At convergence, $x[k + 1] = x[k] = x^*$, and so (6.7) implies that $x^* = ax^* + b$, which implies that $x^* = \frac{b}{1-a}$. (This holds assuming $|a| < 1$.) For example, for (6.6), $x^* = \frac{4\mu}{1-(1-2\mu)} = 2$, which is indeed the minimum.

Thus, both simulation and analysis suggest that the iteration (6.6) is a viable way to find the minimum of the function $J(x)$, as long as μ is suitably small. As will become clearer in later sections, such solutions to optimization problems are almost always possible—as long as the function $J(x)$ is differentiable. Similarly, it is usually quite straightforward to simulate the algorithm to examine its behavior in specific cases, though it is not always so easy to carry out a theoretical analysis.

By their nature, steepest descent and hill climbing methods use only local information. This is because the update from a point $x[k]$ depends only on the value of $x[k]$ and on the value of its derivative evaluated at that point. This can be a problem, since if the objective function has many minima, the steepest descent algorithm may become "trapped" at a minimum that is not (globally) the smallest. These are called local minima. To see how this can happen, consider the problem of finding the value of x that minimizes the function

$$J(x) = e^{-0.1|x|} \sin(x). \tag{6.8}$$

Applying the chain rule, the derivative is

$$e^{-0.1|x|} \cos(x) - 0.1e^{-0.1|x|} \sin(x) \operatorname{sign}(x),$$

where

$$\operatorname{sign}(x) = \left[\begin{array}{cc} 1 & x > 0 \\ -1 & x < 0 \end{array} \right.$$

is the formal derivative of $|x|$. Solving directly for the minimum point is nontrivial (try it!). Yet implementing a steepest descent search for the minimum can be done in a straightforward manner using the iteration

$$x[k + 1] = x[k] - \mu e^{-0.1|x[k]|}(\cos(x[k]) - 0.1 \sin(x[k]) \operatorname{sign}(x)). \tag{6.9}$$

To be concrete, replace the update equation in `polyconverge.m` with

```
x(k+1)=x(k)-mu*exp(-0.1*abs(x(k)))*(cos(x(k))-0.1* sin(x(k))*sign(x(k)));
```

Problem

6.16. Implement the steepest descent strategy to find the minimum of $J(x)$ in (6.8), modelling the program after `polyconverge.m`. Run the program for different values of `mu`, `N`, and `x(1)`, and answer the same questions as in Problem 6.15.

One way to understand the behavior of steepest descent algorithms is to plot the *error surface*, which is basically a plot of the objective as a function of the variable that is being optimized. Figure 6.10(a) displays clearly the single global minimum of the objective function (6.4) while Figure 6.10(b) shows the many minima of the objective function defined by (6.8). As will be clear to anyone who has attempted Problem 6.16, initializing within any one of the valleys causes the algorithm to descend to the bottom of that valley. Although true steepest descent algorithms can never climb over a peak to enter another valley (even if the minimum there is lower) it can sometimes happen in practice when there is a significant amount of noise in the measurement of the downhill direction.

Essentially, the algorithm gradually descends the error surface by moving in the (locally) downhill direction, and different initial estimates may lead to different minima. This underscores one of the limitations of steepest descent methods—if there are many minima, then it is important to initialize near an acceptable one. In some problems such prior information may easily be obtained, while in others it may be truly unknown.

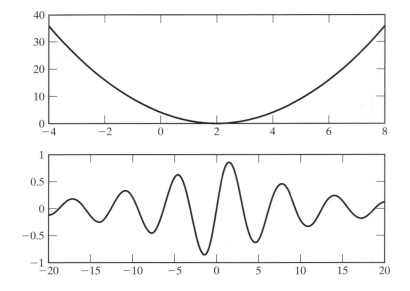

FIGURE 6.10 Error surfaces corresponding to (a) the objective function (6.4) and (b) the objective function (6.8).

The examples of this section are somewhat simple because they involve static functions. Most applications in communication systems deal with signals that evolve over time, and the next section applies the steepest descent idea in a dynamic setting to the problem of Automatic Gain Control (AGC). The AGC provides a simple setting in which all three of the major issues in optimization must be addressed: setting the goal, choosing a method of solution, and verifying that the method is successful.

6.7 Automatic Gain Control

Any receiver is designed to handle signals of a certain average magnitude most effectively. The goal of an AGC is to amplify weak signals and to attenuate strong signals so that they remain (as much as possible) within the normal operating range of the receiver. Typically, the rate at which the gain varies is slow compared with the data rate, though it may be fast by human standards.

The power in a received signal depends on many things: the strength of the broadcast, the distance from the transmitter to the receiver, the direction in which the antenna is pointed, and whether there are any geographic features such as mountains (or tall buildings) that block, reflect, or absorb the signal. While more power is generally better from the point of view of trying to decipher the transmitted message, there are always limits to the power handling capabilities of the receiver. Hence if the received signal is too large (on average), it must be attenuated. Similarly, if the received signal is weak (on average), then it must be amplified.

Figure 6.11 shows the two extremes that the AGC is designed to avoid. In part (a), the signal is much larger than the levels of the sampling device (indicated by the horizontal lines). The gain must be made smaller. In part (b), the signal is much too small to be captured effectively, and the gain must increased.

There are two basic approaches to an AGC. The traditional approach uses analog circuitry to adjust the gain before the sampling. The more modern approach uses the

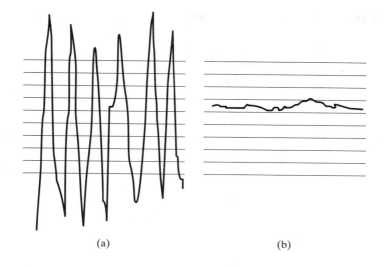

FIGURE 6.11 The goal of the AGC is to maintain the dynamic range of the signal by attenuating it when it is too large (as in (a)) and by increasing it when it is too small (as in (b)).

(a)   (b)

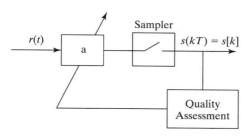

FIGURE 6.12 An automatic gain control must adjust the gain parameter a so that the average energy at the output remains (roughly) fixed, despite fluctuations in the average received energy.

output of the sampler to adjust the gain. The advantage of the analog method is that the two blocks (the gain and the sampling) are separate and do not interact. The advantage of the digital adjustment is that less additional hardware is required since the DSP processing is already present for other tasks.

A simple digital system for AGC gain adjustment is shown in Figure 6.12. The input $r(t)$ is multiplied by the gain a to give the normalized signal $s(t)$. This is then sampled to give the output $s[k]$. The assessment block measures $s[k]$ and determines whether a must be increased or decreased. How can a be adjusted?

The goal is to choose a so that the power (or average energy) of $s(t)$ is approximately equal to some specified d^2. Since

$$a^2 \, \text{avg}\{r^2(t)\}\big|_{t=kT} \approx \text{avg}\{s^2(kT)\} \approx \text{avg}\{s^2[k]\},$$

it would be ideal to choose

$$a^2 \approx \frac{d^2}{\text{avg}\{r^2(kT)\}}, \qquad (6.10)$$

because this would imply that $\text{avg}\{s^2(kT)\} \approx d^2$. The averaging operation (in this case a moving average over a block of data of size N) is defined by

$$\text{avg}\{x[k]\} = \frac{1}{N} \sum_{i=k-N+1}^{k} x[i]$$

and is discussed in Appendix G in amazing detail. Unfortunately, neither the analog input $r(t)$ nor its power are directly available to the assessment block in the DSP portion of the receiver, so it is not possible to directly implement (6.10).

Is there an adaptive element that can accomplish this task? As suggested in the beginning of Section 6.5, there are three steps to the creation of a viable optimization approach: setting a goal, choosing a solution method, and testing. As in any real life engineering task, a proper mathematical statement of the goal can be tricky, and this section proposes two (slightly different) possibilities for the AGC. By comparing the resulting algorithms (essentially, alternative forms for the AGC design), it may be possible to trade off among various design considerations.

One sensible goal is to try to minimize a simple function of the difference between the power of the sampled signal $s[k]$ and the desired power d^2. For instance, the averaged squared error in the powers of s and d,

$$J_{LS}(a) = \text{avg}\left\{\frac{1}{4}(s^2[k] - d^2)^2\right\} = \frac{1}{4}\text{avg}\{(a^2r^2(kT) - d^2)^2\}, \tag{6.11}$$

penalizes values of a which cause $s^2[k]$ to deviate from d^2. This formally mimics the parabolic form of the objective (6.4) in the polynomial minimization example of the previous section. Applying the steepest descent strategy yields

$$a[k+1] = a[k] - \mu \left.\frac{dJ_{LS}(a)}{da}\right|_{a=a[k]}, \tag{6.12}$$

which is the same as (6.5), except that the name of the parameter has changed from x to a. To find the exact form of (6.12) requires the derivative of $J_{LS}(a)$ with respect to the unknown parameter a. This can be approximated by swapping the derivative and the averaging operations, as formalized in (G.13) to give

$$\frac{dJ_{LS}(a)}{da} = \frac{1}{4}\frac{d\text{avg}\{(a^2r^2(kT) - d^2)^2\}}{da}$$

$$\approx \frac{1}{4}\text{avg}\left\{\frac{d(a^2r^2(kT) - d^2)^2}{da}\right\}$$

$$= \text{avg}\{(a^2r^2(kT) - d^2)ar^2(kT)\}.$$

The term $a^2r^2(kT)$ inside the parentheses is equal to $s^2[k]$. The term $ar^2(kT)$ outside the parentheses is not directly available to the assessment mechanism, though it can reasonably be approximated by $\frac{s^2[k]}{a}$. Substituting the derivative into (6.12) and evaluating at $a = a[k]$ gives the algorithm

$$a[k+1] = a[k] - \mu\text{avg}\left\{(s^2[k] - d^2)\frac{s^2[k]}{a[k]}\right\}. \tag{6.13}$$

Care must be taken when implementing (6.13) that $a[k]$ does not approach zero.

Of course, $J_{LS}(a)$ of (6.11) is not the only possible goal for the AGC problem. What is important is not the exact form of the performance function, but where the performance function has its optimal points. Another performance function that has a similar error surface (peek ahead to Figure 6.14) is

$$J_N(a) = \text{avg}\{|a|(\frac{s^2[k]}{3} - d^2)\} = \text{avg}\{|a|(\frac{a^2r^2(kT)}{3} - d^2)\}. \qquad (6.14)$$

Taking the derivative gives

$$\frac{dJ_N(a)}{da} = \frac{d\text{avg}\{|a|(\frac{a^2r^2(kT)}{3} - d^2)\}}{da}$$

$$\approx \text{avg}\{\frac{d|a|(\frac{a^2r^2(kT)}{3} - d^2)}{da}\}$$

$$= \text{avg}\{\text{sgn}(a[k])(s^2[k] - d^2)\},$$

where the approximation arises from swapping the order of the differentiation and the averaging (recall (G.13)) and where the derivative of $|\cdot|$ is the signum or sign function, which holds as long as the argument is nonzero. Evaluating this at $a = a[k]$ and substituting into (6.12) gives another AGC algorithm

$$a[k + 1] = a[k] - \mu \; \text{avg}\{\text{sgn}(a[k])(s^2[k] - d^2)\}. \qquad (6.15)$$

Consider the "logic" of this algorithm. Suppose that a is positive. Since d is fixed,

$$\text{avg}\{\text{sgn}(a[k])(s^2[k] - d^2)\} = \text{avg}\{(s^2[k] - d^2)\} = \text{avg}\{s^2[k]\} - d^2.$$

Thus, if the average energy in $s[k]$ exceeds d^2, a is decreased. If the average energy in $s[k]$ is less than d^2, a is increased. The update ceases when $\text{avg}\{s^2[k]\} \approx d^2$, that is, where $a^2 \approx \frac{d^2}{r^2}$, as desired. (An analogous logic applies when a is negative.)

The two performance functions (6.11) and (6.14) define the updates for the two adaptive elements in (6.13) and (6.15). $J_{LS}(a)$ minimizes the square of the deviation of the power in $s[k]$ from the desired power d^2. This is a kind of "least square" performance function (hence the subscript LS). Such squared-error objectives are common, and will reappear in phase tracking algorithms in Chapter 10, in clock recovery algorithms in Chapter 12, and in equalization algorithms in Chapter 13. On the other hand, the algorithm resulting from $J_N(a)$ has a clear logical interpretation (the N stands for 'naive'), and the update is simpler, since (6.15) has fewer terms and no divisions.

To experiment concretely with these algorithms, `agcgrad.m` provides an implementation in MATLAB. It is easy to control the rate at which $a[k]$ changes by choice of stepsize: a larger μ allows $a[k]$ to change faster, while a smaller μ allows greater smoothing. Thus, μ can be chosen by the system designer to trade off the bandwidth of $a[k]$ (the speed at which $a[k]$ can track variations in the energy levels of the incoming signal) versus the amount of jitter or noise. Similarly, the length over which the averaging is done (specified by the parameter `lenavg`) will also affect the speed of adaptation; longer averages imply slower moving, smoother estimates while shorter averages imply faster moving, more jittery estimates.

---

agcgrad.m: minimize $J(a) = \text{avg}\{|a|((1/3)a^2r^2 - ds)\}$ by choice of a

---

```
n=10000;                              % number of steps in simulation
vr=1.0;                               % power of the input
r=sqrt(vr)*randn(size(1:n));          % generate random inputs
ds=.15;                               % desired power of output = d^2
mu=.001;                              % algorithm stepsize
lenavg=10;                            % length over which to average
a=zeros(size(1:n)); a(1)=1;           % initialize AGC parameter
s=zeros(size(1:n));                   % initialize outputs
avec=zeros(1,lenavg);                 % vector to store terms for averaging
for k=1:n-1
  s(k)=a(k)*r(k);                     % normalize by a(k)
  avec=[sign(a(k))*(s(k)^2-ds),avec(1:end-1)]; % incorporate new update into avec
  a(k+1)=a(k)-mu*mean(avec);          % average adaptive update of a(k)
end
```

Typical output of `agcgrad.m` is shown in Figure 6.13. The gain parameter a adjusts automatically to make the overall power of the output s roughly equal to the specified parameter ds. Using the default values above, where the average power of r is approximately 1, we find that a converges to about 0.38 since $0.38^2 \approx 0.15 = d^2$.

The objective $J_{LS}(a)$ can be implemented similarly by replacing the `avec` calculation inside the `for` loop with

```
avec=[(s(k)^2-ds)*(s(k)^2)/a(k),avec(1:end-1)];
```

In this case, with the default values, a converges to about 0.22, which is the value that minimizes the least square objective $J_{LS}(a)$. Thus, the answer which minimizes $J_{LS}(a)$ is different from the answer which minimizes $J_N(a)$! More on this later.

As it is easy to see when playing with the parameters in `agcgrad.m`, the size of the averaging parameter `lenavg` is relatively unimportant. Even with `lenavg=1`, the algorithms converge and perform approximately the same! This is because the algorithm updates are themselves in the form of a lowpass filter. (See Appendix G for a discussion of the similarity between averagers and lowpass filters.) Removing the averaging from the update gives the simpler form for $J_N(a)$

```
a(k+1)=a(k)-mu*sign(a(k))*(s(k)^2-ds);
```

or, for $J_{LS}(a)$,

```
a(k+1)=a(k)-mu*(s(k)^2-ds)*(s(k)^2)/a(k);
```

Try them!

Perhaps the best way to formally describe how the algorithms work is to plot the performance functions. But it is not possible to directly plot $J_{LS}(a)$ or $J_N(a)$, since they depend on the data sequence $s[k]$. What is possible (and often leads to useful insights) is to plot the performance function averaged over a number of data points (also called the *error surface*). As long as the stepsize is small enough and the average is long enough, the mean behavior of the algorithm will be dictated by the shape of the error surface in the same way that the objective function of the exact steepest descent algorithm (for

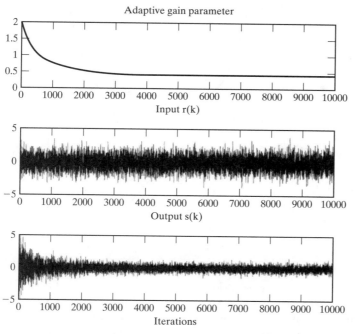

FIGURE 6.13 An automatic gain control adjusts the parameter *a* (in the top panel) automatically to achieve the desired output power.

instance, the objectives (6.4) and (6.8)) dictate the evolution of the algorithms (6.6) and (6.9).

The following code `agcerrorsurf.m` shows how to calculate the error surface for $J_N(a)$: The variable n specifies the number of terms to average over, and `tot` sums up the behavior of the algorithm for all *n* updates at each possible parameter value a. The average of these (`tot/n`) is a close (numerical) approximation to $J_N(a)$ of (6.14). Plotting over all *a* gives the error surface.

---

agcerrorsurf.m: draw error surface

```
n=10000;                        % number of steps in simulation
r=randn(size(1:n));             % generate random inputs
ds=.15;                         % desired power of output = d^2
Jagc=[]; all=-0.7:0.02:0.7;     % all specifies range of values of a
for a=all                       % for each value a
 tot=0;
 for i=1:n
  tot=tot+abs(a)*((1/3)*a^2*r(i)^2-ds); % total cost over all possibilities
 end
 Jagc=[Jagc, tot/n];            % take average value, and save
end
```

Similarly, the error surface for $J_{LS}(a)$ can be plotted by using

```
tot=tot+0.25*(a^2*r(i)^2-ds)^2;  % error surface for JLS
```

The output of `agcerrorsurf.m` for both objective functions is shown in Figure 6.14. Observe that zero (which is a critical point of the error surface) is a local maximum in

FIGURE 6.14 The error surface for the AGC objective functions (6.11) and (6.14) each have two minima. As long as a can be initialized with the correct (positive) sign, there is little danger of converging to the wrong minimum.

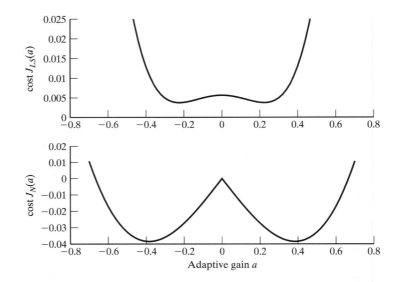

both cases. The final converged answers ($a \approx 0.38$ for $J_N(a)$ and $a \approx 0.22$ for $J_{LS}(a)$) occur at minima. Were the algorithm to be initialized improperly to a negative value, then it would converge to the negative of these values. As with the algorithms in Figure 6.10, examination of the error surfaces shows *why* the algorithms converge as they do. The parameter a descends the error surface until it can go no further.

But why do the two algorithms converge to different places? The facile answer is that they are different because they minimize different performance functions. Indeed, the error surfaces in Figure 6.14 show minima in different locations. The convergent value of $a \approx 0.38$ for $J_N(a)$ is explicable because $0.38^2 \approx 0.15 = d^2$. The convergent value of $a = 0.22$ for $J_{LS}(a)$ is calculated in closed form in Problem 6.18, and this value does a good job minimizing its cost, but it has not necessarily solved the problem of making a^2 close to d^2. Rather, $J_{LS}(a)$ calculates a smaller gain value. After all, $J_{LS}(a)$ contains a fourth power, while $J_N(a)$ contains a third power. Hence the minima are different. The moral is this: Beware your performance functions—they may do what you ask.

Problems

6.17. Use `agcgrad.m` to investigate the AGC algorithm.
   (a) What range of stepsize `mu` works? Can the stepsize be too small? Can the stepsize be too large?
   (b) How does the stepsize `mu` effect the convergence rate?
   (c) How does the variance of the input effect the convergent value of a?
   (d) What range of averages `lenavg` works? Can `lenavg` be too small? Can `lenavg` be too large?
   (e) How does `lenavg` effect the convergence rate?

6.18. Show that the value of a that achieves the minimum of $J_{LS}(a)$ can be expressed as

$$\pm\sqrt{\frac{d^2 \sum_k r_k^2}{\sum_k r_k^4}}.$$

Is there a way to use this (closed form) solution to replace the iteration (6.13)?

6.19. Consider the alternative objective function $J(a) = \frac{1}{2}a^2(\frac{1}{2}\frac{s^2[k]}{3} - d^2)$. Calculate the derivative and implement a variation of the AGC algorithm that minimizes this objective. How does this version compare to the algorithms (6.13) and (6.15)? Draw the error surface for this algorithm. Which version is preferable?

6.20. Try initializing the estimate `a(1)=-2` in `agcgrad.m`. Which minimum does the algorithm find? What happens to the data record?

6.21. Create your own objective function $J(a)$ for the AGC problem. Calculate the derivative and implement a variation of the AGC algorithm that minimizes this objective. How does this version compare to the algorithms (6.13) and (6.15)? Draw the error surface for your algorithm. Which version do you prefer?

6.22. Investigate how the error surface depends on the input signal. Replace `randn` with `rand` in `agcerrorsurf.m` and draw the error surfaces for both $J_N(a)$ and $J_{LS}(a)$.

6.8 Using an AGC to Combat Fading

One of the impairments encountered in transmission systems is the degradation due to fading, when the strength of the received signal changes in response to changes in the transmission path. (Recall the discussion in Section 4.1.5 on page 64.) This section shows how an AGC can be used to counteract the fading, assuming the rate of the fading is slow, and provided the signal does not disappear completely.

Suppose that the input consists of a random sequence undulating slowly up and down in magnitude, as in the top plot of Figure 6.15. The adaptive AGC compensates for the amplitude variations, growing small when the power of the input is large, and large when the power of the input is small. This is shown in the middle graph. The resulting output is of roughly constant amplitude, as shown in the bottom plot of Figure 6.15.

This figure was generated using the following code:

---

agcvsfading.m: compensating for fading with an AGC

```
n=50000;                              % number of steps in simulation
r=randn(1,n);                         % generate raw random inputs
env=0.75+abs(sin(2*pi*(1:n)/n));      % the fading profile
r=r.*env;                             % apply profile to raw input r[k]
ds=.5;                                % desired power of output = d^2
a=zeros(size(1:n)); a(1)=1;           % initialize AGC parameter
s=zeros(size(1:n));                   % initialize outputs
mu=.01;                               % algorithm stepsize
for k=1:n-1
  s(k)=a(k)*r(k);                     % normalize by a(k) to get s[k]
  a(k+1)=a(k)-mu*(s(k)^2-ds);         % adaptive update of a(k)
end
```

The "fading profile" defined by the vector `env` is slow compared with the rate at which the adaptive gain moves, which allows the gain to track the changes. Also, the

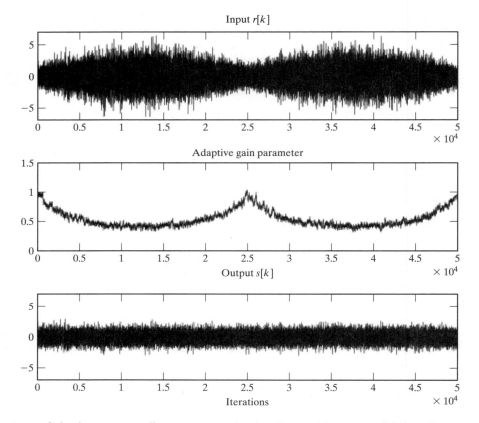

power of the input never dies away completely. The problems that follow ask you to investigate what happens in more extreme situations.

Problems

6.23. Mimic the code in `agcvsfading.m` to investigate what happens when the input signal dies away. (Try removing the `abs` command from the fading profile variable.) Can you explain what you see?

6.24. Mimic the code in `agcvsfading.m` to investigate what happens when the power of the input signal varies rapidly. What happens if the sign of gain estimate is incorrect?

6.25. Would the answers to the previous two problems change if using algorithm (6.13) instead of (6.15)?

6.9 Summary

Sampling transforms a continuous-time analog signal into a discrete-time digital signal. In the time domain, this can be viewed as a multiplication by a train of pulses. In the frequency domain this corresponds to a replication of the spectrum. As long as the sampling rate is fast enough that the replicated spectra do not overlap, the sampling process is reversible; that is, the original analog signal can be reconstructed from the samples.

An AGC can be used to make sure that the power of the analog signal remains in the region where the sampling device operates effectively. The same AGC, when adaptive, can also provide a protection against signal fades. The AGC can be designed using a steepest descent (optimization) algorithm that updates the adaptive parameter by moving in the direction of the negative of the derivative. This steepest descent approach to the solution of optimization problems will be used throughout *Telecommunication Breakdown*.

6.10 For Further Reading

Details about resampling procedures are available in the published works of

- Smith, J. O. "Bandlimited interpolation—interpretation and algorithm," 1993,

which is available at his website at http://ccrma-www.stanford.edu/~jos/resample/.

A general introduction to adaptive algorithms centered around the steepest descent approach can be found in

- B. Widrow and S. D. Stearns, *Adaptive Signal Processing,* Prentice-Hall, 1985.

One of our favorite discussions of adaptive methods is

- C. R. Johnson Jr., *Lectures on Adaptive Parameter Estimation,* Prentice-Hall, 1988.

This whole book can be found in .pdf form on the CD accompanying this book.

7 Digital Filtering and the DFT

Digital filtering is not simply converting from analog to digital filters;
it is a fundamentally different way of thinking about the topic of signal
processing, and many of the ideas and limitations of the analog method
have no counterpart in digital form
—R. W. Hamming, *Digital Filters*, 3d ed., Prentice Hall 1989

Once the received signal is sampled, the real story of the digital receiver begins.

An analog bandpass filter at the front end of the receiver removes extraneous signals (for instance, it removes television frequency signals from a radio receiver) but some portion of the signal from other FDM users may remain. While it would be conceptually possible to remove all but the desired user at the start, accurate retunable analog filters are complicated and expensive to implement. Digital filters, on the other hand, are easy to design, inexpensive (once the appropriate DSP hardware is present) and easy to retune. The job of cleaning up out-of-band interferences left over by the analog BPF can be left to the digital portion of the receiver.

Of course, there are many other uses for digital filters in the receiver, and this chapter focuses on how to "build" digital filters. The discussion begins by considering the digital impulse response and the related notion of discrete-time convolution. Conceptually, this closely parallels the discussion of linear systems in Chapter 4. The meaning of the DFT (discrete Fourier transform) closely parallels the meaning of the Fourier transform, and several examples encourage fluency in the spectral analysis of discrete data signals. The final section on practical filtering shows how to design digital filters with (more or less) any desired frequency response by using special MATLAB commands.

7.1 Discrete Time and Discrete Frequency

The study of discrete-time (digital) signals and systems parallels that of continuous-time (analog) signals and systems. Many digital processes are fundamentally simpler than their analog counterparts, though there are a few subtleties unique to discrete-time implementations. This section begins with a brief overview and comparison, and then proceeds to discuss the DFT, which is the discrete counterpart of the Fourier transform.

Just as the impulse function $\delta(t)$ plays a key role in defining signals and systems in continuous time, the discrete pulse

$$\delta[k] = \begin{cases} 1 & k = 0 \\ 0 & k \neq 0 \end{cases} \tag{7.1}$$

can be used to decompose discrete signals and to characterize discrete-time systems.[1]
Any discrete-time signal can be written as a linear combination of discrete impulses. For
instance, if the signal $w[k]$ is the repeating pattern $\{-1, 1, 2, 1, -1, 1, 2, 1, \ldots\}$, it can
be written

$$w[k] = -\delta[k] + \delta[k-1] + 2\delta[k-2] + \delta[k-3] - \delta[k-4] + \delta[k-5] + 2\delta[k-6] + \delta[k-7]\ldots$$

In general, the discrete time signal $w[k]$ can be written

$$w[k] = \sum_{j=-\infty}^{\infty} w[j]\, \delta[k-j].$$

This is the discrete analog of the sifting property (4.4); simply replace the integral with
a sum, and replace $\delta(t)$ with $\delta[k]$.

Like their continuous-time counterparts, discrete-time systems map input signals into
output signals. Discrete-time linear systems are characterized by an impulse response
$h[k]$, which is the output of the system when the input is an impulse, though, of course,
(7.1) is used instead of (4.2). When an input $x[k]$ is more complicated than a single pulse,
the output $y[k]$ can be calculated by summing all the responses to all the individual terms,
and this leads directly to the definition of discrete-time convolution:

$$y[k] = \sum_{j=-\infty}^{\infty} x[j]\, h[k-j]$$
$$\equiv x[k] * h[k]. \tag{7.2}$$

Observe that the convolution of discrete-time sequences appears in the reconstruction
formula (6.3), and that (7.2) parallels continuous-time convolution in (4.8) with the
integral replaced by a sum and the impulse response $h(t)$ replaced by $h[k]$.

The discrete-time counterpart of the Fourier transform is the discrete Fourier transform
(DFT). Like the Fourier transform, the DFT decomposes signals into their constituent
sinusoidal components. Like the Fourier transform, the DFT provides an elegant way to
understand the behavior of linear systems by looking at the frequency response (which
is equal to the DFT of the impulse response). Like the Fourier transform, the DFT is an
invertible, information preserving transformation.

The DFT differs from the Fourier transform in three useful ways. First, it applies
to discrete-time sequences, which can be stored and manipulated directly in computers
(rather than to analog waveforms, which cannot be directly stored in digital computers).
Second, it is a sum rather than an integral, and so is easy to implement in either hardware
or software. Third, it operates on a finite data record, rather than an integration over all
time. Given a data record (or vector) $w[k]$ of length N, the DFT is defined by

$$W[n] = \sum_{k=0}^{N-1} w[k] e^{-j(2\pi/N)nk} \quad n = 0, 1, 2, \ldots, N-1. \tag{7.3}$$

---

[1] The pulse in discrete time is considerably more straightforward than the implicit definition of the continuous-
time impulse function in (4.2) and (4.3).

For each value n, (7.3) multiplies each term of the data by a complex exponential, and then sums. Compare this to the Fourier transform; for each frequency f, (2.1) multiplies each point of the waveform by a complex exponential, and then integrates. Thus $W[n]$ is a kind of frequency function in the same way that $W(f)$ is a function of frequency. The next section will make this relationship explicit by showing how $e^{-j(2\pi/N)nk}$ can be viewed as a discrete time sinusoid with frequency proportional to n. Just as a plot of the frequency function $W(f)$ is called the spectrum of the signal $w(t)$, plots of the frequency function $W[n]$ are called the (discrete) spectrum of the signal $w[k]$. One source of confusion is that the frequency f in the Fourier transform can take on any value while the frequencies present in (7.3) are all integer multiples n of a single fundamental with frequency $2\pi/N$. This fundamental is precisely the sine wave with period equal to the length N of the window over which the DFT is taken. Thus, the frequencies in (7.3) are constrained to a discrete set; these are the "discrete frequencies" of the section title.

The most common implementation of the DFT is called the *fast Fourier transform* (FFT), which is an elegant way to rearrange the calculations in (7.3) so that it is computationally efficient. For all purposes other than numerical efficiency, the DFT and the FFT are synonymous.

Like the Fourier transform, the DFT is invertible. Its inverse, the IDFT, is defined by

$$w[k] = \frac{1}{N} \sum_{n=0}^{N-1} W[n]e^{j(2\pi/N)nk} \quad k = 0, 1, 2, \ldots, N-1. \tag{7.4}$$

The IDFT takes each point of the frequency function $W[n]$, multiplies by a complex exponential, and sums. Compare this with the IFT; (D.2) takes each point of the frequency function $W(f)$, multiplies by a complex exponential, and integrates. Thus, the Fourier transform and the DFT translate from the time domain into the frequency domain, while the inverse Fourier transform and the IDFT translate from frequency back into time.

Many other aspects of continuous-time signals and systems have analogs in discrete time. Following are some that will be useful in later chapters:

- Symmetry—If the time signal $w[k]$ is real, then $W^*[n] = W[N-n]$. This is analogous to (A.35).

- Parseval's theorem holds in discrete time—$\sum_k w^2[k] = \frac{1}{N} \sum_n |W[n]|^2$. This is analogous to (A.43).

- The frequency response $H[n]$ of a linear system is the DFT of the impulse response $h[k]$. This is analogous to the continuous-time result that the frequency response $H(f)$ is the Fourier transform of the impulse response $h(t)$.

- Time delay property in discrete time—$w[k-l] \Leftrightarrow W[n]e^{-j(2\pi/N)l}$. This is analogous to (A.38).

- Modulation property—This frequency shifting property is analogous to (A.34).

- If $w[k] = \sin(\frac{2\pi f k}{T})$ is a periodic sine wave, then the spectrum is a sum of two delta impulses. This is analogous to the result in Example 4.1.

- Convolution[2] in (discrete) time is the same as multiplication in (discrete) frequency. This is analogous to (4.10).

- Multiplication in (discrete) time is the same as convolution in (discrete) frequency. This is analogous to (4.11).

- The transfer function of a linear system is the ratio of the DFT of the output and the DFT of the input. This is analogous to (4.12).

Problems

7.1. Show why Parseval's theorem is true in discrete time.

7.2. Suppose a filter has impulse response $h[k]$. When the input is $x[k]$, the output is $y[k]$. Show that, if the input is $x_d[k] = x[k] - x[k-1]$, then the output is $y_d[k] = y[k] - y[k-1]$. Compare this result with Problem 4.13.

7.3. Let $w[k] = \sin(2\pi k/N)$ for $k = 1, 2, \ldots, N-1$. Use the definitions (7.3) and (7.4) to find the corresponding values of $W[n]$.

7.1.1 Understanding the DFT

Define a vector $\mathbf{W}$ containing all N frequency values $W[n]$, $n = 1, 2, \ldots, N-1$, and a vector $\mathbf{w}$ containing all N time values $w[k]$, $k = 1, 2, \ldots, N-1$. Then the IDFT equation (7.4) can be rewritten as a matrix multiplication

$$
\mathbf{w} = \begin{bmatrix} w[0] \\ w[1] \\ w[2] \\ w[3] \\ \vdots \\ w[N-1] \end{bmatrix}
$$

$$
= \frac{1}{N} \begin{bmatrix} 1 & 1 & 1 & 1 & \cdots & 1 \\ 1 & e^{j2\pi/N} & e^{j4\pi/N} & e^{j6\pi/N} & \cdots & e^{j2\pi(N-1)/N} \\ 1 & e^{j4\pi/N} & e^{j8\pi/N} & e^{j12\pi/N} & \cdots & e^{j4\pi(N-1)/N} \\ 1 & e^{j6\pi/N} & e^{j12\pi/N} & e^{j18\pi/N} & \cdots & e^{j6\pi(N-1)/N} \\ \vdots & \vdots & \vdots & \vdots & & \vdots \\ 1 & e^{j2(N-1)\pi/N} & e^{j4(N-1)\pi/N} & e^{j6(N-1)\pi/N} & \cdots & e^{j2(N-1)^2\pi/N} \end{bmatrix} \begin{bmatrix} W[0] \\ W[1] \\ W[2] \\ W[3] \\ \vdots \\ W[N-1] \end{bmatrix}
$$

$$
\equiv \frac{1}{N} M^{-1} \mathbf{W}, \tag{7.5}
$$

where the matrix $\frac{1}{N} M^{-1}$ (a matrix of columns of complex exponentials) defines the IDFT operation. The DFT is defined similarly by

$$
\mathbf{W} = NM\mathbf{w}. \tag{7.6}
$$

---

[2] To be precise, this should be *circular convolution*. However, for the purposes of designing a workable receiver, this distinction is not essential. The interested reader can explore the relationship of discrete-time convolution in the time and frequency domains in a concrete way using `waystofilt.m` on page 133.

Since the inverse of an orthonormal matrix is equal to its own complex conjugate transpose, M in (7.6) is the same as M^{-1} in (7.5) with the signs on all the exponents flipped.

The matrix M^{-1} is highly structured. Letting C_n be the nth column of M^{-1} and multiplying both sides by N, (7.5) can be rewritten as

$$
N\mathbf{w} = W[0]\begin{bmatrix} 1 \\ 1 \\ 1 \\ 1 \\ \vdots \\ 1 \end{bmatrix} + W[1]\begin{bmatrix} 1 \\ e^{j2\pi/N} \\ e^{j4\pi/N} \\ e^{j6\pi/N} \\ \vdots \\ e^{j2\pi(N-1)/N} \end{bmatrix} + \ldots + W[N-1]\begin{bmatrix} 1 \\ e^{j2(N-1)\pi/N} \\ e^{j4(N-1)\pi/N} \\ e^{j6(N-1)\pi/N} \\ \vdots \\ e^{j2(N-1)^2\pi/N} \end{bmatrix}
$$

$$
= W[0]\, C_0 + W[1]\, C_1 + \ldots + W[N-1]\, C_{N-1} \tag{7.7}
$$

$$
= \sum_{n=0}^{N-1} W[n]C_n.
$$

This form displays the time vector $\mathbf{w}$ as a linear combination[3] of the columns C_n. What are these columns? They are vectors of discrete (complex valued) sinusoids, each at a different frequency. Accordingly, the DFT reexpresses the time vector as a linear combination of these sinusoids. The complex scaling factors $W[n]$ define how much of each sinusoid is present in the original signal $w[k]$.

To see how this works, consider the first few columns. C_0 is a vector of all ones; it is the zero frequency sinusoid, or DC. C_1 is more interesting. The ith element of C_1 is $e^{j2i\pi/N}$, which means that as i goes from 0 to $N-1$, the exponential assumes N uniformly spaced points around the unit circle. This is clearer in polar coordinates, where the magnitude is always unity and the angle is $2i\pi/N$ radians. Thus, C_1 is the lowest frequency sinusoid that can be represented (other than DC); it is the sinusoid that fits exactly one period in the time interval NT_s, where T_s is the distance in time between adjacent samples. C_2 is similar, except that the ith element is $e^{j4i\pi/N}$. Again, the magnitude is unity and the phase is $4i\pi/N$ radians. Thus, as i goes from 0 to $N-1$, the elements are N uniformly spaced points which go around the circle twice. Thus, C_2 has frequency twice that of C_1, and it represents a complex sinusoid that fits exactly two periods into the time interval NT_s. Similarly, C_n represents a complex sinusoid of frequency n times that of C_1; it orbits the circle n times and is the sinusoid that fits exactly n periods in the time interval NT_s.

One subtlety that can cause confusion is that the sinusoids in C_i are complex valued, yet, most signals of interest are real. Recall from Euler's identities (2.3) and (A.3) that the real-valued sine and cosine can each be written as a sum of two complex valued exponentials that have exponents with opposite signs. The DFT handles this elegantly. Consider C_{N-1}. This is

$$
\begin{bmatrix} 1, & e^{j2(N-1)\pi/N}, & e^{j4(N-1)\pi/N}, & e^{j6(N-1)\pi/N}, \ldots, & e^{j2(N-1)^2\pi/N} \end{bmatrix}^T,
$$

---

[3] Those familiar with advanced linear algebra will recognize that M^{-1} can be thought of as a change of basis that reexpresses $\mathbf{w}$ in a basis defined by the columns of M^{-1}.

which can be rewritten as

$$\left[1, \ e^{-j2\pi/N}, \ e^{-j4\pi/N}, \ e^{-j6\pi/N}, \ldots, \ e^{-j2\pi(N-1)/N}\right]^T,$$

since $e^{-j2\pi} = 1$. Thus, the elements of C_{N-1} are identical to the elements of C_1, except that the exponents have the opposite sign, implying that the angle of the ith entry in C_{N-1} is $-2i\pi/N$ radians. Thus, as i goes from 0 to $N - 1$, the exponential assumes N uniformly spaced points around the unit circle, in the opposite direction from C_1. This is the meaning of what might be interpreted as "negative frequencies" that show up when taking the DFT. The complex exponential proceeds in a (negative) clockwise manner around the unit circle, rather than in a (positive) counterclockwise direction. But it takes *both* to make a real valued sine or cosine, as Euler's formula shows. For real valued sinusoids of frequency $2\pi n/N$, both $W[n]$ and $W[N - n]$ are nonzero and equal in magnitude.[4]

Problem

7.4. Which column C_i represents the highest possible frequency in the DFT? What do the elements of this column look like? Hint: Look at $C_{N/2}$ and think of a square wave. This "square wave" is the highest frequency that can be represented by the DFT, and occurs at exactly the Nyquist rate.

7.1.2 Using the DFT

Fortunately, MATLAB makes it easy to do spectral analysis with the DFT by providing a number of simple commands that carry out the required calculations and manipulations. It is not necessary to program the sum (7.3) or the matrix multiplication (7.5). The single line commands **W** = fft(**w**) and **w** = ifft(**W**) invoke efficient FFT (and IFFT) routines when possible, and relatively inefficient DFT (and IDFT) calculations otherwise. The numerical idiosyncrasies are completely transparent, with one annoying exception. In MATLAB, all vectors, including **W** and **w**, must be indexed from 1 to N instead of from 0 to $N - 1$.

While the FFT/IFFT commands are easy to invoke, their meaning is not always instantly transparent. The intent of this section is to provide some examples that show how to interpret (and how *not* to interpret) the frequency analysis commands in MATLAB.

Begin with a simple sine wave of frequency f sampled every T_s seconds, as is familiar from previous programs such as speccos.m. The first step in any frequency analysis is to define the window over which the analysis will take place, since the FFT/DFT must operate on a finite data record. The program specsin0.m defines the length of the analysis with the variable N (powers of two make for fast calculations), and then analyzes the first N samples of w. It is tempting to simply invoke the MATLAB commands fft and to plot the results. Typing plot(fft(w(1:N))) gives a meaningless answer (try it!) because the output of the fft command is a vector of complex numbers. When MATLAB plots complex numbers, it plots the real vs. the imaginary parts. In order to view the magnitude spectrum, first use the abs command, as shown in specsin0.m.

---

[4] Since $W[n] = W^*[N - n]$ by the discrete version of the symmetry property (A.35), the magnitudes are equal but the phases have opposite signs.

---

specsin0.m: naive and deceptive spectrum of a sine wave via the FFT

```
f=100; Ts=1/1000; time=5.0;        % freq, sampling interval, time
t=Ts:Ts:time;                      % define a time vector
w=sin(2*pi*f*t);                   % define the sinusoid
N=2^10;                            % size of analysis window
fw=abs(fft(w(1:N)));               % find magnitude of DFT/FFT
plot(fw)                           % plot the waveform
```

Running this program results in a plot of the magnitude of the output of the FFT analysis of the waveform w. The top plot in Figure 7.1 shows two large spikes, one near "100" and one near "900". What do these mean? Try a simple experiment. Change the value of N from 2^{10} to 2^{11}. This is shown in the bottom plot of Figure 7.1, where the two spikes now occur at about "200" and at about "1850". But the frequency of the sine wave hasn't changed! It does not seem reasonable that the window over which the analysis is done should change the frequencies in the signal.

There are two problems. First, specsin0.m plots the magnitude data against the index of the vector fw, and this index (by itself) is meaningless. The discussion surrounding (7.7) shows that each element of $W[n]$ represents a scaling of the complex sinusoid with frequency $e^{j2\pi n/N}$. Hence, these indices must be scaled by the time over which the analysis is conducted, which involves both the sampling interval and the number of points in the FFT analysis. The second problem is the ordering of the frequencies. Like

FIGURE 7.1 Naive and deceptive plots of the spectrum of a sine wave in which the frequency of the analyzed wave appears to depend on the size N of the analysis window. The top figure has $N = 2^{10}$, while the bottom uses $N = 2^{11}$.

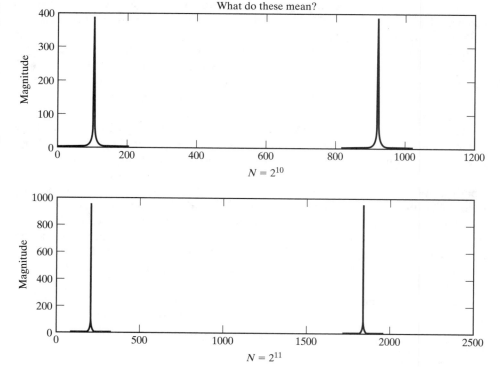

the columns C_n of the DFT matrix M (7.6), the frequencies represented by the $W[N - n]$ are the negative of the frequencies represented by $W[n]$.

There are two solutions. The first is appropriate only when the original signal is real valued. In this case, the $W[n]$'s are symmetric and there is no extra information contained in the negative frequencies. This suggests plotting only the positive frequencies, a strategy that is followed in `specsin1.m`.

---

specsin1.m: spectrum of a sine wave via the FFT/DFT

```
f=100; Ts=1/1000; time=5.0;        % freq, sampling interval, time
t=Ts:Ts:time;                      % define a time vector
w=sin(2*pi*f*t);                   % define the sinusoid
N=2^10;                            % size of analysis window
ssf=(0:N/2-1)/(Ts*N);              % frequency vector
fw=abs(fft(w(1:N)));               % find magnitude of DFT/FFT
plot(ssf,fw(1:N/2))                % plot for positive freq. only
```

The output of `specsin1.m` is shown in the top plot of Figure 7.2. The magnitude spectrum shows a single spike at 100 Hz, as is expected. Change f to other values, and observe that the location of the peak in frequency moves accordingly. Change the width and location of the analysis window N and verify that the location of the peak does not change. Change the sampling interval Ts and verify that the analyzed peak remains at the same frequency.

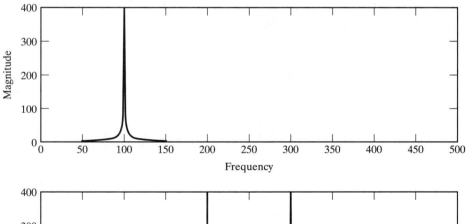

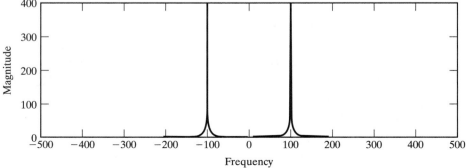

FIGURE 7.2 Proper use the FFT command can be done as in `specsin1.m` (the top graph), which plots only the positive frequencies, or as in `specsin2.m` (the bottom graph), which shows the full magnitude spectrum symmetric about $f = 0$.

The second solution requires more bookkeeping of indices, but gives plots that more closely accord with continuous-time intuition and graphs. `specsin2.m` exploits the built in function `fftshift`, which shuffles the output of the FFT command so that the negative frequencies occur on the left, the positive frequencies on the right, and DC in the middle.

---

specsin2.m: spectrum of a sine wave via the FFT/DFT

```
f=100; Ts=1/1000; time=10.0;       % freq, sampling interval, time
t=Ts:Ts:time;                      % define a time vector
w=sin(2*pi*f*t);                   % define the sinusoid
N=2^10;                            % size of analysis window
ssf=(-N/2:N/2-1)/(Ts*N);           % frequency vector
fw=fft(w(1:N));                    % do DFT/FFT
fws=fftshift(fw);                  % shift it for plotting
plot(ssf,abs(fws))                 % plot magnitude spectrum
```

Running this program results in the bottom plot of Figure 7.2, which shows the complete magnitude spectrum for both positive and negative frequencies. It is also easy to plot the phase spectrum by substituting `phase` for `abs` in either of the preceding two programs.

Problems

7.5. Explore the limits of the FFT/DFT technique by choosing extreme values. What happens when:

   (a) `f` becomes too large? Try $f = 200, 300, 450, 550, 600, 800, 2200$ Hz. Comment on the relationship between `f` and `Ts`.

   (b) `Ts` becomes to large? Try $Ts = 1/500, 1/250, 1/50$. Comment on the relationship between `f` and `Ts`. (You may have to increase `time` in order to have enough samples to operate on.)

   (c) `N` becomes too large or too small? What happens to the location in the peak of the magnitude spectrum when $N = 2^{11}, 2^{14}, 2^8, 2^4, 2^2, 2^{20}$? What happens to the width of the peak in each of these cases? (You may have to increase `time` in order to have enough samples to operate on).

7.6. Replace the `sin` function with $\sin^2$. Use

```
w=sin(2*pi*f*t).^2
```

What is the spectrum of $\sin^2$? What is the spectrum of $\sin^3$? Consider $\sin^k$. What is the largest k for which the results make sense? Explain what limitations there are.

7.7. Replace the sin function with sinc. What is the spectrum of the sinc function? What is the spectrum of sinc^2?

7.8. Plot the spectrum of $w(t) = sin(t) + je^{-t}$. Should you use the technique of `specsin1.m` or of `specsin2.m`? Hint: Think symmetry.

7.9. The FFT of a real sequence is typically complex, and sometimes it is important to look at the phase (as well as the magnitude).

(a) Let `w=sin(2*pi*f*t+phi)`. For `phi=0, 0.2, 0.4, 0.8, 1.5, 3.14`, find the phase of the FFT output at the frequencies $\pm f$.

(b) Find the phase of the output of the FFT when

```
w=sin(2*pi*f*t+phi).^2
```

These are all examples of "simple" functions, which can be investigated (in principle, anyway) analytically. The greatest strength of the FFT/DFT is that it can also be used for the analysis of data when no functional form is known. There is a data file on the CD called `gong.wav`, which is a sound recording of an Indonesian gong (a large struck metal plate). The following code reads in the waveform and analyzes its spectrum using the FFT. Make sure that the file `gong.wav` is in an active MATLAB path, or you will get a "file not found" error. If there is a sound card (and speakers) attached, the `sound` command plays the `.wav` file at the sampling rate $f_s = 1/T_s$.

---

specgong.m find spectrum of the "gong" sound

```
filename='gong.wav';                  % name of wave file goes here
[x,sr]=wavread(filename);             % read in wavefile
Ts=1/sr; siz=length(x);               % sample interval and # of samples
N=2^16; x=x(1:N)';                    % length for analysis
sound(x,1/Ts)                         % play sound, if sound card installed
time=Ts*(0:length(x)-1);             % establish time base for plotting
subplot(2,1,1), plot(time,x)          % and plot top figure
magx=abs(fft(x));                     % take FFT magnitude
ssf=(0:N/2-1)/(Ts*N);                 % establish freq base for plotting
subplot(2,1,2), plot(ssf,magx(1:N/2)) % plot mag spectrum
```

Running `specgong.m` results in the plot shown in Figure 7.3. The top figure shows the time behavior of the sound as it rises very quickly (when the gong is struck) and then slowly decays over about 1.5 seconds. The variable `N` defines the window over which the frequency analysis occurs. The middle plot shows the complete spectrum, and the bottom plot zooms in on the low frequency portion where the largest spikes occur. This sound consists primarily of three major frequencies, at about 520, 630, and 660 Hz. Physically, these represent the three largest resonant modes of the vibrating plate.

With $N = 2^{16}$, `specgong.m` analyzes approximately 1.5 seconds (`Ts*N` seconds, to be precise). It is reasonable to suppose that the gong might undergo important transients during the first few milliseconds. This can be investigated by decreasing `N` and applying the DFT to different segments of the data record.

Problems

7.10. Determine the spectrum of the gong sound during the first 0.1 seconds. What value of `N` is needed? Compare this to the spectrum of a 0.1 second segment chosen from the middle of the sound. How do they differ?

FIGURE 7.3 Time and frequency plots of the gong waveform. The top figure shows the decay of the signal over 1.5 seconds. The middle figure shows the magnitude spectrum, and the bottom figure zooms in on the low frequency portion so that the frequencies are more legible.

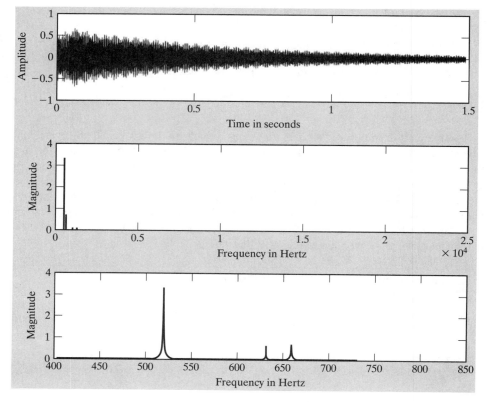

7.11. A common practice when taking FFTs is to plot the magnitude on a log scale. This can be done in MATLAB by replacing the `plot` command with `semilogy`. Try it in `specgong.m`. What extra details can you see?

7.12. The waveform of another, much larger gong is given in `gong2.wav` on the CD. Conduct a thorough analysis of this sound, looking at the spectrum for a variety of analysis windows (values of N) and at a variety of times within the waveform.

7.13. Choose a .wav file from the CD (in the Sounds folder) or download a .wav file of a song from the Internet. Conduct a FFT analysis of the first few seconds of sound, and then another analysis in the middle of the song. How do the two compare? Can you correlate the FFT analysis with the pitch of the material? With the rhythm? With the sound quality?

The key factors in a DFT or FFT based frequency analysis are as follows:

- The sampling interval T_s is the time resolution, the shortest time over which any event can be observed. The sampling rate $f_s = \frac{1}{T_s}$ is inversely proportional.
- The total time is $T = NT_s$ where N is the number of samples in the analysis.
- The frequency resolution is $\frac{1}{T} = \frac{1}{NT_s} = \frac{f_s}{N}$. Sinusoids closer together (in frequency) than this value are indistinguishable.

For instance, in the analysis of the gong conducted in `specgong.m`, the sampling inter-val $T_s = \frac{1}{44100}$ is defined by the recording. With $N = 2^{16}$, the total time is $NT_s = 1.48$ seconds, and the frequency resolution is $\frac{1}{NT_s} = 0.67$ Hz.

Sometimes the total absolute time T is fixed. Sampling faster decreases T_s and increases N, but cannot give better resolution in frequency. Sometimes it is possible to increase the total time. Assuming a fixed T_s, this implies an increase in N and bet-ter frequency resolution. Assuming a fixed N, this implies an increase in T_s and worse resolution in time. Thus, better resolution in frequency means worse resolution in time. Conversely, better resolution in time means worse resolution in frequency. If this is still confusing, or if you would like to see it from a different perspective, check out Appendix D.

The DFT is a key tool in analyzing and understanding the behavior of communications systems. Whenever data flows through a system, it is a good idea to plot it as a function of time, and also to plot it as a function of frequency; that is, to look at it in the time domain and in the frequency domain. Often, aspects of the data that are clearer in time are hard to see in frequency, and aspects that are obvious in frequency are obscure in time. Using both points of view is common sense.

7.2 Practical Filtering

Filtering can be viewed as the process of emphasizing or attenuating certain frequen-cies within a signal. Linear filters are common because they are easy to understand and straightforward to implement. Whether in discrete or continuous time, a linear filter is characterized by its impulse response (i.e., its output when the input is an impulse). The process of convolution aggregates the impulse responses from all the input instants into a formula for the output. It is hard to visualize the action of convolution directly in the time domain, making analysis in the frequency domain an important concep-tual tool. The Fourier transform (or the DFT in discrete time) of the impulse response gives the frequency response, which is easily interpreted as a plot that shows how much gain or attenuation (or phase shift) each frequency undergoes by the filtering operation. Thus, while implementing the filter in the time domain as a convolution, it is normal to specify, design, and understand it in the frequency domain as a point-by-point multiplication of the spectrum of the input and the frequency response of the filter.

In principle, this provides a method not only of understanding the action of a fil-ter, but also of designing a filter. Suppose that a particular frequency response is desired, say one that removes certain frequencies, while leaving others unchanged. For example, if the noise is known to lie in one frequency band while the impor-tant signal lies in another frequency band, then it is natural to design a filter that removes the noisy frequencies and passes the signal frequencies. This intuitive notion translates directly into a mathematical specification for the frequency response. The impulse response can then be calculated directly by taking the inverse transform, and this impulse response defines the desired filter. While this is the basic principle of filter design, there are a number of subtleties that can arise, and sophisticated routines are available in MATLAB that make the filter design process flexible, even if they are not foolproof.

Filters can be classified in several ways:

- Lowpass filters (LPF) try to pass all frequencies below some cutoff frequency and remove all frequencies above.

- Highpass filters try to pass all frequencies above some specified value and remove all frequencies below.

- Notch (or bandstop) filters try to remove particular frequencies (usually in a narrow band) and to pass all others.

- Bandpass filters try to pass all frequencies in a particular range and to reject all others.

The region of frequencies allowed to pass through a filter is called the *passband*, while the region of frequencies removed is called the *stopband*. Sometimes there is a region between where it is relatively less important what happens, and this is called the *transition band*.

By linearity, more complex filter specifications can be implemented as sums and concatenations of the above basic filter types. For instance, if $h_1[k]$ is the impulse response of a bandpass filter that passes only frequencies between 100 and 200 Hz, and $h_2[k]$ is the impulse response of a bandpass filter that passes only frequencies between 500 and 600 Hz, then $h[k] = h_1[k] + h_2[k]$ passes only frequencies between 100 and 200 Hz or between 500 and 600 Hz. Similarly, if $h_l[k]$ is the impulse response of a lowpass filter that passes all frequencies below 600 Hz, and $h_h[k]$ is the impulse response of a highpass filter that passes all frequencies above 500 Hz, then $h[k] = h_l[k] * h_h[k]$ is a bandpass filter that passes only frequencies between 500 and 600 Hz.

For the most part, *Telecommunication Breakdown* talks about filters in which the passband is flat because these are the most common filters in a typical receiver. Other filter profiles are possible, and the techniques of filter design are not restricted to flat passbands.

The next section shows how such (digital) filters can be implemented in MATLAB. The succeeding sections shows how to design filters, and how they behave on a number of test signals.

7.2.1 Implementing Filters

Suppose that the impulse response of a discrete-time filter is $h[i]$, $i = 0, 1, 2, \ldots, N - 1$. If the input to the filter is the sequence $x[i]$, $i = 0, 1, \ldots, M - 1$, then the output is given by the convolution Equation (7.2). There are four ways to implement this filtering in MATLAB:

- `conv` directly implements the convolution equation and outputs a vector of length $N + M - 1$.

- `filter` implements the convolution so as to supply one output value for each input value; the output is of length M.

- In the frequency domain, take the FFT of the input, the FFT of the output, multiply the two, and take the IFFT to return to the time domain.

- In the time domain, pass through the input data, at each time multiplying by the impulse response and summing the result.

Probably the easiest way to see the differences is to play with the four methods.

---

waystofilt.m "conv" vs. "filter" vs. "freq domain" vs. "time domain"

```
h=[1 -1 2 -2 3 -3];                        % impulse response h[k]
x=[1 2 3 4 5 6 -5 -4 -3 -2 -1];            % input data x[k]
yconv=conv(h,x)                            % convolve x[k]*h[k]
yfilt=filter(h,1,x)                        % filter x[k] with h[k]
n=length(h)+length(x)-1;                   % pad length for FFT
ffth=fft([h zeros(1,n-length(h))]);        % FFT of impulse response = H[n]
fftx=fft([x, zeros(1,n-length(x))]);       % FFT of input = X[n]
ffty=ffth.*fftx;                           % product of H[n] and X[n]
yfreq=real(ifft(ffty))                     % IFFT of product gives y[k]
z=[zeros(1,length(h)-1),x];                % initial state in filter = 0
for k=1:length(x)                          % time domain method
  ytim(k)=fliplr(h)*z(k:k+length(h)-1)';   % iterates once for each x[k]
end                                        % to directly calculate y[k]
```

Observe that the first M terms of yconv, yfilt, yfreq, and ytim are the same, but that both yconv and yfreq have N-1 extra values at the end. For both the time domain method and the filter command, the output values are aligned in time with the input values, one output for each input. Effectively, the filter command is a single line implementation of the time domain for loop.

For the FFT method, the two vectors (input and convolution) must both have length N+M-1. The raw output has complex values due to numerical roundoff, and the command real is used to strip away the imaginary parts. Thus, the FFT based method requires more MATLAB commands to implement. Observe also that conv(h,x) and conv(x,h) are the same, whereas filter(h,1,x) is not the same as filter(x,1,h).

To view the frequency response of the filter h, MATLAB provides the command freqz, which automatically zero pads[5] the impulse response and then plots both the magnitude and the phase. Type

```
freqz(h)
```

to see that the filter with impulse response h=[1, 1, 1, 1, 1] is a (poor) lowpass filter with two dips at 0.4 and 0.8 of the Nyquist frequency as shown in Figure 7.4. The command freqz always normalizes the frequency axis so that "1.0" corresponds to the Nyquist frequency $f_s/2$. The passband of this filter (all frequencies less than the point where the magnitude drops 3 dB below the maximum) ends just below 0.2. The maximum magnitude in the stopband occurs at about 0.6, where it is about 12 dB down from the peak at zero. Better (i.e., closer to the ideal) lowpass filters would attenuate more in the stopband, would be flatter across the passband, and would have narrower transition bands.

---

[5] By default, the MATLAB command freqz creates a length 512 vector containing the specified impulse response followed by zeros. The FFT of this elongated vector is used for the magnitude and phase plots, giving the plots a smoother appearance than when taking the FFT of the raw impulse response.

FIGURE 7.4 The frequency response of the filter with impulse response h=[1, 1, 1, 1, 1] has a poor lowpass character. It is easier to see this in the frequency domain than directly in the time domain.

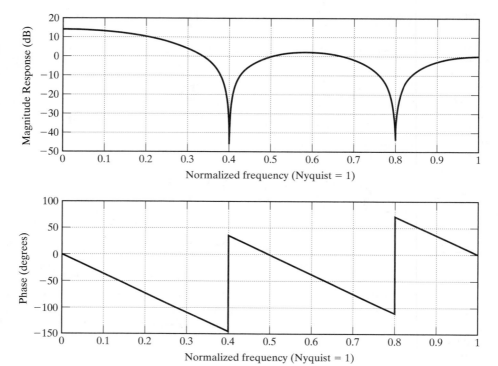

7.2.2 Filter Design

This section gives an extended explanation of how to use MATLAB to design a bandpass filter to fit a specified frequency response with a flat passband. The same procedure (with suitable modification) also works for the design of any of the other basic filter types.

A bandpass filter is intended to scale, but not distort, signals with frequencies that fall within the passband, and to reject signals with frequencies in the stopband. An ideal, distortionless response for the passband would be perfectly flat in magnitude, and would have linear phase (corresponding to a delay). The transition band from the passband to the stopband should be as narrow as possible. In the stopband, the frequency response magnitude should be sufficiently small and the phase is of no concern. These objectives are captured in Figure 7.5. Recall (from (A.35)) for a real $w(t)$ that $|W(f)|$ is even and $\angle W(f)$ is odd, as illustrated in Figure 7.5.

MATLAB has several commands that carry out filter design. The remez command provides a linear phase impulse response (with real, symmetric coefficients $h[k]$) that has the best approximation of a specified (piecewise flat) frequency response.[6] The syntax of the remez command for the design of a bandpass filter as in Figure 7.5 is

```
b = remez(fl,fbe,damps)
```

which has inputs fl, fbe, and damps, and output b.

---

[6] There are many possible meanings of the word "best"; for the remez algorithm, "best" is defined in terms of maintaining an equal ripple in the flat portions.

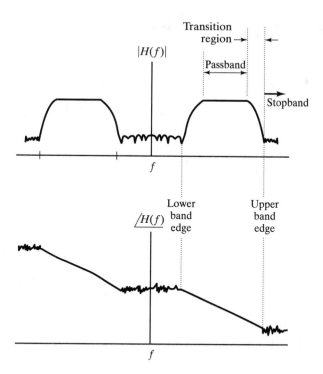

FIGURE 7.5 Specification of a bandpass filter in terms of magnitude and phase spectra.

- f1 specifies (one less than) the number of terms in the impulse response of the desired filter. Generally, more is better in terms of meeting the design specifications. However, larger f1 are also more costly in terms of computation and in terms of the total throughput delay, so a compromise is usually made.

- fbe is a vector of frequency band edge values as a fraction of the prevailing Nyquist frequency. For example, the filter specified in Figure 7.5 needs six values: the bottom of the stopband (presumably zero), the top edge of the lower stopband (which is also the lower edge of the lower transition band), the lower edge of the passband, the upper edge of the passband, the lower edge of the upper stopband, and the upper edge of the upper stopband (generally the last value will be 1). The transition bands must have some nonzero width (the upper edge of the lower stopband cannot equal the lower passband edge) or MATLAB produces an error message.

- damps is the vector of desired amplitudes of the frequency response at each band edge. The length of damps must match the length of f1.

- b is the output vector containing the impulse response of the specified filter.

The following MATLAB script designs a filter to the specifications of Figure 7.5:

FIGURE 7.6 Bandpass
filter frequency
response.

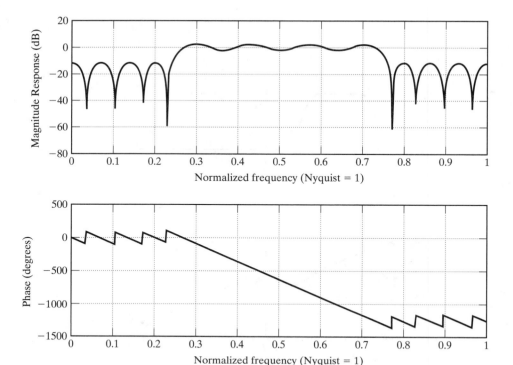

---

bandex.m design a bandpass filter and plot frequency response

---

```
fbe=[0 0.24 0.26 0.74 0.76 1];    % frequency band edges as
                                  % fraction of Nyquist frequency
damps=[0 0 1 1 0 0];              % desired amplitudes at band edges
fl=30;                            % filter size
b=remez(fl,fbe,damps);           % b is the designed impulse response
freqz(b)                          % plot frequency response to check design
```

The frequency response of the resulting finite impulse response (FIR) filter is shown in
Figure 7.6. Observe that the stopband is about 14 dB lower than the passband, a marginal
improvement over the naive lowpass filter of Figure 7.4, but the design is much flatter in
the pass band. The "equiripple" nature of this filter is apparent in the slow undulations
of the magnitude in the passband.

While commands such as remez make filter design easy, be warned—strange things
can happen, even to nice people. Always check to make sure that the output of the
design is a filter that behaves as expected. There are many other ways to design linear
filters, and MATLAB includes several commands that design filter coefficients: cremez,
firls, fir1, fir2, butter, cheby1, cheby2, and ellip. The subject of filter
design is vast, and each of these is useful in certain applications. For simplicity, we
have chosen to present all examples throughout *Telecommunication Breakdown* by using
remez.

Problems

7.14. Rerun `bandex.m` with very narrow transition regions, for instance `fbe = [0 0.24 0.2401 0.6 0.601 1]`. What happens to the ripple in the passband? Compare the minimum magnitude in the passband with the maximum value in the stopband.

7.15. Returning to the filter specified in Figure 7.5, try using different numbers of terms in the impulse response, `fl=5, 10, 100, 500, 1000`. Comment on the resulting designs in terms of flatness of the frequency response in the passband, attenuation from the passband to the stopband, and the width of the transition band.

7.16. Specify and design a lowpass filter with cutoff at 0.15. What values of `fl`, `fbe`, and `damps` work best?

7.17. Specify and design a filter that has two passbands, one between [0.2, 0.3] and another between [0.5 0.6]. What values of `fl`, `fbe`, and `damps` work best?

7.18. Use the filter designed in `bandex.m` to filter a white noise sequence (i.e., `message=randn(1,1000)`) using the time domain method from `waystofilt.m`.

The preceding filter designs do not explicitly require the sampling rate of the signal. However, since the sampling rate determines the Nyquist rate, it is used implicitly. The next exercise asks that you familiarize yourself with "real" units of frequency in the filter design task.

Problems

7.19. In Problem 7.10, the program `specgong.m` was used to analyze the sound of an Indonesian gong. The three most prominent partials (or narrowband components) were found to be at about 520, 630, and 660 Hz.
  (a) Design a filter using `remez` that will remove the two highest partials from this sound without affecting the lowest partial.
  (b) Use the `filter` command to process the `gong.wav` file with your filter.
  (c) Take the FFT of the resulting signal (the output of your filter) and verify that the partial at 520 remains while the others are removed.
  (d) If a sound card is attached to your computer, compare the sound of the raw and the filtered gong sound by using MATLAB's `sound` command. Comment on what you hear.

The next set of problems examines how accurate digital filters really are.

Problems

7.20. With a sampling rate of 44100 Hz, let $x[k]$ be a sinusoid of frequency 3000 Hz. Design a lowpass filter with a cutoff frequency `fl` of 1500 Hz, and let $y[k] = LPF\{x[k]\}$ be the output of the filter.
  (a) How much does the filter attenuate the signal? (Express your answer as the ratio of the power in the output $y[k]$ to the power in the input $x[k]$.)

FIGURE 7.7 The linear system $y[k + 1] = y[k] + \mu x[k]$, with input $x[k]$ and output $y[k]$, effectively adds up all the input values. This is often called a *summer*, or, by analogy with continuous time, an *integrator*. It can be drawn more concisely in a single block.

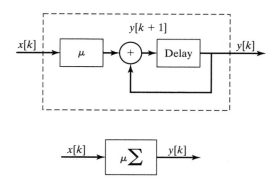

(b) Now use a LPF with a cutoff of 2500 Hz. How much does the filter attenuate the signal?

(c) Now use a LPF with a cutoff of 2900 Hz. How much does the filter attenuate the signal?

7.21. Repeat Problem 7.20 without using the `filter` command (implement the filtering, using the time domain method in `waystofilt.m`).

7.22. With the same setup as in Problem 7.20, generate $x[k]$ as a bandlimited noise signal containing frequencies between 3000 Hz and the Nyquist rate.

(a) Use a LPF with cutoff frequency `f1` of 1500 Hz. How much does the filter attenuate the signal?

(b) Now use a LPF with a cutoff of 2500 Hz. How much does the filter attenuate the signal?

(c) Now use a LPF with a cutoff of 3100 Hz. How much does the filter attenuate the signal?

(d) Now use a LPF with a cutoff of 4000 Hz. How much does the filter attenuate the signal?

7.23. Let $f_1 < f_2 < f_3$. Suppose $x[k]$ has no frequencies above f_1 Hz, while $z[k]$ has no frequencies below f_3. If a LPF has cutoff frequency f_2. In principle,

$$LPF\{x[k] + z[k]\} = LPF\{x[k]\} + LPF\{z[k]\} = x[k] + 0 = x[k].$$

Explain how this is (and is not) consistent with the results of Problems 7.20 and 7.22.

7.24. Let the output $y[k]$ of a linear system be created from the input $x[k]$ according to the formula

$$y[k + 1] = y[k] + \mu x[k],$$

where μ is a small constant. This is drawn in Figure 7.7.

(a) What is the impulse response of this filter?

(b) What is the frequency response of this filter?

(c) Would you call this filter lowpass, highpass, or bandpass?

7.25. Using one of the alternative filter design routines (`cremez`, `firls`, `fir1`, `fir2`, `butter`, `cheby1`, `cheby2`, or `ellip`), repeat Problems 7.14–7.19. Comment on the subtle (and the not-so-subtle) differences in the resulting designs.

7.26. The effect of bandpass filtering can be accomplished by

1. modulating to DC,
2. lowpass filtering, and
3. modulating back.

Repeat the task given in Problem 7.19 (the Indonesian gong filter design problem) by modulating with a 520 Hz cosine, lowpass filtering, and then remodulating. Compare the final output of this method with the direct bandpass filter design.

7.3 For Further Reading

Here are some of our favorite books on signal processing:

- K. Steiglitz, *A Digital Signal Processing Primer*, Addison-Wesley, Pubs, 1996.
- J. H. McClellan, R. W. Schafer, M. A. Yoder, *DSP First: A Multimedia Approach*, Prentice Hall, 1998.
- C. S. Burrus and T. W. Parks, *DFT/FFT and Convolution Algorithms: Theory and Implementation*, Wiley-Interscience, 1985.

8 *Bits to Symbols to Signals*

> How much will two bits be worth in the digital marketplace?
> —Hal Varian, *Scientific American*, Sept. 1995

Any message, whether analog or digital, can be translated into a string of binary digits. In order to transmit or store these digits, they are often clustered or encoded into a more convenient representation whose elements are the symbols of an *alphabet*. In order to utilize bandwidth efficiently, these symbols are then translated (again!) into short analog waveforms called *pulse shapes* that are combined to form the actual transmitted signal.

The receiver must undo each of these translations. First, it examines the received analog waveform and decodes the symbols. Then it translates the symbols back into binary digits, from which the original message can (hopefully) be reconstructed.

This chapter briefly examines each of these translations, and the tools needed to make the receiver work. One of the key ideas is *correlation* which can be used as a kind of pattern matching tool for discovering key locations within the signal stream. Section 8.3 shows how correlation can be viewed as a kind of linear filter, and hence its properties can be readily understood in both the time and frequency domains.

8.1 Bits to Symbols

The information that is to be transmitted by a communication system comes in many forms: a pressure wave in the air, a flow of electrons in a wire, a digitized image or sound file, the text in a book. If the information is in analog form, then it can be sampled (as in Chapter 6). For instance, an analog-to-digital converter can transform the output of a microphone into a stream of numbers representing the pressure wave in the air, or it can turn measurements of the current in the wire into a sequence of numbers that are proportional to the electron flow. The sound file, which is already digital, contains a long list of numbers that correspond to the instantaneous amplitude of the sound. Similarly, the picture file contains a list of numbers that describe the intensity and color of the pixels in the image. The text can be transformed into a numerical list using the ASCII code. In all these cases, the raw data represent the information that must be transmitted by the communication system. The receiver, in turn, must ultimately translate the received signal back into the data.

Once the information is encoded into a sequence of numbers, it can be reexpressed as a string of binary digits 0 and 1. This is discussed at length in Chapter 14. But the binary

0–1 representation is not usually very convenient from the point of view of efficient and reliable data transmission. For example, directly modulating a binary string with a cosine wave would result in a small piece of the cosine wave for each 1 and nothing (the zero waveform) for each 0. It would be very hard to tell the difference between a message that contained a string of zeroes, and no message at all!

The simplest solution is to recode the binary 0, 1 into binary ± 1. This can be accomplished using either the linear operation $2x - 1$ (which maps 0 into -1, and 1 into 1), or by $-2x + 1$ (which maps 0 into 1, and 1 into -1). This "binary" ± 1 is an example of a two-element symbol set. There are many other common symbol sets. In *multilevel signaling*, the binary terms are gathered into groups. Regrouping in pairs, for instance, recodes the information into a four-level signal. For example, the binary sequence might be paired thus:

$$\dots 000010110101 \dots \rightarrow \dots 00\ 00\ 10\ 11\ 01\ 01 \dots . \tag{8.1}$$

Then the pairs might be encoded as

$$\begin{array}{ccc} 11 & \rightarrow & +3 \\ 10 & \rightarrow & +1 \\ 01 & \rightarrow & -1 \\ 00 & \rightarrow & -3 \end{array} \tag{8.2}$$

to produce the symbol sequence

$$\dots 00\ 00\ 10\ 11\ 01\ 01 \dots \rightarrow \dots -3,\ -3,\ +1,\ +3 - 1,\ -1 \dots .$$

Of course, there are many ways that such a mapping between bits and symbols might be made, and Problem 8.2 explores one simple alternative called the Grey code. The binary sequence may be grouped in many ways: into triplets for an 8-level signal, into quadruplets for a 16-level scheme, into "in-phase" and "quadrature" parts for transmission through a quadrature system. The values assigned to the groups ($\pm 1, \pm 3$ in (8.2)) are called the *alphabet* of the given system.

Example 8.1

Text is commonly encoded using ASCII, and MATLAB automatically represents any string file as a list of ASCII numbers. For instance, let `str='I am text'` be a text string. This can be viewed in its internal form by typing `real(str)`, which returns the vector 73 32 97 109 32 116 101 120 116, which is the (decimal) ASCII representation of this string. This can be viewed in binary using `dec2base(str,2,8)`, which returns the binary (base 2) representation of the decimal numbers, each with 8 digits.

The MATLAB function `letters2pam`, provided on the CD, changes a text string into the 4-level alphabet $\pm 1, \pm 3$. Each letter is represented by a sequence of 4 elements, for instance the letter *I* is $-1\ -3\ 1\ -1$. The function is invoked with the syntax `letters2pam(str)`. The inverse operation is `pam2letters`. Thus `pam2letters(letters2pam(str))` returns the original string.

One complication in the decoding procedure is that the receiver must figure out when the groups begin in order to parse the digits properly. For example, if the first element of the sequence in (8.1) was lost, then the message would be mistranslated as

$$\ldots 00010110101 \ldots \rightarrow \ldots 00\ 01\ 01\ 10\ 10 \ldots \rightarrow \ldots -3,\ -1,\ -1,\ 1,\ 1, \ldots.$$

Similar parsing problems occur whenever messages start or stop. For example, if the message consists of pixel values for a television image, it is important that the decoder be able to determine precisely when the image scan begins. These kinds of synchronization issues are typically handled by sending a special "start of frame" sequence that is known to both the transmitter and the receiver. The decoder then searches for the start sequence, usually using some kind of correlation (pattern matching) technique. This is discussed in detail in Section 8.3.

Example 8.2

There are many ways to translate data into binary equivalents. Example 8.1 showed one way to convert text into 4-PAM and then into binary. Another way exploits the MATLAB function `text2bin.m` and its inverse `bin2text.m`, which use the 7-bit version of the ASCII code (rather than the 8-bit version). This representation is more efficient, since each pair of text letters can be represented by 14 bits (or seven 4-PAM symbols) rather than 16 bits (or eight 4-PAM symbols). On the other hand, the 7-bit version can encode only half as many characters as the 8-bit version. Again, it is important to be able to correctly identify the start of each letter when decoding.

Problems

8.1. The MATLAB code in `naivecode.m`, which is on the CD, implements the translation from binary to 4-PAM (and back again) suggested in (8.2). Examine the resiliency of this translation to noise by plotting the number of errors as a function of the noise variance v. What is the largest variance for which no errors occur? At what variance are the errors near 50%?

8.2. A Grey code has the property that the binary representation for each symbol differs from its neighbors by exactly one bit. A Grey code for the translation of binary into 4-PAM is

$$
\begin{aligned}
01 &\rightarrow +3 \\
11 &\rightarrow +1 \\
10 &\rightarrow -1 \\
00 &\rightarrow -3
\end{aligned}
$$

Mimic the code in `naivecode.m` to implement this alternative and plot the number of errors as a function of the noise variance v. Compare your answer with Problem 8.1. Which code is better?

8.2 Symbols to Signals

Even though the original message is translated into the desired alphabet, it is not yet ready for transmission: it must be turned into an analog waveform. In the binary case, a simple method is to use a rectangular pulse of duration T seconds to represent $+1$, and the same rectangular pulse inverted (i.e., multiplied by -1) to represent the element -1. This is called a polar non-return-to-zero line code. The problem with such simple codes is that they use bandwidth inefficiently. Recall that the Fourier transform of the rectangular pulse in time is the $\text{sinc}(f)$ function in frequency (A.20), which dies away slowly as f increases. Thus, simple codes like the non-return-to-zero are compact in time, but wide in frequency, limiting the number of simultaneous nonoverlapping users in a given spectral band.

More generally, consider the four-level signal of (8.2). This can be turned into an analog signal for transmission by choosing a pulse shape $p(t)$ (that is not necessarily rectangular and not necessarily of duration T) and then transmitting

$$
\begin{array}{ll}
p(t-kT) & \text{if the } k\text{th symbol is } 1 \\
-p(t-kT) & \text{if the } k\text{th symbol is } -1 \\
3p(t-kT) & \text{if the } k\text{th symbol is } 3 \\
-3p(t-kT) & \text{if the } k\text{th symbol is } -3
\end{array}
$$

Thus, the sequence is translated into an analog waveform by initiating a scaled pulse at the symbol time kT, where the amplitude scaling is proportional to the associated symbol value. Ideally, the pulse would be chosen so that

- the value of the message at time k does not interfere with the value of the message at other sample times (the pulse shape causes no *intersymbol interference*),
- the transmission makes efficient use of bandwidth, and
- the system is resilient to noise.

Unfortunately, these three requirements cannot all be optimized simultaneously, and so the design of the pulse shape must consider carefully the tradeoffs that are needed. The focus in Chapter 11 is on how to design the pulse shape $p(t)$, and the consequences of that choice in terms of possible interference between adjacent symbols and in terms of the signal-to-noise properties of the transmission.

For now, to see concretely how pulse shaping works, let's pick a simple nonrectangular shape and proceed without worrying about optimality. Let $p(t)$ be the symmetrical blip shape shown in the top part of Figure 8.1, and defined in pulseshape0.m by the hamming command. The text string in str is changed into a 4-level signal as in Example 8.1, and then the complete transmitted waveform is assembled by assigning an appropriately scaled pulse shape to each data value. The output appears in the bottom of Figure 8.1. Looking at this closely, observe that the first letter T is represented by the four values $-1 -1 -1 -3$, which corresponds exactly to the first four negative blips, three small and one large.

The program pulseshape0.m represents the "continuous-time" or analog signal by oversampling both the data sequence and the pulse shape by a factor of M. This technique was discussed in Section 6.3, where an "analog" sine wave sine100hzsamp.m was represented digitally at two sampling intervals, a slow digital interval T_s and a faster

FIGURE 8.1 The process of pulse shaping replaces each symbol of the alphabet (in this case, ±1, ±3) with an analog pulse (in this case, the short blip function shown in the top panel).

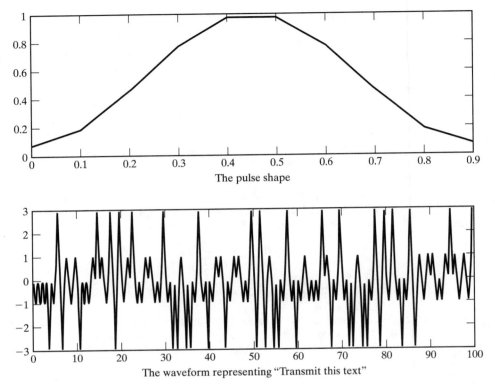

rate (shorter interval) T_s/M representing the underlying analog signal. The pulse shaping itself is carried out by the `filter` command which convolves the pulse shape with the data sequence.

---

pulseshape0.m: applying a pulse shape to a text string

```
str='Transmit this text string';        % message to be transmitted
m=letters2pam(str); N=length(m);         % 4-level signal of length N
M=10; mup=zeros(1,N*M); mup(1:M:end)=m;  % oversample by M
ps=hamming(M);                           % blip pulse of width M
x=filter(ps,1,mup);                      % convolve pulse shape with data
```

Problems

8.3. For $T = 0.1$, plot the spectrum of the output x. What is the bandwidth of this signal?

8.4. For $T = 0.1$, plot the spectrum of the output x when the pulse shape is changed to a rectangular pulse. (Change the definition of `ps` in the next to last line of `pulseshape0.m`.) What is the bandwidth of this signal?

8.5. Can you think of a pulse shape that will have a narrower bandwidth than either of the above but that will still be time limited by T? Implement it by changing the definition of ps, and check to see if you are correct.

Thus the raw message, the samples, are prepared for transmission by

- encoding into an alphabet (in this case $\pm 1, \pm 3$), and then
- pulse shaping the elements of the alphabet using $p(t)$.

The receiver must undo these two operations; it must examine the received signal and recover the elements of the alphabet, and then decode these to reconstruct the message. Both of these tasks are made easier using correlation, which is discussed in the next section. The actual decoding processes used in the receiver are then discussed in Section 8.4.

8.3 Correlation

Suppose there are two signals or sequences. Are they similar, or are they different? If one is just shifted in time relative to the other, how can the time shift be determined? The approach called correlation shifts one of the sequences in time, and calculates how well they match (by multiplying point by point and summing) at each shift. When the sum is small, they are not much alike; when the sum is large, many terms are similar. Thus, correlation is a simple form of pattern matching, which is useful in communication systems for aligning signals in time. This can be applied at the level of symbols when it is necessary to find appropriate sampling times, and it can be applied at the "frame" level when it is necessary to find the start of a message (for instance, the beginning of each frame of a television signal). This section discusses various techniques of cross-correlation and autocorrelation, which can be viewed in either the time domain or the frequency domain.

In discrete time, cross-correlation is a function of the time shift j between two sequences $w[k]$ and $v[k + j]$:

$$R_{wv}(j) = \lim_{T \to \infty} \frac{1}{T} \sum_{k=-T/2}^{T/2} w[k]v[k + j]. \tag{8.3}$$

For finite data records, the sum need only be accumulated over the nonzero elements, and the normalization by $1/T$ is often ignored. (This is how MATLAB's xcorr function works.) While this may look like the convolution Equation (7.2), it is not identical since the indices are different (in convolution, the index of $v(\cdot)$ is $j - k$ instead of $k + j$). The operation and meaning of the two processes are also not identical: convolution represents the manner in which the impulse response of a linear system acts on its inputs to give the outputs, while cross-correlation quantifies the similarity of two signals.

In many communication systems, each message is parcelled into segments or frames, each having a predefined header. As the receiver decodes the transmitted message, it must determine where the message segments start. The following code simulates this in a simple setting in which the header is a predefined binary string and the data consist of a much longer binary string that contains the header hidden somewhere inside. After performing the correlation, the index with the largest value is taken as the most likely location of the header.

---

correx.m: correlation can locate the header within the data

```
header=[1 -1 1 -1 -1 1 1 1 -1 -1];          % header is a predefined string
l=30; r=25;                                  % place header l=30 from start
data=[sign(randn(1,l)) header sign(randn(1,r))]; % generate signal
sd=0.25; data=data+sd*randn(size(data));     % add noise
y=xcorr(header, data);                        % do cross-correlation
[m,ind]=max(y);                               % location of largest correlation...
headstart=length(data)-ind;                   % ...gives place where header starts
```

Running `correx.m` results in a trio of figures much like those in Figure 8.2. (Details will differ each time it is run, because the actual "data" are randomly generated with MATLAB's `randn` function.) The top plot in Figure 8.2 shows the 10-sample binary header. The data vector is constructed to contain $l=30$ data values followed by the header (with noise added), and then $r = 25$ more data points, for a total block of 65 points. It is plotted in the middle of Figure 8.2. Observe that it is difficult to "see" where the header lies among the noisy data record. The correlation between the data and the header is calculated and plotted in the bottom of Figure 8.2 as a function of the lag index. The index where the correlation attains its largest value defines where the best match between the data and the header occurs. Most likely this will be at index `ind=35` (as in Figure 8.2). Because of the way MATLAB orders its output, the calculations represent

FIGURE 8.2 The correlation can be used to locate a known header within a long signal. The predefined header is shown in the top graph. The data consist of a random binary string with the header embedded and noise added. The bottom plot shows the correlation. The location of the header is determined by the peak occurring at 35.

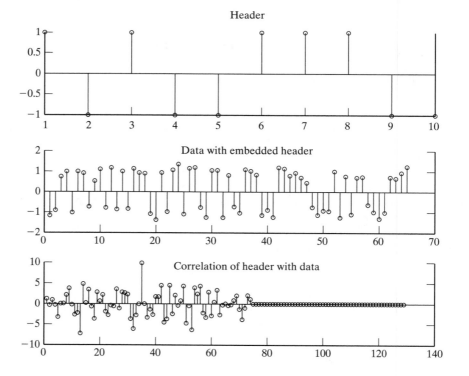

sliding the first vector (the header), term by term, across the second vector (the data). The long string of zeroes at the end[1] occurs because the two vectors are of different lengths. MATLAB computes `xcorr` over a window twice the length of the longest vector (which in this case is the length of the vector `data`). Hence the start of the header is given by `length(data)-ind`.

One way that the correlation might fail to find the correct location of the header is if the header string accidently occurred in the data values. If this happened, then the correlation would be as large at the "accidental" location as at the intended location. This becomes increasingly unlikely as the header is made longer, though a longer header also wastes bandwidth. Another way to decrease the likelihood of false hits is to average over several headers.

Problems

8.6. Rerun `correx.m` with different length data vectors (try `l=100`, `r=100` and `l=10`, `r=10`). Observe how the location of the peak changes.

8.7. Rerun `correx.m` with different length headers. Does the peak in the correlation become more or less distinct as the number of terms in the header increases?

8.8. Rerun `correx.m` with different amounts of noise. Try `sd=0`, `.1`, `.3`, `.5`, `1`, `2`. How large can the noise be made if the correlation is still to find the true location of the header?

8.9. The code in `corrvsconv.m` explores the relationship between the correlation and convolution. The convolution of two sequences is the same as the cross-correlation of the time-reversed signal, though the correlation is padded with extra zeroes. (The MATLAB function `fliplr` carries out the time reversal.) If `h` is made longer than `x`, what needs to be changed so that `yconv` and `ycorr` remain equal?

---

corrvsconv.m: "correlation" vs "convolution"

```
h=[1 -1 2 -2 3 -3];                  % define sequence h[k]
x=[1 2 3 4 5 6 -5 -4 -3 -2 -1];      % define sequence x[k]
yconv=conv(x,h)                      % convolve x[k]*h[k]
ycorr=xcorr(fliplr(x),h)             % correlation of flipped x and h
```

8.4 Receive Filtering: From Signals to Symbols

Suppose that a message has been coded into its alphabet, pulse shaped into an analog signal, and transmitted. The receiver must then "un–pulse-shape" the analog signal back into the alphabet, which requires finding where in the received signal the pulse shapes are located. Correlation can be used to accomplish this task, because it is effectively the task of locating a known sequence (in this case the sampled pulse shape) within a longer sequence (the sampled received signal). This is analogous to the problem of finding the header within the received signal, although some of the details have

---

[1] Some versions of MATLAB use a different convention with the `xcorr` command. If you find that the string of zeros occurs at the beginning, then reverse the order of the arguments.

changed. While optimizing this procedure is somewhat involved (and is therefore postponed until Chapter 11), the gist of the method is reasonably straightforward, and is shown by continuing the example begun in `pulseshape0.m`.

The code in `rectfilt.m` below begins by repeating the pulse shaping code from `pulseshape0.m`, using the pulse shape ps defined in the top plot of Figure 8.1. This creates an "analog" signal x that is oversampled by a factor M. The receiver begins by correlating the pulse shape with the received signal, using the xcorr function.[2] After appropriate scaling, this is downsampled to the symbol rate by choosing one out of each M (regularly spaced) samples. These values are then quantized to the nearest element of the alphabet using the function quantalph (which was introduced in Problem 3.19). The function quantalph has two vector arguments; the elements of the first vector are quantized to the nearest elements of the second vector (in this case quantizing z to the nearest elements of $[-3, -1, 1, 3]$).

If all has gone well, the quantized output mprime should be identical to the original message string. The function pam2letters rebuilds the message from the received signal. The final line of the program calculates how many symbol errors have occurred (how many of the ± 1, ± 3 differ between the message m and the reconstructed message mprime).

---

recfilt.m: undo pulse shaping using correlation

```
% first run pulseshape0.m to create the transmitted signal x
y=xcorr(x,p);                        % correlate pulse with received signal
z=y(N*M:M:end)/(pow(ps)*M);          % downsample to symbol rate and normalize
mprime=quantalph(z,[-3,-1,1,3])';    % quantize to +/-1 and +/-3 alphabet
pam2letters(mprime)                  % reconstruct message
sum(abs(sign(mprime-m)))             % calculate number of errors
```

In essence, `pulseshape0.m` from page 144 is a transmitter, and `recfilt.m` is the corresponding receiver. Many of the details of this simulation can be changed and the message will still arrive intact. The following exercises encourage exploration of some of the options.

Problems

8.10. Other pulse shapes may be used. Try
   (a) a sinusoidal shaped pulse `ps=sin(0.1*pi*(0:M-1));`
   (b) a sinusoidal shaped pulse `ps=cos(0.1*pi*(0:M-1));`
   (c) a rectangular pulse shape `ps=ones(1,M);`

8.11. What happens if the pulse shape used at the transmitter differs from the pulse shape used at the receiver? Try using the original pulse shape from `pulse-shape0.m` at the transmitter, but using
   (a) `ps=sin(0.1*pi*(0:M-1));` at the receiver. What percentage errors occur?

---

[2] Because of the connections between cross-correlation, convolution, and filtering, this process is often called *pulse-matched filtering* because the impulse response of the filter is matched to the shape of the pulse.

(b) `ps=cos(0.1*pi*(0:M-1));` at the receiver. What percentage errors occur?

8.12. The received signal may not always arrive at the receiver unchanged. Simulate a noisy channel by including the command `x=x+1.0*randn(size(x))` before the `xcorr` command in `recfilt.m`. What percentage errors occur? What happens as you increase or decrease the amount of noise (by changing the `1.0` to a larger or smaller number)?

8.5 Frame Synchronization: From Symbols to Bits

In many communication systems, the data in the transmitted signal is separated into chunks called frames. In order to correctly decode the text at the receiver, it is necessary to locate the boundary (the start) of each chunk. This was done by fiat in the receiver of `recfilt.m` by correctly indexing into the received signal y. Since this starting point will not generally be known beforehand, it must somehow be located. This is an ideal job for correlation and a marker sequence.

The marker is a set of predefined symbols embedded at some specified location within each frame. The receiver can locate the marker by cross-correlating it with the incoming signal stream. What makes a good marker sequence? This section shows that not all markers are created equally.

Consider the binary data sequence

$$\ldots + 1, \ -1, \ +1, \ +1, \ -1, \ -1, \ -1, \ +1, \ \text{marker}, \ +1, \ -1, \ +1, \ \ldots, \qquad (8.4)$$

where the marker is used to indicate a frame transition. A seven-symbol marker is to be used. Consider two candidates:

- marker A: 1, 1, 1, 1, 1, 1, 1
- marker B: 1, 1, 1, −1, −1, 1, −1

The correlation of the signal with each of the markers can be performed as indicated in Figure 8.3.

For marker A, correlation corresponds to a simple sum of the last seven values. Starting at the location of the seventh value available to us in the data sequence (two data points before the marker), marker A produces the sequence

$$-1, \ -1, \ 1, \ 1, \ 1, \ 3, \ 6, \ 7, \ 7, \ 7, \ 5, \ 5.$$

For marker B, starting at the same point in the data sequence and performing the associated moving weighted sum, produces

$$1, \ 1, \ 3, \ -1, \ -5, \ -1, \ -1, \ 1, \ 7, \ -1, \ 1, \ -3.$$

With the two correlator output sequences shown, started two values prior to the start of the seven-symbol marker, we want the flag indicating a frame start to occur with point number 9 in the correlator sequences shown. Clearly, the correlator output for marker B has a much sharper peak at its ninth value than the correlator output of marker A. This should enhance the robustness of the use of marker B relative to that of marker A against the unavoidable presence of noise.

FIGURE 8.3 Correlation diagram

Sum products of adjacent values

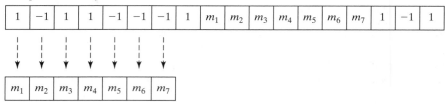

Shift marker to right and repeat

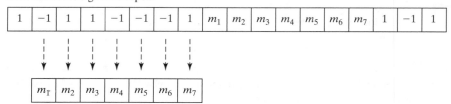

Marker B is a "maximum-length pseudonoise (PN)" sequence. One property of a maximum-length PN sequence $\{c_i\}$ of plus and minus ones is that its autocorrelation is quite peaked:

$$R_c(k) = \frac{1}{N} \sum_{n=0}^{N-1} c_n c_{n+k} = \begin{cases} 1, & k = \ell N \\ \frac{-1}{N}, & k \neq \ell N \end{cases}.$$

Another technique that involves the chunking of data and the need to locate boundaries between chunks is called *scrambling*. Scrambling is used to "whiten" a message sequence (to make its spectrum flatter) by decorrelating the message. The transmitter and receiver agree on a binary scrambling sequence s that is repeated over and over to form a periodic string S that is the same size as the message. S is then added (using modulo 2 arithmetic) bit by bit to the message m at the transmitter, and then S is added bit by bit again at the receiver. Since both $1 + 1 = 0$ and $0 + 0 = 0$,

$$m + S + S = m$$

and the message is recaptured after the two summing operations. The scrambling sequence must be aligned so that the additions at the receiver correspond to the appropriate additions at the transmitter. The alignment can be accomplished using correlation.

Problems

8.13. Redo the example of this section, using MATLAB.

8.14. Add a channel with impulse response $1, 0, 0, a, 0, 0, 0, b$ to this example. (Convolve the impulse response of the channel with the data sequence.)
    (a) For $a = 0.1$ and $b = 0.4$, how does the channel change the likelihood that the correlation correctly locates the marker? Try using both markers A and B.
    (b) Answer the same question for $a = 0.5$ and $b = 0.9$.
    (c) Answer the same question for $a = 1.2$ and $b = 0.4$.

8.15. Generate a long sequence of binary random data with the marker embedded every 25 points. Check that marker A is less robust (on average) than marker B by counting the number of times marker A misses the frame start compared with the number of times marker B misses the frame start.

8.16. Create your own marker sequence, and repeat the previous problem. Can you find one that does better than marker B?

8.17. Use the 4-PAM alphabet with symbols $\pm 1, \pm 3$. Create a marker sequence, and embed it in a long sequence of random 4-PAM data. Check to make sure it is possible to correctly locate the markers.

8.18. Add a channel with impulse response $1, 0, 0, a, 0, 0, 0, b$ to this 4-PAM example.
   (a) For $a = 0.1$ and $b = 0.4$, how does the channel change the likelihood that the correlation correctly locates the marker?
   (b) Answer the same question for $a = 0.5$ and $b = 0.9$.

8.19. Choose a binary scrambling sequence s that is 17 bits long. Create a message that is 170 bits long, and scramble it using bit-by-bit mod 2 addition.
   (a) Assuming the receiver knows where the scrambling begins, add S to the scrambled data and verify that the output is the same as the original message.
   (b) Embed a marker sequence in your message. Use correlation to find the marker and to automatically align the start of the scrambling.

9 *Stuff Happens*

This practical guide leads the reader through solving the problem from start to finish. You will learn to: define a problem clearly, organize your problem solving project, analyze the problem to identify the root causes, solve the problem by taking corrective action, and prove the problem is really solved by measuring the results.

—Jeanne Sawyer, *When Stuff Happens: A Practical Guide to Solving Problems Permanently*, Sawyer Publishing Group, 2001

There is nothing new in this chapter. Really. By peeling away the outer, most accessible layers of the communication system, the previous chapters have provided all of the pieces needed to build an idealized digital communication system, and this chapter just shows how to combine the pieces into a functioning system. Then we get to play with the system a bit, asking a series of "what if" questions.

In outline, the idealized system consists of two parts, rather than three, since the channel is assumed to be noiseless and disturbance free.

The Transmitter

- codes a message (in the form of a character string) into a sequence of symbols,

- transforms the symbol sequence into an analog signal using a pulse shape, and

- modulates the scaled pulses up to the passband.

The Digital Receiver

- samples the received signal,

- demodulates to baseband,

- filters the signal to remove unwanted portions of the spectrum,

- correlates with the pulse shape to help emphasize the "peaks" of the pulse train,

- downsamples to the symbol rate, and

- decodes the symbols back into the character string.

Each of these procedures is familiar from earlier chapters, and you may have already written MATLAB code to perform them. It is time to combine the elements into a full simulation of a transmitter and receiver pair that can function successfully in an ideal setting.

9.1 An Ideal Digital Communication System

The system is illustrated in the block diagram of Figure 9.1. This system is described in great detail in Section 9.2, which also provides a MATLAB version of the transmitter and receiver. Once everything is pieced together, it is easy to verify that messages can be sent reliably from transmitter to receiver.

Unfortunately, some of the assumptions made in the ideal setting are unlikely to hold in practice; for example, the presumption that there is no interference from other transmitters, that there is no noise, that the gain of the channel is always unity, that the signal leaving the transmitter is exactly the same as the signal at the input to the digital receiver. All of these assumptions will almost certainly be violated in practice. Stuff happens!

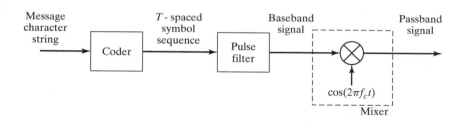

(a) Transmitter

FIGURE 9.1 Ideal communication system.

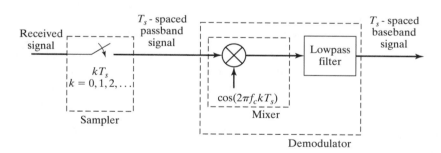

Demodulator

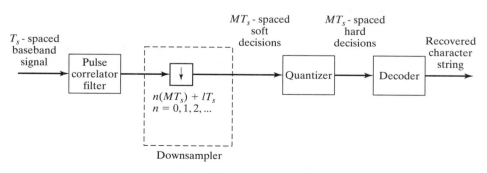

(b) Receiver

Section 9.3 begins to accommodate some of the nonidealities encountered in real systems by addressing the possibility that the channel gain might vary with time. For example, a large metal truck might abruptly move between a cell phone and the antenna at the base station, causing the channel gain to drop precipitously. If the receiver cannot react to such a change, it may suffer debilitating errors when reconstructing the message. Section 9.3 examines the effectiveness of incorporating an automatic gain control (AGC) adaptive element (as described in Section 6.8) at the front-end of the receiver. With care, the AGC can accommodate the varying gain. The success of the AGC is encouraging. Perhaps there are simple ways to compensate for other common impairments.

Section 9.4 presents a series of "what if" questions concerning the various assumptions made in the construction of the ideal system, focusing on performance degradations caused by synchronization loss and various kinds of distortions:

- What if there is channel noise? (The ideal system is noise free.)

- What if the channel has multipath interference? (There are no reflections or echoes in the ideal system.)

- What if the phase of the oscillator at the transmitter is unknown (or guessed incorrectly) at the receiver? (The ideal system knows the phase exactly.)

- What if the frequency of the oscillator at the transmitter is off just a bit from its specification? (In the ideal system, the frequency is known exactly.)

- What if the sample instant associated with the arrival of top-dead-center of the leading pulse is inaccurate so that the receiver samples at the "wrong" times? (The sampler in the ideal system is never fooled.)

- What if the number of samples between symbols assumed by the receiver is different from that used at the transmitter? (These are the same in the ideal case.)

These questions are investigated via a series of experiments that require only modest modification of the ideal system simulation. These simulations will show (as with the time-varying channel gain) that small violations of the idealized assumptions can often be tolerated. However, as the operational conditions become more severe (as more stuff happens), the receiver must be made more robust.

Of course, its not possible to fix all these problems in one chapter. That's what the rest of the book is for!

- Chapter 10 deals with methods to acquire and track changes in the carrier phase and frequency.

- Chapter 11 describes better pulse shapes and corresponding receive filters that perform well in the presence of channel noise.

- Chapter 12 discusses techniques for tracking the symbol clock so that the samples can be taken at the best possible times.

- Chapter 13 designs a symbol-spaced filter that undoes multipath interference and can reject certain kinds of in-band interference.

- Chapter 14 describes simple coding schemes that provide protection against channel noise.

9.2 Simulating the Ideal System

The simulation of the digital communication system in Figure 9.1 divides into two parts just as the figure does. The first part creates the analog transmitted signal, and the second part implements the discrete-time receiver.

The message consists of the character string

```
01234 I wish I were an Oscar Meyer wiener 56789
```

In order to transmit this important message, it is first translated into the 4-PAM symbol set $\pm 1, \pm 3$ (which is designated $m[i]$ for $i = 1, 2, \ldots, N$) using the subroutine `let-ters2pam`. This can be represented formally as the analog pulse train $\sum_{i=0}^{N-1} m[i]\delta(t - iT)$, where T is the time interval between symbols. The simulation operates with an oversampling factor M, which is the speed at which the "analog" portion of the system evolves. The pulse train enters a filter with pulse shape $p(t)$. By the sifting property (A.56), the output of the pulse shaping filter is the analog signal $\sum_{i=0}^{N-1} m[i]p(t - iT)$, which is then modulated (by multiplication with a cosine at the carrier frequency f_c) to form the transmitted signal

$$\sum_{i=0}^{N-1} m[i]p(t - iT)\cos(2\pi f_c t).$$

Since the channel is assumed to be ideal, this is equal to the received signal $r(t)$. This ideal transmitter is simulated in the first part of `idsys.m`.

---

idsys.m: (part 1) idealized transmission system - the transmitter

---

```
% encode text string as T-spaced PAM (+/-1, +/-3) sequence
str='01234 I wish I were an Oscar Meyer wiener 56789';
m=letters2pam(str); N=length(m);    % 4-level signal of length N
% zero pad T-spaced symbol sequence to create upsampled T/M-spaced
% sequence of scaled T-spaced pulses (with T = 1 time unit)
M=100; mup=zeros(1,N*M); mup(1:M:end)=m; % oversampling factor
% Hamming pulse filter with T/M-spaced impulse response
p=hamming(M);                        % blip pulse of width M
x=filter(p,1,mup);                   % convolve pulse shape with data
figure(1), plotspec(x,1/M)           % baseband signal spectrum
                                     % am modulation
t=1/M:1/M:length(x)/M;               % T/M-spaced time vector
fc=20;                               % carrier frequency
c=cos(2*pi*fc*t);                    % carrier
r=c.*x;                              % modulate message with carrier
```

Since MATLAB cannot deal directly with analog signals, the "analog" signal $r(t)$ is sampled at M times the symbol rate, and $r(t)|_{t=kT_s}$ (the signal $r(t)$ sampled at times $t = kT_s$) is the vector r in the MATLAB code. The vector r is also the input to the digital portion of the receiver. Thus, the first sampling block in the receiver of Figure 9.1 is implicit in the way MATLAB emulates the analog signal. To be specific, k can be represented as the sum of an integer multiple of M and some positive integer ρ smaller

Stuff Happens

than M such that

$$kT_s = (iM + \rho)T_s.$$

Since $T = MT_s$,

$$kT_s = iT + \rho T_s.$$

Thus, the received signal sampled at $t = kT_s$ is

$$r(t)|_{t=kT_s} = \sum_{i=0}^{N-1} m[i]p(t - iT)\cos(2\pi f_c t)|_{t=kT_s=iT+\rho T_s}$$

$$= \sum_{i=0}^{N-1} m[i]p(kT_s - iT)\cos(2\pi f_c kT_s).$$

The receiver performs downconversion in the second part of idsys.m with a mixer that uses a synchronized cosine wave, followed by a lowpass filter that removes out-of-band signals. A quantizer makes hard decisions that are then decoded back from symbols to the characters of the message. When all goes well, the reconstructed message is the same as the original.

---

idsys.m: (part 2) idealized transmission system - the receiver

---

```
% am demodulation of received signal sequence r
c2=cos(2*pi*fc*t);                  % synchronized cosine for mixing
x2=r.*c2;                           % demod received signal
fl=50;                              % LPF length
fbe=[0 0.5 0.6 1]; damps=[1 1 0 0]; % design of LPF parameters
b=remez(fl,fbe,damps);              % create LPF impulse response
x3=2*filter(b,1,x2);                % LPF and scale downconverted signal
% extract upsampled pulses using correlation implemented as a convolving filter
y=filter(fliplr(p)/(pow(p)*M),1,x3); % filter rec'd sig with pulse; normalize
% set delay to first symbol-sample and increment by M
z=y(0.5*fl+M:M:end);                % downsample to symbol rate
figure(2), plot([1:length(z)],z,'.') % soft decisions
% decision device and symbol matching performance assessment
mprime=quantalph(z,[-3,-1,1,3])';   % quantize to +/-1 and +/-3 alphabet
cluster_variance=(mprime-z)*(mprime-z)'/length(mprime), % cluster variance
lmp=length(mprime);
percentage_symbol_errors=100*sum(abs(sign(mprime-m(1:lmp))))/lmp, % symb err
% decode decision device output to text string
reconstructed_message=pam2letters(mprime)    % reconstruct message
```

This ideal system simulation is composed primarily of code recycled from previous chapters. The transformation from a character string to a 4-level T-spaced sequence to an upsampled (T/M-spaced) T-wide (Hamming) pulse shape filter output sequence mimics pulseshape0.m from Section 8.2. This sequence of T/M-spaced pulse filter outputs and its magnitude spectrum are shown in Figure 9.2 (type plotspec(x,1/M) after running idsys.m).

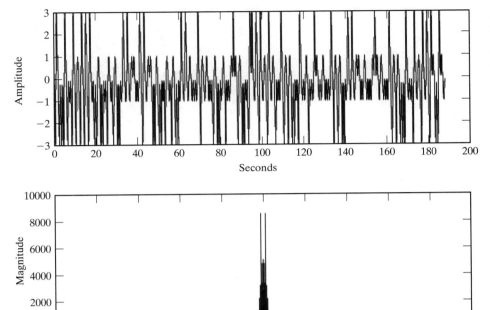

FIGURE 9.2 The transmitter creates the signal in the top plot, which has the magnitude spectrum shown in the bottom.

Each pulse is 1 time unit long, so successive pulses can be initiated without any overlap. The unit duration of the pulse could be a millisecond (for a pulse frequency of 1 kHz) or a microsecond (for a pulse frequency of 1 MHz). The magnitude spectrum in Figure 9.2 has little apparent energy outside bandwidth 2 (the meaning of 2 in Hz is dependent on the units of time).

This oversampled waveform is upconverted by multiplication with a sinusoid. This is familiar from `AM.m` of Section 5.2. The transmitted passband signal and its spectrum (created using `plotspec(v,1/M)`) are shown in Figure 9.3. The default carrier frequency is `fc=20`. Nyquist sampling of the received signal occurs as long as the sample frequency $1/(T/M) = M$ for $T = 1$ is twice the highest frequency in the received signal, which will be the carrier frequency plus the baseband signal bandwidth of approximately 2. Thus, M should be greater than 44 to prevent aliasing of the received signal. This allows reconstruction of the analog received signal at any desired point, which could prove valuable if the times at which the samples were taken were not synchronized with the received pulses.

The transmitted signal reaches the receiver portion of the ideal system in Figure 9.1. Downconversion is accomplished by multiplying the samples of the received signal by an oscillator that (miraculously) has the same frequency and phase as was used in the transmitter. This produces a signal with the spectrum shown in Figure 9.4 (type `plotspec(x2,1/M)` after running `idsys.m`), a spectrum that has substantial nonzero components (that must be removed) at about twice the carrier frequency.

FIGURE 9.3 The signal and its spectrum after upconversion.

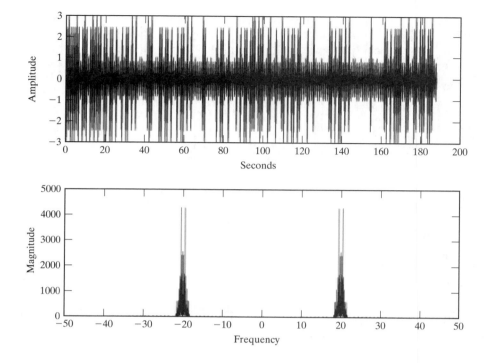

FIGURE 9.4 The received signal and spectrum after down-conversion (mixing).

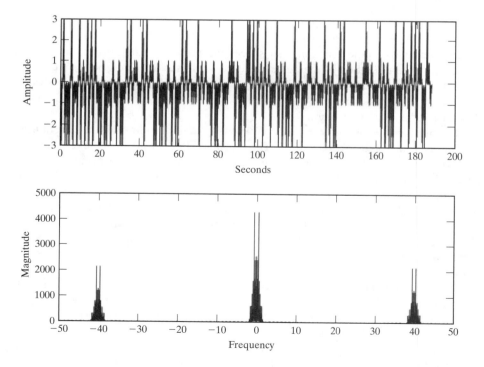

To suppress the components centered around ±40 in Figure 9.4 and to pass the baseband component without alteration (except for possibly a delay), the lowpass filter is designed with a cutoff between 25 and 30. For $M = 100$, the Nyquist frequency is 50. (Section 7.2.2 details the use of `remez` for filter design.) The frequency response of the resulting FIR filter (from `freqz(b)`) is shown in Figure 9.5. To make sense of the horizontal axes, observe that the "1" in Figure 9.5 corresponds to the "50" in Figure 9.4. Thus the cutoff between normalized frequencies 0.5 to 0.6 corresponds to unnormalized frequencies of 25 and 30, as desired.

The output of the lowpass filter in the demodulator is a signal with the spectrum shown in Figure 9.6 (drawn using `plotspec(x3,1/M)`). The spectrum in Figure 9.6 should compare quite closely to that of the transmitter baseband in Figure 9.2, as indeed it does. It is easy to check the effectiveness of the lowpass filter design by attempting to use a variety of different lowpass filters, as suggested in Problem 9.4.

Recall the discussion in Section 8.4 of matching two scaled pulse shapes. Viewing the pulse shape as a kind of marker, it is reasonable to correlate the pulse shape with the received signal in order to locate the pulses. (More justification for this procedure is forthcoming in Chapter 11.) This appears in Figure 9.1 as the block labelled "pulse correlation filter." The code in `idsys.m` implements this using the `filter` command to carry out the correlation (rather than the `xcorr` function), though the choice was a matter of convenience. (Refer to `corrvsconv.m` in Problem 8.9 to see how the two functions are related.)

The first $4M$ samples of the resulting signal y are plotted in Figure 9.7 (via `plot(y(1:4*M))`). The first three symbols of the message (i.e., `m(1:3)`) are −3, 3, and −3, and Figure 9.7 shows why it is best to take the samples at indices $125 + kM$.

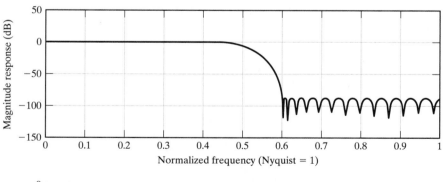

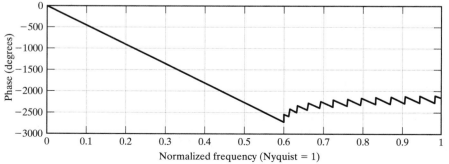

FIGURE 9.5 Frequency response of the lowpass filter.

FIGURE 9.6 Signal and spectrum after the demodulation and lowpass filtering. Compare to the baseband transmitted signal (and spectrum) in Figure 9.2.

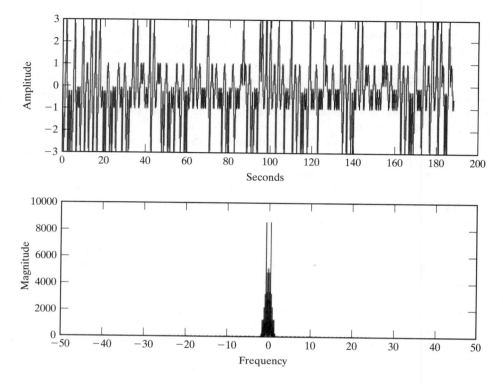

FIGURE 9.7 The first four symbol periods (recall the oversampling factor was $M = 100$) of the signal at the receiver (after the demodulation, LPF, and pulse correlator filter). The first three symbol values are -3, $+3$, -3, which can be deciphered from the signal assuming the delay can be selected appropriately.

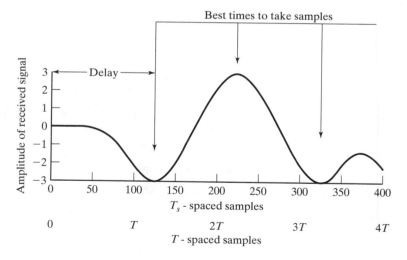

The initial delay of 125 corresponds to half the length of the lowpass filter (`0.5*fl`) plus half the length of the correlator filter (`0.5*M`) plus half a symbol period (`0.5*M`), which accounts for the delay from the start of each pulse to its peak.

Selecting this delay and the associated downsampling are accomplished in the code

```
z=y(0.5*fl+M:M:end);            % downsample to symbol rate
```

in `idsys.m`, which recovers the T-spaced samples z. With reference to Figure 9.1, the parameter l in the downsampling block is 125.

A revealing extension of Figure 9.7 is to plot the oversampled waveform y for the complete transmission in order to see if the subsequent peaks of the pulses occur at regular intervals precisely on source alphabet symbol values, as we would hope. However, even for small messages (such as the wiener jingle), squeezing such a figure onto one graph makes a detailed visual examination of the peaks fruitless. This is precisely why we plotted Figure 9.7—to see the detailed timing information for the first few symbols.

One idea is to plot the next four symbol periods on top of the first four by shifting the start of the second block to time zero. Continuing this approach throughout the data record mimics the behavior of a well-adjusted oscilloscope that triggers at the same point in each symbol group. This operation can be implemented in MATLAB by first determining the maximum number of groups of `4*M` samples that fit inside the vector y from the `l`th sample on. Let

```
ul=floor((length(y)-l-1)/(4*M));
```

then the `reshape` command can be used to form a matrix with `4*M` rows and `ul` columns. This is easily plotted using

```
plot(reshape(y(1:ul*4*M+124),4*M,ul))
```

and the result is shown in Figure 9.8. Note that the first element plotted in Figure 9.8 is the `l`th element of Figure 9.7. This type of figure, called an *eye diagram*, is commonly used in practice as an aid in troubleshooting. Eye diagrams will also be used routinely in succeeding chapters.

Four is an interesting grouping size for this particular problem because four symbols are used to represent each character in the coding and decoding implemented in `letters2pam` and `pam2letters`. One idiosyncrasy is that each character starts off with

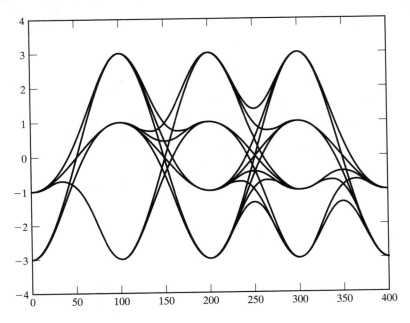

FIGURE 9.8 Repeatedly overlaying a time width of four symbols yields an *eye diagram*.

FIGURE 9.9 Recon-
structed symbols,
called the *soft deci-
sions*, are plotted in
this *constellation dia-
gram* time history.

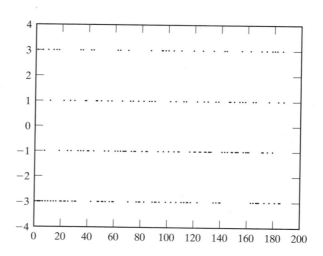

a negative symbol. Another is that the second symbol in each character is never −1 in
our chosen message. These are not generic effects; they are a consequence of the particular coding and message used in idsys.m. Had we chosen to implement a scrambling
scheme (recall Problem 8.19) the received signal would be whitened and these particular
peculiarities would not occur.

The vector z contains estimates of the decoded symbols, and the command
plot([1:length(z)],z,'.') produces a time history of the output of the down-
sampler, as shown in Figure 9.9. This is called the time history of a *constellation diagram*
in which all the dots are meant to lie near the allowable symbol values. Indeed, the points
in Figure 9.9 cluster tightly about the alphabet values ±1 and ±3. How tightly they cluster can be quantified using the *cluster variance*, which is the average of the square of the
difference between the decoded symbol values (the *soft* decisions) in z and the nearest
member of the alphabet (the final *hard* decisions).

The MATLAB function quantalph.m is used in idsys.m to calculate the
hard decisions, which are then converted back into a text character string using
pam2letters. If all goes well, this reproduces the original message. The only flaw
is that the last symbol of the message has been lost due to the inherent delay of the
lowpass filtering and the pulse shape correlation. Because four symbols are needed to
decode a single character, the loss of the last symbol also results in the loss of the last
character. The function pam2letters provides a friendly reminder in the MATLAB
command window that this has happened.

The problems that follow give you a few more ways to explore the behavior of the
ideal system.

Problem

9.1. Using idsys.m, examine the effect of using different carrier frequencies. Try
fc=50, 30, 3, 1, 0.5. What are the limiting factors that cause some to
work and others to fail?

9.2. Using `idsys.m`, examine the effect of using different oversampling frequencies. Try `M=1000, 25, 10`. What are the limiting factors that cause some to work and others to fail?

9.3. What happens if the LPF at the beginning of the receiver is removed? What do you think will happen if there are other users present? Try adding in "another user" at `fc = 30`.

9.4. What are the limits to the LPF design at the beginning of the receiver? What is the lowest cutoff frequency that works? The highest?

9.5. Using the same specifications (`fbe=[0 0.1 0.2 1]; damps=[1 1 0 0 ];`), how short can you make the LPF? Explain.

9.3 Flat Fading: A Simple Impairment and a Simple Fix

Unfortunately, a number of the assumptions made in the simulation of the ideal system `idsys.m` are routinely violated in practice. The designer of a receiver must somehow compensate by improving the receiver. This section presents an impairment (flat fading) for which we have already developed a fix (an AGC). Later sections describe misbehavior due to a wider variety of common impairments that we will spend the rest of the book combating.

Flat fading occurs when there are obstacles moving in the path between the transmitter and receiver or when the transmitter and receiver are moving with respect to each other. It is most commonly modelled as a time-varying channel gain that attenuates the received signal. The modifier "flat" implies that the loss in gain is uniform over all frequencies.[1] This section begins by studying the loss of performance caused by a time-varying channel gain (using a modified version of `idsys.m`) and then examines the ability of an adaptive element (the automatic gain control, AGC) to make things right.

In the ideal system of the preceding section, the gain between the transmitter and the receiver was implicitly assumed to be unity. What happens when this assumption is violated, when flat fading is experienced in midmessage? To examine this question, suppose that the channel gain is unity for the first 20% of the transmission, but that for the last 80% it drops by half. This flat fade can easily be studied by inserting the following code between the transmitter and the receiver parts of `idsys.m`.

---

modification of idsys.m for time-varying fading channel

```
lr=length(r);                    % length of transmitted signal vector
fp=[ones(1,floor(0.2*lr)),0.5*ones(1,lr-floor(0.2*lr))]; % flat fading profile
r=r.*fp;                         % apply profile to transmitted signal vector
```

The resulting plot of the soft decisions in Figure 9.10 (via `plot([1:length(z)], z,'.')`) shows the effect of the fade in the latter 80% of the response. Shrinking the magnitude of the symbols ±3 by half puts it in the decision region for ±1, which

---

[1] In communications jargon, it is not *frequency selective*.

FIGURE 9.10 Soft
decisions with uncom-
pensated flat fading.

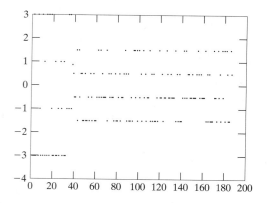

generates a large number of symbol errors. Indeed, the recovered message looks nothing
like the original.

Section 6.8 has already introduced an adaptive element designed to compensate for
flat fading: the automatic gain control, which acts to maintain the power of a signal at a
constant known level. Stripping out the AGC code from `agcvsfading.m` on page 117
and combining it with the fading channel just discussed creates a simulation in which
the fade occurs, but in which the AGC can actively work to restore the power of the
received signal to its desired nominal value ds≈ 1.

---

further modification of idsys.m: fading plus automatic gain control

```
ds=pow(r);                              % desired average power of signal
lr=length(r);                           % length of transmitted signal vector
fp=[ones(1,floor(0.2*lr)),0.5*ones(1,lr-floor(0.2*lr))]; % flat fading profile
r=r.*fp;                                % apply profile to transmitted signal vector
g=zeros(1,lr); g(1)=1;                  % initialize gain
nr=zeros(1,lr);
mu=0.0003;                              % stepsize
for i=1:lr-1                            % adaptive AGC element
 nr(i)=g(i)*r(i);                       % AGC output
 g(i+1)=g(i)-mu*(nr(i)^2-ds);          % adapt gain
end
r=nr;                                   % received signal is still called r
```

Inserting this segment into `idsys.m` (immediately after the time-varying fading channel modification) results in only a small number of errors that occur right at the time of
the fade. Very quickly, the AGC kicks in to restore the received power. The resulting
plot of the soft decisions (via `plot([1:length(z)],z,'.')`) in Figure 9.11 shows
how quickly after the abrupt fade the soft decisions return to the appropriate sector. (Look
for where the larger soft decisions exceed a magnitude of 2.)

Figure 9.12 plots the trajectory of the AGC gain g as it moves from the vicinity of
unity to the vicinity of 2 (just what is needed to counteract a 50% fade). Increasing the
stepsize `mu` can speed up this transition, but also increases the range of variability in
the gain as it responds to short periods when the square of the received signal does not
closely match its long-term average.

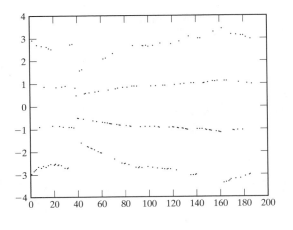

FIGURE 9.11 Soft decisions with an AGC compensating for an abrupt flat fade.

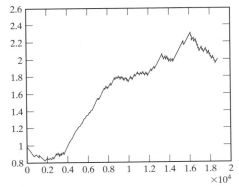

FIGURE 9.12 Trajectory of the AGC gain parameter as it moves to compensate for the fade.

Problem

9.6. Another idealized assumption made in `idsys.m` is that the receiver knows the start of each frame; that is, it knows where each four-symbol group begins. This is a kind of "frame synchronization" problem and was absorbed into the specification of the parameter l. (With the default settings, l was 125.) This problem poses the question. "What if this is not known, and how can it be fixed?"

(a) Verify, using `idsys.m`, that the message becomes scrambled if the receiver is mistaken about the start of each group of four. Add a random number of 4-PAM symbols before the message sequence, but do not "tell" the receiver that you have done so (i.e., do not change l). What value of l would fix the problem? Can l really be known before hand?

(b) Section 8.5 proposed the insertion of a marker sequence as way to synchronize the frame. Add a seven-symbol marker sequence just prior to the first character of the text. In the receiver, implement a correlator that searches for the known marker. Demonstrate the success of this modification by adding random symbols at the start of the transmission. Where in the receiver have you chosen to put the correlation procedure? Why?

(c) One quirk of the system (observed in the eye diagram in Figure 9.8) is that each group of four begins with a negative number. Use this feature (rather than a separate marker sequence) to create a correlator in the receiver that can be used to find the start of the frames.

(d) The previous two exercises showed two possible solutions to the frame synchronization problem. Explain the pros and cons of each method, and argue which is a "better" solution.

9.4 Other Impairments: More "What Ifs"

Of course, a fading channel is not the only thing that can go wrong in a telecommunication system. (Think back to the "what if" questions in the first section of this chapter.) This section considers a range of synchronization and interference impairments that violate the assumptions of the idealized system. Though each impairment is studied separately (i.e., assuming that everything functions ideally except for the particular impairment of interest), a single program is written to simulate any of the impairments. The program impsys.m leaves both the transmitter and the basic operation of the receiver unchanged; the primary impairments are to the sampled sequence that is delivered to the receiver.

The rest of this chapter conducts a series of experiments dealing with stuff that can happen to the system. Interference is added to the received signal as additive gaussian channel noise and as multipath interference. The oscillator at the transmitter is no longer presumed to be synchronized with the oscillator at the receiver. The best sample times are no longer presumed to be known exactly in either phase or period.

---

impsys.m: transmission system with uncompensated impairments

```
% specification of impairments
cng=input('channel noise gain: try 0, 0.6 or 2 :: ');
cdi=input('channel multipath: 0 for none, 1 for mild or 2 for harsh :: ');
fo=input('transmitter mixer freq offset in %: try 0 or 0.01 :: ');
po=input('transmitter mixer phase offset in rad: try 0, 0.7 or 0.9 :: ');
toper=input('baud timing offset as % of symb period: try 0, 20 or 30 :: ');
so=input('symbol period offset: try 0 or 1 :: ');

% INSERT TRANSMITTER CODE (FROM IDSYS.M) HERE

if cdi < 0.5,               % channel ISI
  mc=[1 0 0];               % distortion-free channel
elseif cdi<1.5,
  mc=[1 zeros(1,M) 0.28 zeros(1,2.3*M) 0.11];  % mild multipath channel
else
  mc=[1 zeros(1,M) 0.28 zeros(1,1.8*M) 0.44];  % harsh multipath channel
end
mc=mc/(sqrt(mc*mc'));       % normalize channel power
dv=filter(mc,1,r);          % filter transmitted signal through channel
nv=dv+cng*(randn(size(dv)));  % add Gaussian channel noise
to=floor(0.01*toper*M);     % fractional period delay in  sampler
rnv=nv(1+to:end);           % delay in on-symbol designation
rt=(1+to)/M:1/M:length(nv)/M; % modified time vector with delayed message start
rM=M+so;                    % receiver sampler timing offset

% INSERT RECEIVER CODE (FROM IDSYS.M) HERE
```

The first few lines of `impsys.m` prompt the user for parameters that define the impairments. The channel noise gain parameter `cng` is a gain factor associated with a Gaussian noise that is added to the received signal. The suggested values of 0, 0.6, and 2 represent no impairment, mild impairment (that only rarely causes symbol recovery errors), and a harsh impairment (that causes multiple symbol errors), respectively.

The second prompt selects the multipath interference: none, mild, or harsh. In the mild and harsh cases, three copies of the transmitted signal are summed at the receiver, each with a different delay and amplitude. This is implemented by passing the transmitted signal through a filter whose impulse response is specified by the variable `mc`. As occurs in practice, the transmission delays are not necessarily integer multiples of the symbol interval. Each of the multipath models has its largest tap first. If the largest path gain were not first, this could be interpreted as a delay between the receipt of the first sample of the first pulse of the message and the optimal sampling instant.

The next pair of prompts concern the transmitter and receiver oscillators. The receiver assumes that the phase of the oscillator at the transmitter is zero at the time of arrival of the first sample of the message. In the ideal system, this assumption was correct. In `impsys.m`, however, the receiver makes this same assumption, but it may no longer be correct. Mismatch between the phase of the oscillator at the transmitter and the phase of the oscillator at the receiver is an inescapable impairment (unless there is also a separate communication link or added signal such as an embedded pilot tone that synchronizes the oscillators). The user is prompted for a carrier phase offset in radians (the variable `po`) that is added to the phase of the oscillator at the transmitter, but not at the receiver. Similarly, the frequencies of the oscillators at the transmitter and receiver may differ by a small amount. The user specifies the frequency offset in the variable `fo` as a percentage of the carrier frequency. This is used to scale the carrier frequency of the transmitter, but not of the receiver. This represents a difference between the nominal values used by the receiver and the actual values achieved by the transmitter.

Just as the receiver oscillator need not be fully synchronized with the transmitter oscillator, the symbol clock at the receiver need not be properly synchronized with the transmitter symbol period clock. Effectively, the receiver must choose when to sample the received signal based on its best guess as to the phase and frequency of the symbol clock at the transmitter. In the ideal case, the delay between the receipt of the start of the signal and the first sample time was readily calculated using the parameter l. But l cannot be known in a real system because the "first sample" depends, for instance, on when the receiver is turned on. Thus, the phase of the symbol clock is unknown at the receiver. This impairment is simulated in `impsys.m` using the timing offset parameter `toper`, which is specified as a percentage of the symbol period. Subsequent samples are taken at positive integer multiples of the presumed sampling interval. If this interval is incorrect, then the subsequent sample times will also be incorrect. The final impairment is specified by the "symbol period offset," which sets the symbol period at the transmitter to `so` less than that at the receiver.

Using `impsys.m`, it is now easy to investigate how each impairment degrades the performance of the system.

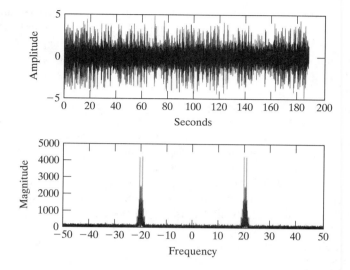

FIGURE 9.13 When
noise is added, the
received signal
appears jittery. The
spectrum has a notice-
able noise floor.

9.4.1 Additive Channel Noise

Whenever the channel noise is greater than half the gap between two adjacent symbols in the source constellation, a symbol error may occur. For the constellation of ±1s and ±3s, if a noise sample has magnitude larger than 1, then the output of the quantizer may be erroneous.

Suppose that a white, broadband noise is added to the transmitted signal. The spectrum of the received signal, which is plotted in Figure 9.13 (via `plotspec(nv,1/rM)`), shows a nonzero noise floor compared with the ideal (noise-free) spectrum in Figure 9.3. A noise gain factor of `cng=0.6` leads to a cluster variance of about 0.02 and no symbol errors. A noise gain of `cng=2` leads to a cluster variance of about 0.2 and results in approximately 2% symbol errors. When there are 10% symbol errors, the reconstructed text becomes undecipherable (for the particular coding used in `letters2pam` and `pam2letters`). Thus, as should be expected, the performance of the system degrades as the noise is increased. It is worthwhile taking a closer look to see exactly what goes wrong.

The eye diagram for the noisy received signal is shown in Figure 9.14, which should be compared with the noise-free eye diagram in Figure 9.8. This is plotted using the MATLAB commands:

```
ul=floor((length(x3)-124)/(4*rM));
plot(reshape(x3(125:ul*4*rM+124),4*rM,ul)).
```

Hopefully, it is clear from the noisy eye diagram that it would be very difficult to correctly decode the symbols directly from this signal. Fortunately, the correlation filter reduces the noise significantly, as shown in the eye diagram in Figure 9.15. (This is plotted as before, substituting `y` for `x3`.) Comparing Figures 9.14 and 9.15 closely, observe that the whole of the latter is shifted over in time by about 50 samples. This is the effect of the time delay of the correlator filter, which is half the length of the filter. Clearly, it is much easier to correctly decode using `y` than using `x3`, though the pulse shapes of Figure 9.15 are still blurred when compared with the ideal pulse shapes in Figure 9.8.

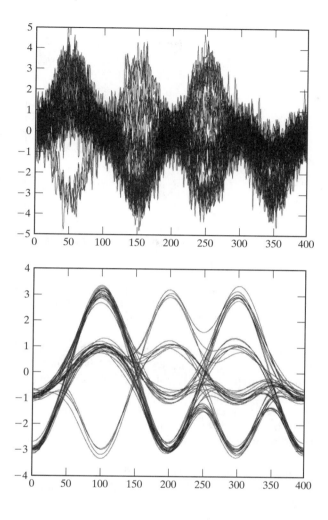

FIGURE 9.14 The eye diagram of the received signal x3 repeatedly overlays four-symbol wide segments. The channel noise is not insignificant.

FIGURE 9.15 The eye diagram of the received signal y after the correlation filter. The noise is reduced significantly.

Problem

9.7. The correlation filter in `impsys.m` is a lowpass filter with impulse response given by the pulse shape p.

    (a) Plot the frequency response of this filter. What is the bandwidth of this filter?

    (b) Design a lowpass filter using `remez` that has the same length and the same bandwidth as the correlation filter.

    (c) Use your new filter in place of the correlation filter in `impsys.m`. Has the performance improved or worsened? Explain in detail what tests you have used.

No peeking ahead to Chapter 11.

9.4.2 Multipath Interference

The next impairment is interference caused by a multipath channel, which occurs whenever there is more than one route between the transmitter and the receiver. Because

FIGURE 9.16 With mild multipath interference, the soft decisions can readily be segregated into four stripes that correspond to the four symbol values.

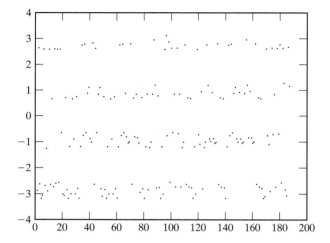

these paths experience different delays and attenuations, multipath interference can be modelled as a linear filter. Since filters can have complicated frequency responses, some frequencies may be attenuated more than others, and so this is called *frequency-selective fading*.

The "mild" multipath interference in `impsys.m` has three (nonzero) paths between the transmitter and the receiver. Its frequency response has numerous dips and bumps that vary in magnitude from about $+2$ to -4 dB. (Verify this using `freqz`.) A plot of the soft decisions is shown in Figure 9.16 (from `plot([1:length(z)],z,'.')`), which should be compared with the ideal constellation diagram in Figure 9.9. The effect of the mild multipath interference is to smear the lines into stripes. As long as the stripes remain separated, the quantizer is able to recover the symbols, and hence the message, without errors.

The "harsh" multipath channel in `impsys.m` also has three paths between the transmitter and receiver, but the later reflections are larger than in the mild case. The frequency response of this channel has peaks up to about $+4$ dB and down to about -8 dB, so its effects are considerably more severe. The effect of this channel can be seen directly by looking at the constellation diagram of the soft decisions in Figure 9.17. The constellation diagram is smeared, and it is no longer possible to visually distinguish the four stripes that represent the four symbol values. It is no surprise that the message becomes garbled. As the output shows, there are about 10% symbol errors, and a majority of the recovered characters are wrong.

9.4.3 Carrier Phase Offset

For the receiver in Figure 9.1, the difference between the phase of the modulating sinusoid at the transmitter and the phase of the demodulating sinusoid at the receiver is the carrier phase offset. The effect of a nonzero offset is to scale the received signal by a factor equal to the cosine of the offset, as was shown in (5.4) of Section 5.2. Once the phase offset is large enough, the demodulated signal contracts so that its maximum magnitude is less than 2. When this happens, the quantizer always produces a ± 1. Symbol errors abound.

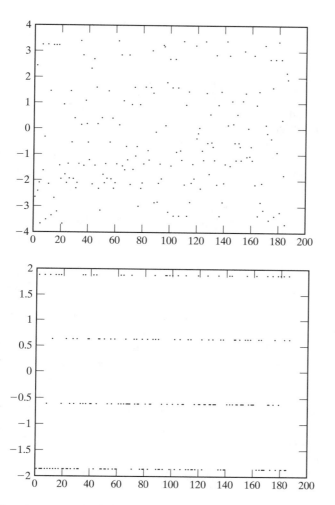

FIGURE 9.17 With harsher multipath interference, the soft decisions smear and it is no longer possible to see which points correspond to which of the four symbol values.

FIGURE 9.18 Soft decisions for harsh carrier phase offset are never greater than two. The quantizer finds no ±3's and many symbol errors occur.

When running `impsys.m`, there are two suggested nonzero choices for the phase offset parameter po. With `po=0.9`, $\cos(0.9) = 0.62$, and $3\cos(0.9) < 2$. This is shown in the plot of the soft decision errors in Figure 9.18. For the milder carrier phase offset (`po=0.7`), the soft decisions result in no symbol errors, because the quantizer will still decode values at $\pm 3\cos(0.7) = \pm 2.3$ as ± 3.

As long as the constellation diagram retains distinct horizontal stripes, all is not lost. In Figure 9.18, even though the maximum magnitude is less than two, there are still four distinct stripes. If the quantizer could be scaled properly, the symbols could be decoded successfully. Such a scaling might be accomplished, for instance, by another AGC, but such scaling would not improve the signal-to-noise ratio. A better approach is to identify the 'unknown' phase offset, as discussed in Chapter 10.

Problem

9.8. Using `impsys.m` as a basis, implement an AGC-style adaptive element to compensate for a phase offset. Verify that your method works for a phase offset of 0.9

Stuff
Happens

and for a phase offset of 1.2. Show that the method fails when the phase offset is $\pi/2$.

9.4.4 Carrier Frequency Offset

The receiver in Figure 9.1 has a carrier frequency offset when the frequency of the carrier at the transmitter differs from the assumed frequency of the carrier at the receiver. As was shown in (5.5) in Section 5.2, this impairment is like a modulation by a sinusoid with frequency equal to the offset. This modulating effect is catastrophic when the low frequency modulator approaches a zero crossing, since then the gain of the signal approaches zero. This effect is apparent for a 0.01% frequency offset in `impsys.m` in the plot of the soft decisions (via `plot([1:length(z)],z,'.')`) in Figure 9.19. This experiment suggests that receiver mixer frequency must be adjusted to track that of the transmitter.

9.4.5 Downsampler Timing Offset

As shown in Figure 9.7, there is a sequence of "best times" at which to downsample. When the starting point is correct and no intersymbol interference (ISI) is present, as in the ideal system, the sample times occur at the top of the pulses. When the starting point is incorrect, all the sample times are shifted away from the top of the pulses. This was set in the ideal simulation using the parameter l, with its default value of 125. The timing offset parameter `toper` in `impsys.m` is used to offset the received signal. Essentially, this means that the best value of l has changed, though the receiver does not know it.

This is easiest to see by drawing the eye diagram. Figure 9.20 shows an overlay of 4-symbol wide segments of the received signal (using the `reshape` command as in the code on page 161). The receiver still thinks the best times to sample are at $l + nT$, but this is clearly no longer true. In fact, whenever the sample time begins between 100 and 140 (and lies in this or any other shaded region), there will be errors when quantizing. For example, all samples taken at 125 lie between ± 1, and hence no symbols will ever be decoded at their ± 3 value. In fact, some even have the wrong sign! This is a far worse

FIGURE 9.19 Soft decisions for 0.01% carrier frequency offset.

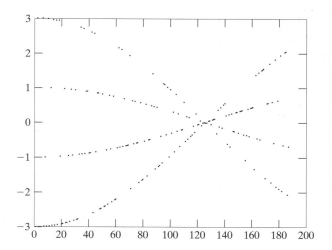

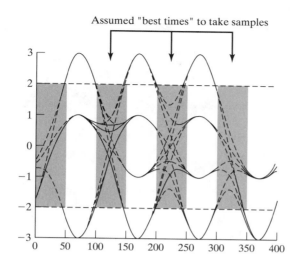

Assumed "best times" to take samples

FIGURE 9.20 Eye diagram with down-sampler timing offset of 50%. Sample times used by the ideal system are no longer valid, and lead to numerous symbol errors.

situation than in the carrier phase impairment because no simple amplitude scaling will help. Rather, a solution must correct the problem; it must slide the times so that they fall in the unshaded regions. Because these unshaded regions are wide open, this is often called the *open eye* region. The goal of an adaptive element designed to fix the timing offset problem is to *open the eye* as wide as possible.

9.4.6 Downsampler Period Offset

When the assumed period of the downsampler is in error, there is no hope. As mentioned in the previous impairment, the receiver believes that the best times to sample are at $l + nT$. When there is a period offset, it means that the value of T used at the receiver differs from the value actually used at the transmitter.

The prompt in `impsys.m` for symbol period offset suggests trying 0 or 1. A response of 1 results in the transmitter creating the signal assuming that there are $M - 1$ samples per symbol period, while the receiver retains the setting of M samples per symbol, which is used to specify the correlator filter and to pick subsequent downsampling instants once the initial sample time is selected. The symptom of a misaligned sample period is a periodic collapse of the constellation, similar to that observed when there is a carrier frequency offset (recall Figure 9.19). For an offset of 1, the soft decisions are plotted in Figure 9.21. Can you connect the value of the period of this periodic collapse to the parameters of the simulated example?

9.4.7 Repairing Impairments

When stuff happens and the receiver continues to operate as if all were well, the trans-mitted message can become unintelligible. The various impairments of the preceding sections point the way to the next onion-like layer of the design by showing the kinds of problems that may arise. Clearly, the receiver must be improved to counteract these impairments.

Coding (Chapter 14) and matched receive filtering (Chapter 11) are intended primarily to counter the effects of noise. Equalization (Chapter 13) compensates for multipath

FIGURE 9.21 When there is a 1% down-sampler period offset, all is lost, as shown by the eye diagram in the top plot and the soft decisions in the bottom.

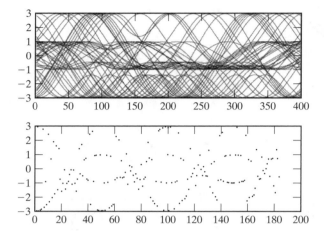

interference, and can reject narrowband interferers. Carrier recovery (Chapter 10) will be used to adjust the phase, and possibly the frequency as well, of the receiver oscillator. Timing recovery (Chapter 12) aims to reduce downsampler timing and period offset. All of these fixes can be viewed as digital signal processing (DSP) solutions to the impairments explored in `impsys.m`.

Each of these fixes will be designed separately, as if the problem it is intended to counter were the only problem in the world. Fortunately, somehow they can all work together simultaneously. Examining possible interactions between the various fixes, which is normally a part of the testing phase of a receiver design, will be part of the receiver design project of Chapter 15.

IV

The Adaptive Component Layer

The current layer describes all the practical fixes that are required in order to create a workable radio. One by one the various pragmatic problems are studied and solutions are proposed, implemented, and tested. These include fixes for additive noise, for timing offset problems, for clock frequency mismatches and jitter, and for multipath reflections. The order in which topics are discussed is the order in which they appear in our receiver.

10 *Carrier Recovery*

A man with one watch knows what time it is. A man with two watches is never sure.

—Segal's law

Figure 10.1 shows a generic transmitter and receiver pair that emphasizes the modulation and corresponding demodulation. Even assuming that the transmission path is ideal (as in Figure 10.1), the signal that arrives at the receiver is a complicated analog waveform that must be downconverted and sampled before the message can be recovered. For the demodulation to be successful, the receiver must be able to figure out both the frequency and phase of the modulating sinusoid used in the transmitter, as was shown in (5.4) and (5.5) and graphically illustrated in Figures 9.18 and 9.19. This chapter discusses a variety of strategies that can be used to estimate the phase and frequency of the carrier and to fix the gain problem (of (5.4) and Figure 9.18) and the undulation problem (of (5.5) and Figure 9.19). This process of estimating the frequency and phase of the carrier is called *carrier recovery*.

Figure 10.1 shows two downconversion steps: one analog and one digital. In a purely analog system, no sampler or digital downconversion would be needed. The problem is that accurate analog downconversion requires highly precise analog components, which can be expensive. In a purely digital receiver, the sampler would directly digitize the received signal, and no analog downconversion would be required. The problem is that sampling this fast can be prohibitively expensive. The happy compromise is to use an inexpensive analog downconverter to translate to some lower intermediate frequency, where it is possible to sample cheaply enough. At the same time, sophisticated digital processing can be used to compensate for inaccuracies in the cheap analog components. Indeed, the same adaptive elements that estimate and remove the unknown phase offset between the transmitter and the receiver automatically compensate for any additional phase inaccuracies in the analog portion of the receiver.

Normally, the transmitter and receiver agree to use a particular frequency for the carrier, and in an ideal world, the frequency of the carrier of the transmitted signal would be known exactly. But even expensive oscillators may drift apart in frequency over time, and cheap (inaccurate) oscillators may be an economic necessity. Thus, there needs to be a way to align the frequency of the oscillator at the transmitter with the frequency of the oscillator at the receiver. Since the goal is to find the frequency and phase of a signal, why not use a Fourier Transform (or, more properly, an FFT)? Section

FIGURE 10.1 Schematic of a communications system emphasizing the need for synchronization of the frequency and phase of the carrier.

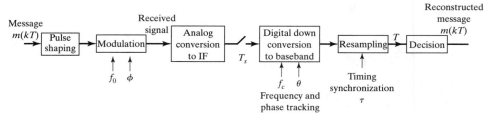

10.1 shows how to isolate a sinusoid that is at twice the frequency of the carrier by squaring and filtering the received signal. The frequency and phase of this sinusoid can then be found in a straightforward manner by using the FFT, and the frequency and phase of the carrier can then be simply deduced. Though feasible, this method is rarely used because of the computational cost.

The strategy of the following sections is to replace the FFT operation with an adaptive element that achieves its optimum value when the phase of an estimated carrier equals the phase of the actual carrier. By moving the estimates in the direction of the gradient of a suitable performance function, the element can recursively hone in on the correct value. By first assuming that the frequency is known, there are a variety of ways to structure adaptive elements that iteratively estimate the unknown phase of a carrier. One such performance function, discussed in Section 10.2, is the square of the difference between the received signal and a locally generated sinusoid. Another performance function leads to the well-known *phase locked loop*, which is discussed in depth in Section 10.3, and yet another performance function leads to the *Costas loop* of Section 10.4. An alternative approach uses the *decision directed* method detailed in Section 10.5. Each of these methods is derived from an appropriate performance function, each is simulated in MATLAB, and each can be understood by looking at the appropriate error surface. This approach should be familiar from Chapter 6, where it was used in the design of the AGC.

Section 10.6 then shows how to modify the adaptive elements to attack the frequency estimation problem. Two ways are shown. The first tries (unsuccessfully) to apply a direct adaptive method, and the reasons for the failure provide a cautionary counterpoint to the indiscriminate application of adaptive elements. The second, a simple indirect method, exploits the relationship between the phase and frequency of a signal and forms the basis for an effective adaptive frequency tracking element that is detailed in Section 10.6.2. Of course, there are other possibilities. A method that uses an integrator in the single phase locked loop (PLL) loop is discussed in the document *Analysis of the Phase Locked Loop*, which can be found on the CD.

10.1 Phase and Frequency Estimation via an FFT

As indicated in Figure 10.1, the received signal consists of a message $m(kT_s)$ modulated by the carrier. In the simplest case, when the modulation is done using AM with large carrier as in Section 5.1, it may be quite easy to locate the carrier and its phase. More generally, however, the carrier will be well hidden within the received signal and some kind of extra processing will be needed to bring it to the foreground.

To see the nature of the carrier recovery problem explicitly, the following code generates two different "received signals": the first is AM modulated with large carrier and the second is AM modulated with suppressed carrier. The phase and frequencies of both signals can be recovered using an FFT, though the suppressed carrier scheme requires additional processing before the FFT can successfully be applied.

Drawing on the code in `pulseshape0.m` on page 144, and modulating with the carrier c, `pulrecsig.m` creates the two different received signals. The `pam` command creates a random sequence of symbols drawn from the alphabet ± 1, ± 3, and then uses `hamming` to create a pulse shape.[1] The oversampling factor M is used to simulate the "analog" portion of the transmission, and MT_s is equal to the symbol time T.

---

pulrecsig.m: create pulse shaped received signal

```
N=10000; M=20; Ts=.0001;            % no. symbols, oversampling factor
time=Ts*(N*M-1); t=0:Ts:time;       % sampling interval and time vectors
m=pam(N,4,5);                       % 4-level signal of length N
mup=zeros(1,N*M); mup(1:M:end)=m;   % oversample by integer length M
ps=hamming(M);                      % blip pulse of width M
s=filter(ps,1,mup);                 % convolve pulse shape with data
f0=1000; phoff=-1.0;                % carrier freq. and phase
c=cos(2*pi*f0*t+phoff);             % construct carrier
rsc=s.*c;                           % modulated signal (small carrier)
rlc=(s+1).*c;                       % modulated signal (large carrier)
```

Figure 10.2 plots the spectra of both the large and suppressed carrier signals `rlc` and `rsc`. The carrier itself is clearly visible in the top plot, and its frequency and phase can readily be found by locating the maximum value in the FFT:

```
fftrlc=fft(rlc);                    % spectrum of rlc
[m,imax]=max(abs(fftrlc(1:end/2))); % index of max peak
ssf=(0:length(t))/(Ts*length(t));  % frequency vector
freqL=ssf(imax)                     % freq at the peak
phaseL=angle(fftrlc(imax))          % phase at the peak
```

Changing the default phase offset `phoff` changes the `phaseL` variable accordingly. Changing the frequency `f0` of the carrier changes the frequency `freqL` at which the maximum occurs. Note that the `max` function used in this fashion returns both the maximum value m and the index `imax` at which the maximum occurs.

On the other hand, applying the same code to the FFT of the suppressed carrier signal does not recover the phase offset. In fact, the maximum often occurs at frequencies other than the carrier, and the phase values reported bear no resemblance to the desired phase offset `phoff`. There needs to be a way to process the received signal to emphasize the carrier.

A common scheme uses a squaring nonlinearity followed by a bandpass filter, as shown in Figure 10.3. When the received signal $r(t)$ consists of the pulse modulated data signal $s(t)$ times the carrier $\cos(2\pi f_0 t + \phi)$, the output of the squaring block is

$$r^2(t) = s^2(t)\cos^2(2\pi f_0 t + \phi). \tag{10.1}$$

---

[1] This is not a common (or a particularly useful) pulse shape. It is just easy to use. Good pulse shapes are considered in detail in Chapter 11.

FIGURE 10.2 The magnitude spectrum of the received signal of a system using AM with large carrier has a prominent spike at the frequency of the carrier, as shown in the top plot. When using the suppressed carrier method in the middle plot, the carrier is not clearly visible. After preprocessing of the suppressed carrier signal using the scheme in Figure 10.3, a spike is clearly visible at twice the desired frequency (and with twice the desired phase). In time, this corresponds to an undulating sine wave.

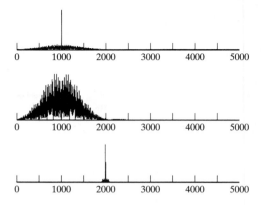

This can be rewritten using the identity $2\cos^2(x) = 1 + \cos(2x)$ in (A.4) to produce

$$r^2(t) = (1/2)s^2(t)[1 + \cos(4\pi f_0 t + 2\phi)].$$

Rewriting $s^2(t)$ as the sum of its (positive) average value and the variation about this average yields

$$s^2(t) = s_{avg}^2 + v(t).$$

Thus,

$$r^2(t) = (1/2)[s_{avg}^2 + v(t) + s_{avg}^2\cos(4\pi f_0 t + 2\phi) + v(t)\cos(4\pi f_0 t + 2\phi)].$$

A narrow bandpass filter centered around $2f_0$ passes the pure cosine term in r^2, and suppresses the DC component, the (presumably) lowpass $v(t)$, and the upconverted $v(t)$. The output of the bandpass filter is approximately

$$r_p(t) = BPF\{r^2(t)\} \approx \frac{1}{2}s_{avg}^2\cos(4\pi f_0 t + 2\phi + \psi), \qquad (10.2)$$

where ψ is the phase shift added by the BPF at frequency $2f_0$. Since ψ is known at the receiver, $r_p(t)$ can be used to find the frequency and phase of the carrier. Of course, the primary component in $r_p(t)$ is at twice the frequency of the carrier, the phase is twice the original unknown phase, and it is necessary to take ψ into account. Thus some extra bookkeeping is needed. The amplitude of $r_p(t)$ undulates slowly as s_{avg}^2 changes.

The following MATLAB code carries out the preprocessing of Figure 10.3. First, run `pulrecsig.m` to generate the suppressed carrier signal `rsc`.

---

pllpreprocess.m: send received signal through square and BPF

```
r=rsc;                            % r generated with suppressed carrier
q=r.^2;                           % square nonlinearity
fl=500; ff=[0 .38 .39 .41 .42 1]; % BPF center frequency at .4
fa=[0 0 1 1 0 0];                 % which is twice f_0
h=remez(fl,ff,fa);                % BPF design via remez
rp=filter(h,1,q);                 % filter to give preprocessed r
```

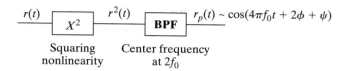

FIGURE 10.3 Prepro-
cessing the input to
a PLL via a squaring
nonlinearity and BPF
results in a sinusoidal
signal at twice the
frequency and with a
phase offset of twice
the original.

Then the phase and frequency of rp can be found directly by using the FFT.

---

pllpreprocess.m: recover "unknown" freq and phase using FFT

```
fftrBPF=fft(rp);                      % spectrum of rBPF
[m,imax]=max(abs(fftrBPF(1:end/2)));  % find frequency of max peak
ssf=(0:length(rp))/(Ts*length(rp));   % frequency vector
freqS=ssf(imax)                       % freq at the peak
phasep=angle(fftrBPF(imax));          % phase at the peak
[IR,f]=freqz(h,1,length(rp),1/Ts);    % frequency response of filter
[mi,im]=min(abs(f-freqS));            % at freq where peak occurs
phaseBPF=angle(IR(im));               % angle of BPF at peak freq
phaseS=mod(phasep-phaseBPF,pi)        % estimated angle
```

Observe that both freqS and phaseS are twice the nominal values of f0 and phoff, though there may be a π ambiguity (as will occur in any phase estimation).

The intent of this section is to clearly depict the problem of recovering the frequency and phase of the carrier even when it is buried within the data modulated signal. The method used to solve the problem (application of the FFT) is not common, primarily because of the numerical complexity. Most practical receivers use some kind of adaptive element to iteratively locate and track the frequency and phase of the carrier. Such elements are explored in the remainder of this chapter.

Problems

10.1. The squaring nonlinearity is only one possibility in the pllpreprocess.m routine.
   (a) Try replacing the $r^2(t)$ with $|r(t)|$. Does this result in a viable method of emphasizing the carrier?
   (b) Try replacing the $r^2(t)$ with $r^3(t)$. Does this result in a viable method of emphasizing the carrier?
   (c) Can you think of other functions that will result in viable methods of emphasizing the carrier?
   (d) Will a linear function work? Why or why not?

10.2. Determine the phase shift ψ of the BPF when
   (a) fl=490, 496, 502.
   (b) Ts=0.0001, 0.000101.
   (c) M=19, 20, 21.
   Explain why ψ should depend on fl, Ts, and M.

10.2 Squared Difference Loop

The problem of phase tracking is to determine the phase ϕ of the carrier and to follow any changes in ϕ using only the received signal. The frequency f_0 of the carrier is assumed known, though ultimately it too must be estimated. The received signal can be preprocessed (as in the previous section) to create a signal that strips away the data, in essence fabricating a sinusoid which has twice the frequency at twice the phase of the unmodulated carrier. This can be idealized to

$$r_p(t) = \cos(4\pi f_0 t + 2\phi), \tag{10.3}$$

which suppresses[2] the dependence on the known phase shift ψ of the BPF (recall (10.2)). The form of $r_p(t)$ implies that there is an essential ambiguity in the phase since ϕ can be replaced by $\phi + n\pi$ for any integer n without changing the value of (10.3). What can be done to recover ϕ (modulo π) from $r_p(t)$?

Is there some way to use an adaptive element? Section 6.5 suggested that there are three steps to the creation of a good adaptive element: setting a goal, finding a method, and then testing. As a first try, consider the goal of minimizing the average of the squared difference between $r_p(t)$ and a sinusoid generated, using an estimate of the phase; that is, seek to minimize

$$J_{SD}(\theta) = \text{avg}\{e^2(\theta, k)\} = \frac{1}{4}\text{avg}\{(r_p(kT_s) - \cos(4\pi f_0 kT_s + 2\theta))^2\} \tag{10.4}$$

by choice of θ, where $r_p(kT_s)$ is the value of $r_p(t)$ sampled at time kT_s. (The subscript SD stands for squared difference, and is used to distinguish this performance function from others that will appear in this and other chapters.) This goal makes sense because, if θ could be found so that $\theta = \phi + n\pi$, then the value of the performance function would be zero. When $\theta \neq \phi + n\pi$, then $r_p(kT_s) \neq \cos(4\pi f_0 kT_s + 2\theta)$, $e(\theta, k) \neq 0$, and so $J_{SD}(\theta) > 0$. Hence, (10.4) is minimized when θ has correctly identified the phase offset, modulo the inevitable π ambiguity.

While there are many methods of minimizing (10.4), an adaptive element that descends the gradient of the performance function $J_{SD}(\theta)$ leads to the algorithm[3]

$$\theta[k+1] = \theta[k] - \mu \left. \frac{dJ_{SD}(\theta)}{d\theta} \right|_{\theta=\theta[k]}, \tag{10.5}$$

which is the same as (6.5) with the variable changed from x to θ. Using the approximation detailed in (G.13), which holds for small μ, we find that the derivative and the average commute. Thus,

$$
\begin{aligned}
\frac{dJ_{SD}(\theta)}{d\theta} &= \frac{d\text{avg}\{e^2(\theta, k)\}}{d\theta} \\
&\approx \text{avg}\left\{\frac{de^2(\theta, k)}{d\theta}\right\} \\
&= \frac{1}{2}\text{avg}\left\{e(\theta, k)\frac{de(\theta, k)}{d\theta}\right\} \\
&= \text{avg}\{(r_p(kT_s) - \cos(4\pi f_0 kT_s + 2\theta))\sin(4\pi f_0 kT_s + 2\theta)\}.
\end{aligned} \tag{10.6}
$$

---

[2] An example that takes ψ into account is given in Problem 10.8.
[3] Recall the discussion surrounding the AGC elements in Chapter 6.

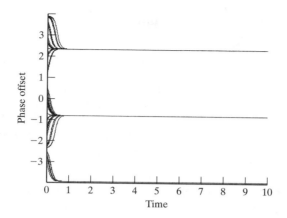

FIGURE 10.4 The phase
tracking algorithm
(10.7) converges to
the correct phase off-
set (in this case -0.8
or to some multiple
$-0.8 + n\pi$) depending
on the initial estimate.

Substituting this into (10.5) and evaluating at $\theta = \theta[k]$ gives[4]

$$\theta[k+1] = \theta[k] - \mu \text{avg}\{(r_p(kT_s) - \cos(4\pi f_0 kT_s + 2\theta[k]))$$

$$\sin(4\pi f_0 kT_s + 2\theta[k])\}. \tag{10.7}$$

This is implemented in `pllsd.m` for a phase offset of `phoff=-0.8` (i.e., ϕ of (10.3) is -0.8, though this value is unknown to the algorithm). Figure 10.4 plots the estimates `theta` for 50 different initial guesses `theta(1)`. Observe that many converge to the correct value at -0.8. Others converge to $-0.8 + \pi$ (about 2.3) and to $-0.8 - \pi$ (about -4).

---

pllsd.m: phase tracking minimizing SD

```
Ts=1/10000; time=10; t=0:Ts:time-Ts;      % time interval and time vector
f0=100; phoff=-0.8;                        % carrier freq. and phase
rp=cos(4*pi*f0*t+2*phoff);                 % simplified received signal
mu=.001;                                   % algorithm stepsize
theta=zeros(1,length(t)); theta(1)=0;      % initialize vector for estimates
fl=25; h=ones(1,fl)/fl;                    % fl averaging coefficients
z=zeros(1,fl);                             % initialize buffers for avg
for k=1:length(t)-1                        % run algorithm
  filtin=(rp(k)-cos(4*pi*f0*t(k)+2*theta(k)))*sin(4*pi*f0*t(k)+2*theta(k));
  z=[z(2:fl), filtin];                     % z's contain fl past inputs
  theta(k+1)=theta(k)-mu*fliplr(h)*z';     % convolve z with h and update
end
```

Observe that the averaging (a kind of lowpass filter, as discussed in Appendix G) is not implemented using the `filter` or `conv` commands because the complete input is not available at the start of the simulation. Instead, the "time domain" method is used, and the code here may be compared to the fourth method in `waystofilt.m` on page 133. At each time k, there is a vector z of past inputs. These are multiplied, point by point, with the impulse response h, which is flipped in time so that the sum properly implements a

---

[4] Recall the convention that $\theta[k] = \theta(kT_s) = \theta(t)|_{t=kT_s}$.

convolution. Because the filter is just a moving average, the impulse response is constant (1/f1) over the length of the filtering.

Problems

10.3. Use the preceding code to "play with" the SD phase tracking algorithm.
   (a) How does the stepsize mu affect the convergence rate?
   (b) What happens if mu is too large (say mu=10)?
   (c) Does the convergence speed depend on the value of the phase offset?
   (d) How does the final converged value depend on the initial estimate theta(1)?

10.4. Investigate these questions by making suitable modifications to pllsd.m.
   (a) What happens if the phase slowly changes over time? Consider a slow, small amplitude undulation in phoff.
   (b) Consider a slow linear drift in phoff.
   (c) What happens if the frequency f0 used in the algorithm is (slightly) different from the frequency used to construct the carrier?
   (d) What happens if the frequency f0 used in the algorithm is greatly different from the frequency used to construct the carrier?

10.5. How much averaging is necessary? Reduce the length of the averaging filter. Can you make the algorithm work with *no* averaging? Why does this work? Hint: Consider the relationship between (10.7) and (G.4).

10.6. Derive (10.6), following the technique used in Example G.3.

The performance function $J_{SD}(\theta)$ of (10.4) provides a mathematical statement of the goal of an adaptive phase tracking element, the method is defined by the algorithm (10.7), and simulations such as pllsd.m suggest that the algorithm can function as desired. But *why* does it work?

One way to understand adaptive elements, as discussed in Section 6.6 and shown in Figure 6.10 on page 110 is to draw the "error surface" for the performance function. But it is not immediately clear what this looks like, since $J_{SD}(\theta)$ depends on the frequency f_0, the time kT_s, and the unknown ϕ (through $r_p(kT_s)$), as well as the estimate θ. Recognizing that the averaging operation acts as a kind of lowpass filter (see Appendix G if this makes you nervous) allows considerable simplification of $J_{SD}(\theta)$. Rewrite (10.4) as

$$J_{SD}(\theta) = \frac{1}{4}\text{LPF}\{(r_p(kT_s) - \cos(4\pi f_0 kT_s + 2\theta))^2\}. \tag{10.8}$$

Substituting $r_p(kT_s)$ from (10.3), this can be rewritten

$$J_{SD}(\theta) = \frac{1}{4}\text{LPF}\{(\cos(4\pi f_0 kT_s + 2\phi) - \cos(4\pi f_0 kT_s + 2\theta))^2\}.$$

Expanding the square gives

$$J_{SD}(\theta) = \frac{1}{4}\text{LPF}\{(\cos^2(4\pi f_0 kT_s + 2\phi) - 2\cos(4\pi f_0 kT_s + 2\phi)\cos(4\pi f_0 kT_s + 2\theta)$$
$$+ \cos^2(4\pi f_0 kT_s + 2\theta)\}.$$

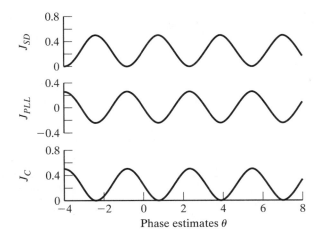

FIGURE 10.5 The error surface (10.9) for the SD phase tracking algorithm is shown in the top plot. Analogous error surfaces for the phase locked loop (10.11) and the Costas loop (10.13) are shown in the middle and bottom plots. All have minima (or maxima) at the desired locations (in this case −0.8) plus $n\pi$ offsets.

Using the trigonometric formula (A.4) for the square of a cosine and the formula (A.13) for the cosine angle sum (i.e., expand $\cos(x + y)$ with $x = 4\pi f_0 k T_s$ and $y = 2\phi$, and then again with $y = 2\theta$) yields

$$J_{SD}(\theta) = \frac{1}{8}\text{LPF}\{2 + \cos(8\pi f_0 k T_s + 4\phi) - 2\cos(2\phi - 2\theta) -$$
$$2\cos(8\pi f_0 k T_s + 2\phi + 2\theta) + \cos(8\pi f_0 k T_s + 4\theta)\}.$$

By the linearity of the LPF,

$$J_{SD}(\theta) = \text{LPF}\left\{\frac{1}{4}\right\} + \frac{1}{8}\text{LPF}\{\cos(8\pi f_0 k T_s + 4\phi)\} - \frac{1}{4}\text{LPF}\{\cos(2\phi - 2\theta)\} -$$
$$\frac{1}{4}\text{LPF}\{\cos(8\pi f_0 k T_s + 2\phi + 2\theta)\} + \frac{1}{8}\cos(8\pi f_0 k T_s + 4\theta)\}.$$

Assuming that the cutoff frequency of the lowpass filter is less than $4 f_0$, this simplifies to

$$J_{SD}(\theta) \approx \frac{1}{4}(1 - \cos(2\phi - 2\theta)), \tag{10.9}$$

which is shown in the top plot of Figure 10.5 for $\phi = -0.8$. The algorithm (10.7) is initialized with $\theta[0]$ at some point on the surface of the undulating sinusoidal curve. At each iteration of the algorithm, it moves downhill. Eventually, it will reach one of the nearby minima, which occur at $\theta = -0.8 \pm n\pi$ for some n. Thus, Figure 10.5 provides evidence that the algorithm can successfully locate the unknown phase, assuming that the preprocessed signal $r_p(t)$ has the form of (10.3).

Figure 10.6 shows the algorithm (10.5) with the averaging operation replaced by the more general LPF. In fact, this provides a concrete answer to problem 10.5; the averaging, the LPF, and the integral block all act as lowpass filters. All that was required of the filtering in order to arrive at (10.9) from (10.8) was that it remove frequencies below $4 f_0$. This mild requirement is accomplished even by the integrator alone.

FIGURE 10.6 A block diagram of the squared difference phase tracking algorithm (10.5). The input $r_p(kT_s)$ is a preprocessed version of the received signal as shown in Figure 10.3. The integrator block Σ has a lowpass character, and is equivalent to a sum and delay as shown in Figure 7.7.

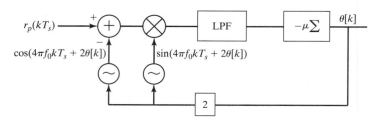

Problems

10.7. The code in `pllsd.m` is simplified in the sense that the received signal `rp` contains just the unmodulated carrier. Implement a more realistic scenario by combining `pulrecsig.m` to include a binary message sequence, `pllpreprocess.m` to create `rp`, and `pllsd.m` to recover the unknown phase offset of the carrier.

10.8. Using the default values in `pulrecsig.m` and `pllpreprocess.m` results in a ψ of zero. Problem 10.2 provided several situations in which $\psi \neq 0$. Modify `pllsd.m` to allow for nonzero ψ, and verify the code for the cases suggested in Problem 10.2.

10.9. Investigate how the SD algorithm performs when the received signal contains pulse shaped 4-PAM data. Can you choose parameters so that $\theta \to \phi$?

10.3 The Phase Locked Loop

Perhaps the best loved method of phase tracking is known as the phase locked loop (PLL). This section shows that the PLL can be derived as an adaptive element ascending the gradient of a simple performance function. The key idea is to modulate the (processed) received signal $r_p(t)$ of Figure 10.3 down to DC, using a cosine of known frequency $2f_0$ and phase $2\theta + \psi$. After filtering to remove the high-frequency components, the magnitude of the DC term can be adjusted by changing the phase. The value of θ that maximizes the DC component is the same as the phase ϕ of $r_p(t)$.

To be specific, let

$$J_{PLL}(\theta) = \frac{1}{2}\text{LPF}\{r_p(kT_s)\cos(4\pi f_0 kT_s + 2\theta + \psi)\}. \qquad (10.10)$$

Using the definition of $r_p(t)$ from (10.3) and the cosine product relationship (A.9), this equation becomes

$$J_{PLL}(\theta) = \frac{1}{2}\text{LPF}\{\cos(4\pi f_0 kT_s + 2\phi + \psi)\cos(4\pi f_0 kT_s + 2\theta + \psi)\}$$

$$= \frac{1}{4}\text{LPF}\{\cos(2\phi - 2\theta) + \cos(8\pi f_0 kT_s + 2\theta + 2\phi + 2\psi)\}$$

$$= \frac{1}{4}\text{LPF}\{\cos(2\phi - 2\theta)\} + \frac{1}{4}\text{LPF}\{\cos(8\pi f_0 kT_s + 2\theta + 2\phi + 2\psi)\}$$

$$\approx \frac{1}{4}\cos(2\phi - 2\theta), \qquad (10.11)$$

assuming that the cutoff frequency of the lowpass filter is well below $4f_0$. This is shown in the middle plot of Figure 10.5 and is the same as $J_{SD}(\theta)$, except for a constant and a sign. The sign change implies that, while $J_{SD}(\theta)$ needs to be minimized to find the correct answer, $J_{PLL}(\theta)$ needs to be maximized. The substantive difference between the SD and the PLL performance functions lies in the way that the signals needed in the algorithm are extracted.

Assuming a small stepsize, the derivative of (10.10) with respect to θ at time k can be approximated (using (G.13)) as

$$\left. \frac{d\text{LPF}\{r_p(kT_s)\cos(4\pi f_0 kT_s + 2\theta + \psi)\}}{d\theta} \right|_{\theta=\theta[k]}$$

$$\approx \text{LPF}\left\{ \left. \frac{dr_p(kT_s)\cos(4\pi f_0 kT_s + 2\theta + \psi)}{d\theta} \right|_{\theta=\theta[k]} \right\}$$

$$= \text{LPF}\{-r_p(kT_s)\sin(4\pi f_0 kT_s + 2\theta[k] + \psi)\}.$$

The corresponding adaptive element,

$$\theta[k+1] = \theta[k] - \mu\text{LPF}\{r_p(kT_s)\sin(4\pi f_0 kT_s + 2\theta[k] + \psi)\}, \qquad (10.12)$$

is shown in Figure 10.7. Observe that the sign of the derivative is preserved in the update (rather than its negative), indicating that the algorithm is searching for a maximum of the error surface rather than a minimum. One difference between the PLL and the SD algorithms is clear from a comparison of Figures 10.6 and 10.7. The PLL requires one less oscillator (and one less addition block). Since the performance functions $J_{SD}(\theta)$ and $J_{PLL}(\theta)$ are effectively the same, the performance characteristics of the two are roughly equivalent.

Suppose that f_0 is the frequency of the transmitter and f_c is the assumed frequency at the receiver (with f_0 close to f_c). The following program simulates (10.12) for time seconds. Note that the remez filter creates an h with a zero phase at the center frequency and so ψ is set to zero.

---

pllconverge.m: simulate Phase Locked Loop

```
Ts=1/10000; time=1; t=Ts:Ts:time;      % time vector
f0=1000; phoff=-0.8;                    % carrier freq. and phase
rp=cos(4*pi*f0*t+2*phoff);              % simplified received signal
fl=10; ff=[0 .01 .02 1]; fa=[1 1 0 0];
h=remez(fl,ff,fa);                      % LPF design
mu=.003;                                % algorithm stepsize
fc=1000;                                % assumed freq. at receiver
theta=zeros(1,length(t)); theta(1)=0;   % initialize vector for estimates
z=zeros(1,fl+1);                        % initialize buffer for LPF
for k=1:length(t)-1                     % z contains past fl+1 inputs
  z=[z(2:fl+1), rp(k)*sin(4*pi*fc*t(k)+2*theta(k))];
  update=fliplr(h)*z';                  % new output of LPF
  theta(k+1)=theta(k)-mu*update;        % algorithm update
end
```

FIGURE 10.7 A block diagram of the phase locked loop algorithm (10.12).

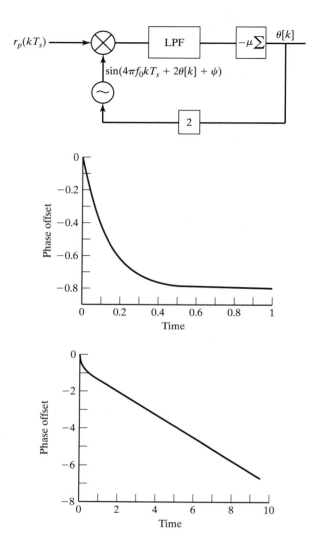

FIGURE 10.8 Using the PLL, the estimates θ converge to a region about the phase offset ϕ, and then oscillate.

FIGURE 10.9 When the frequency estimate is incorrect, θ becomes a "line" whose slope is proportional to the frequency difference.

Figures 10.8 and 10.9 show the output of the program when $f_0 = f_c$ and $f_0 \neq f_c$, respectively. Observe that, when the assumption of equality is fulfilled, θ converges to a region about the correct phase offset ϕ and wiggles about, with a size proportional to the size of μ and dependent on details of the LPF.

When the frequencies are not the same, θ has a definite trend (the simulation in Figure 10.9 used $f_0 = 1000$ Hz and $f_c = 1001$ Hz). Can you figure out how the slope of θ relates to the frequency offset? The caption in Figure 10.9 provides a hint. Can you imagine how the PLL might be used to estimate the frequency as well as to find the phase offset? These questions, and more, will be answered in Section 10.6.

Problems

10.10. Use the preceding code to "play with" the phase locked loop algorithm. How does μ affect the convergence rate? How does μ affect the oscillations in θ?

What happens if μ is too large (say $\mu = 1$)? Does the convergence speed depend on the value of the phase offset?

10.11. In `pllconverge.m`, how much filtering is necessary? Reduce the length of the filter. Does the algorithm still work with *no* LPF? Why? How does your filter affect the convergent value of the algorithm? How does your filter affect the tracking of the estimates when $f_0 \neq f_c$?

10.12. The code in `pllconverge.m` is simplified in the sense that the received signal `rp` contains just the unmodulated carrier. Implement a more realistic scenario by combining `pulrecsig.m` to include a binary message sequence, `pllpre-process.m` to create `rp`, and `pllconverge.m` to recover the unknown phase offset of the carrier.

10.13. Using the default values in `pulrecsig.m` and `pllpreprocess.m` results in a ψ of zero. Problem 10.2 provided several situations in which $\psi \neq 0$. Modify `pllconverge.m` to allow for nonzero ψ, and verify the code on the cases suggested in Problem 10.2.

10.14. Investigate how the PLL algorithm performs when the received signal contains pulse shaped 4-PAM data. Can you choose parameters so that $\theta \to \phi$?

10.15. Many variations on the basic PLL theme are possible. Letting $u(kT_s) = r_p(kT_s) \cos(2\pi kT_s + \theta)$, the preceding PLL corresponds to a performance function of $J_{PLL}(\theta) = \text{LPF}\{u(kT_s)\}$. Consider the alternative $J(\theta) = \text{LPF}\{u^2(kT_s)\}$, which leads directly to the algorithm[5]

$$\theta[k+1] = \theta[k] - \mu\text{LPF}\left\{u(kT_s)\left.\frac{du(kT_s)}{d\theta}\right|_{\theta=\theta[k]}\right\},$$

which is

$$\theta[k+1] = \theta[k] - \mu\text{LPF}\left\{r_p^2(kT_s)\,\sin(4\pi kT_s + 2\theta[k])\,\cos(4\pi kT_s + 2\theta[k])\right\}.$$

(a) Modify the code in `pllconverge.m` to "play with" this variation on the PLL. Try a variety of initial values `theta(1)`. Are the convergent values always the same as with the PLL?

(b) How does μ effect the convergence rate?

(c) How does μ effect the oscillations in θ?

(d) What happens if μ is too large (say $\mu = 1$)?

(e) Does the convergence speed depend on the value of the phase offset?

(f) What happens when the LPF is removed (set equal to unity)?

(g) Can you draw the appropriate error surface?

10.16. Consider the alternative performance function $J(\theta) = |u_k|$. Derive the appropriate adaptive element, and implement it by imitating the code in `pllconverge.m`. In what ways is this algorithm better than the standard PLL? In what ways is it worse?

---

[5] This is sensible because θ that minimize $u^2(kT_s)$ also minimize $u(kT_s)$.

The PLL can be used to identify the phase offset of the carrier. It can be derived as a gradient descent on a particular performance function, and can be investigated via simulation (with variants of `pllconverge.m`, for instance). The CD-ROM also contains a document called *Analysis of the Phase Locked Loop* which goes further, carrying out a linearized analysis of the behavior of the PLL algorithm, and showing how the parameters of the LPF affect the convergence and tracking performance of the loop. Moreover, when the phase offset changes, the PLL can track the changes (up to some maximum rate). Conceptually, tracking a small frequency offset is identical to tracking a changing phase, and Section 10.6 investigates how to use the PLL as a building block for the estimation of frequency offsets.

10.4 The Costas Loop

The PLL and the SD algorithms are two ways of synchronizing the phase at the receiver to the phase at the transmitter. Both require that the received signal be preprocessed (for instance, by a squaring nonlinearity and a BPF as in Figure 10.3) in order to extract a "clean" version of the carrier, albeit at twice the frequency and phase. An alternative approach operates directly on the received signal $r(kT_s) = s(kT_s) \cos(2\pi f_0 kT_s + \phi)$ by reversing the order of the processing: first modulating to DC, then low pass filtering, and finally squaring. This reversal of operations leads to the performance function

$$J_C(\theta) = \text{avg}\{(\text{LPF}\{r(kT_s)\cos(2\pi f_0 kT_s + \theta)\})^2\}. \tag{10.13}$$

The resulting algorithm is called the *Costas loop* after its inventor J. P. Costas. Because of the way that the squaring nonlinearity enters $J_C(\theta)$, it can operate without preprocessing of the received signal as in Figure 10.3. To see why this works, substitute $r(kT_s)$ into (10.13):

$$J_C(\theta) = \text{avg}\{(\text{LPF}\{s(kT_s)\cos(2\pi f_0 kT_s + \phi)\cos(2\pi f_0 kT_s + \theta)\})^2\}.$$

Assuming that the cutoff of the LPF is larger than the absolute bandwidth of $s(kT_s)$, and following the same logic as in (10.11) but with ϕ instead of 2ϕ, θ in place of 2θ and $2\pi f_0 kT_s$ replacing $4\pi f_0 kT_s$ shows that

$$\text{LPF}\{s(kT_s)\cos(2\pi f_0 kT_s + \phi)\cos(2\pi f_0 kT_s + \theta)\} = \frac{1}{2}s(kT_s)\cos(\phi - \theta). \tag{10.14}$$

Substituting (10.14) into (10.13) yields

$$
\begin{aligned}
J_C(\theta) &= \text{avg}\left\{\left(\frac{1}{2}s(kT_s)\cos(\phi - \theta)\right)^2\right\} \\
&= \frac{1}{4}\text{avg}\{s^2(kT_s)\cos^2(\phi - \theta))\} \\
&\approx \frac{1}{4}s_{\text{avg}}^2\cos^2(\phi - \theta),
\end{aligned}
$$

where s_{avg}^2 is the (fixed) average value of the square of the data sequence $s(kT_s)$. Thus $J_C(\theta)$ is proportional to $\cos^2(\phi - \theta)$. This performance function is plotted (for an "unknown" phase offset of $\phi = -0.8$) in the bottom part of Figure 10.5. Like the error surface for the PLL (the middle plot), this achieves a maximum when the estimate θ is equal to ϕ. Other maxima occur at $\phi + n\pi$ for integer n. In fact, except for a scaling and a constant, this is the same as J_{PLL} because $\cos^2(\phi - \theta) = \frac{1}{2}(1 + \cos(2\phi - 2\theta))$, as shown using (A.4).

The Costas loop can be implemented as a standard adaptive element (10.5). The derivative of $J_C(\theta)$ is approximated by swapping the order of the differentiation and the averaging (as in (G.13)), applying the chain rule, and then swapping the derivative with the LPF. In detail, this is

$$\frac{dJ_C(\theta)}{d\theta} \approx \text{avg} \left\{ \frac{d\text{LPF}\{r(kT_s)\cos(2\pi f_0 kT_s + \theta)\}^2}{d\theta} \right\}$$

$$= 2 \, \text{avg} \left\{ \text{LPF}\{r(kT_s)\cos(2\pi f_0 kT_s + \theta)\} \frac{d\text{LPF}\{r(kT_s)\cos(2\pi f_0 kT_s + \theta)\}}{d\theta} \right\}$$

$$\approx 2 \, \text{avg} \left\{ \text{LPF}\{r(kT_s)\cos(2\pi f_0 kT_s + \theta)\} \text{LPF} \frac{\{dr(kT_s)\cos(2\pi f_0 kT_s + \theta)\}}{d\theta} \right\}$$

$$= -2 \, \text{avg} \{\text{LPF}\{r(kT_s)\cos(2\pi f_0 kT_s + \theta)\} \text{LPF}\{r(kT_s)\sin(2\pi f_0 kT_s + \theta)\}\} .$$

Accordingly, an implementable version of the Costas loop can be built as

$$\theta[k + 1] = \theta[k] + \mu \left. \frac{dJ_C(\theta)}{d\theta} \right|_{\theta=\theta[k]}$$

$$= \theta[k] - \mu \, \text{avg}\{\text{LPF}\{r(kT_s)\cos(2\pi f_0 kT_s + \theta[k])\}$$

$$\text{LPF}\{r(kT_s)\sin(2\pi f_0 kT_s + \theta[k])\}\}.$$

This is diagrammed in Figure 10.10, leaving off the first (or outer) averaging operation (as is often done), since it is redundant given the averaging effect of the two LPFs and the averaging effect inherent in the small stepsize update. With this averaging removed, the algorithm is

$$\theta[k + 1] = \theta[k] - \mu\text{LPF}\{r(kT_s)\cos(2\pi f_0 kT_s + \theta[k])\}$$

$$\text{LPF}\{r(kT_s)\sin(2\pi f_0 kT_s + \theta[k])\}.$$

Basically, there are two paths. The upper path modulates by a cosine and then lowpass filters to create (10.14), while the lower path modulates by a sine wave and then lowpass filters to give $-s(kT_s)\sin(\phi - \theta)$. These combine to give the equation update, which is integrated to form the new estimate of the phase. The latest phase estimate is then fed back (this is the "loop" in "Costas loop") into the oscillators, and the recursion proceeds.

Suppose that a 4-PAM transmitted signal `r` is created as in `pulrecsig.m` (from page 179) with carrier frequency `f0=1000`. Then the Costas loop phase tracking method (10.15) can be implemented in much the same way that the PLL was implemented in `pllconverge.m`.

FIGURE 10.10 The
Costas loop is a
phase tracking algo-
rithm based on the
performance func-
tion (10.13).

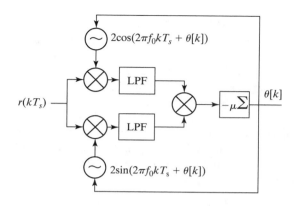

---

costasloop.m: costas loop - input rsc from pulrecsig.m

```
r=rsc;                                    % rsc is from pulrecsig.m
fl=500; ff=[0 .01 .02 1]; fa=[1 1 0 0];
h=remez(fl,ff,fa);                        % LPF design
mu=.003;                                  % algorithm stepsize
fc=1000;                                  % assumed freq. at receiver
theta=zeros(1,length(t)); theta(1)=0;     % initialize estimate vector
zs=zeros(1,fl+1); zc=zeros(1,fl+1);       % initialize buffers for LPFs
for k=1:length(t)-1                       % z's contain past fl+1 inputs
  zs=[zs(2:fl+1), 2*r(k)*sin(2*pi*fc*t(k)+theta(k))];
  zc=[zc(2:fl+1), 2*r(k)*cos(2*pi*fc*t(k)+theta(k))];
  lpfs=fliplr(h)*zs'; lpfc=fliplr(h)*zc'; % new output of filters
  theta(k+1)=theta(k)-mu*lpfs*lpfc;       % algorithm update
end
```

Typical output of `costasloop.m` is shown in Figure 10.11, which shows the evolution of the phase estimates for 50 different starting values `theta(1)`. A number of these converge to $\phi = -0.8$, and a number to nearby π multiples. These stationary points occur at all the minima of the error surface (the bottom plot in Figure 10.5).

When the frequency is not exactly known, the phase estimates of the Costas algorithm try to follow. For example, in Figure 10.12, the frequency of the carrier is $f_0 = 1000$, while the assumed frequency at the receiver was $f_c = 1000.1$. Fifty different starting points were used, and in all cases, the estimates converge to a line. Section 10.6 shows how this linear phase motion can be used to estimate the frequency difference.

Problems

10.17. Use the preceding code to "play with" the Costas loop algorithm.
  (a) How does the stepsize `mu` affect the convergence rate?
  (b) What happens if `mu` is too large (say `mu=1`)?
  (c) Does the convergence speed depend on the value of the phase offset?
  (d) When there is a small frequency offset, what is the relationship between the slope of the phase estimate and the frequency difference?

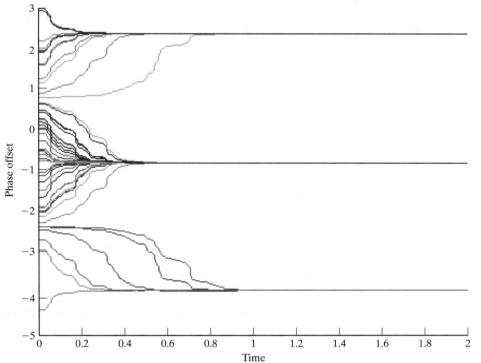

FIGURE 10.11 Depending on where it is initialized, the estimates made by the Costas loop algorithm converge to $\phi \pm n\pi$. For this plot, the "unknown" ϕ was -0.8, and there were 50 different initializations.

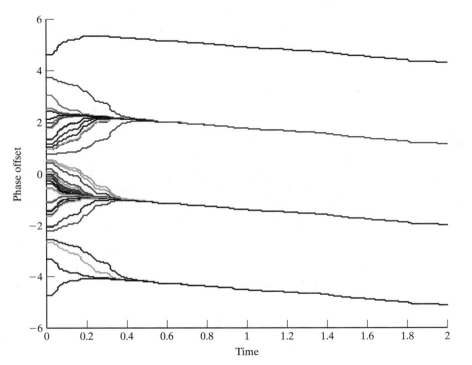

FIGURE 10.12 When the frequency of the carrier is unknown at the receiver, the phase estimates "converge" to a line.

FIGURE 10.13 An alternative implementation of the Costas loop trades off less expensive oscillators for a more complex structure.

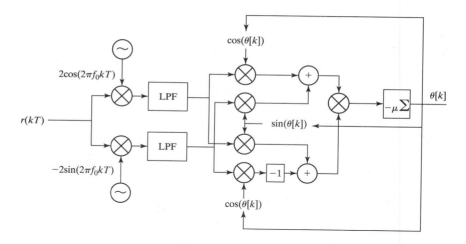

10.18. How does the filter h influence the performance of the Costas loop?
  (a) Try fl=1000, 30, 10, 3.
  (b) Remove the LPFs completely from costasloop.m. How does this affect the convergent values? The tracking performance?

10.19. Oscillators that have the ability to adjust their phase in response to an input signal are more expensive than free running oscillators. Figure 10.13 shows an alternative implementation of the Costas loop.
  (a) Show that this is actually carrying out the same calculations (albeit in a different order) as the implementation in Figure 10.10.
  (b) Write a simulation (or modify costasloop.m) to implement this alternative.

10.20. Reconsider the modified PLL of Problem 10.15. This algorithm also incorporates a squaring operation. Does it require the preprocessing step of Figure 10.3? Why?

In some applications, the Costas loop is considered a better solution than the standard PLL because it can be more robust in the presence of noise.

10.5 Decision Directed Phase Tracking

A method of phase tracking that works only in digital systems exploits the error between the received value and the nearest symbol. For example, suppose that a 0.9 is received in a binary ±1 system, suggesting that a +1 was transmitted. Then the difference between the 0.9 and the nearest symbol +1 provides information that can be used to adjust the phase estimate. This method is called decision directed (DD) because the "decisions" (the choice of the nearest allowable symbol) "direct" (or drive) the adaptation.

To see how this works, let $s(t)$ be a pulse-shaped signal created from a message in which the symbols are chosen from some (finite) alphabet. At the transmitter, $s(t)$ is modulated by a carrier at frequency f_0 with unknown phase ϕ, creating the signal $r(t) = s(t) \cos(2\pi f_0 t + \phi)$. At the receiver, this signal is demodulated by a sinusoid and then lowpass filtered to create

$$x(t) = 2\text{LPF}\{s(t) \cos(2\pi f_0 t + \phi) \cos(2\pi f_c t + \theta)\}. \tag{10.15}$$

As shown in Chapter 5, when the frequencies (f_0 and f_c) and phases (ϕ and θ) are equal, then $x(t) = s(t)$. In particular, $x(kT_s) = s(kT_s)$ at the sample instants $t = kT_s$, where the $s(kT_s)$ are elements of the alphabet. On the other hand, if $\phi \neq \theta$, then $x(kT_s)$ will not be a member of the alphabet. The difference between what $x(kT_s)$ is, and what it should be, can be used to form a performance function and hence a phase tracking algorithm. A quantization function $Q(x)$ is used to find the nearest element of the symbol alphabet.

The performance function for the decision directed method is

$$J_{DD}(\theta) = \frac{1}{4}\text{avg}\{(Q(x[k]) - x[k])^2\}. \tag{10.16}$$

This can be used as the basis of an adaptive element by using the approximation (G.13) to calculate

$$\frac{dJ_{DD}(\theta)}{d\theta} \approx \frac{1}{4}\text{avg}\left\{\frac{d\,(Q(x[k]) - x[k])^2}{d\theta}\right\} = -\frac{1}{2}\text{avg}\left\{(Q(x[k]) - x[k])\frac{dx[k]}{d\theta}\right\},$$

which assumes that the derivative of Q with respect to θ is zero. The derivative of $x[k]$ can similarly be approximated as (recall that $x[k] = x(kT_s) = x(t)|_{t=kT_s}$ is defined in (10.15))

$$\frac{dx[k]}{d\theta} \approx -2\text{LPF}\{r[k]\sin(2\pi f_c kT_s + \theta)\}.$$

Thus, the decision directed algorithm for phase tracking is

$$\theta[k+1] = \theta[k] - \mu\text{avg}\{(Q(x[k]) - x[k])\,\text{LPF}\{r[k]\sin(2\pi f_c kT + \theta[k])\}\}.$$

Suppressing the (redundant) outer averaging operation gives

$$\theta[k+1] = \theta[k] - \mu\,(Q(x[k]) - x[k])\,\text{LPF}\{r[k]\sin(2\pi f_c kT + \theta[k])\}, \tag{10.17}$$

which is shown in block diagram form in Figure 10.14.

Suppose that a 4-PAM transmitted signal r is created as in `pulrecsig.m` (from page 179) with oversampling factor M=20 and carrier frequency f0=1000. Then the DD phase tracking method (10.17) can be simulated.

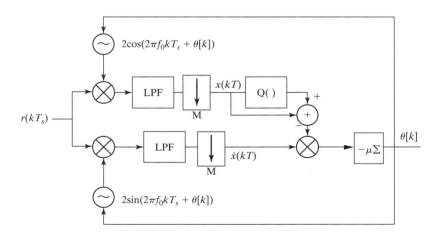

FIGURE 10.14 The decision directed phase tracking algorithm (10.17).

---

plldd.m: decision directed phase tracking

```
fl=100; fbe=[0 .2 .3 1]; damps=[1 1 0 0 ]; % parameters for LPF
h=remez(fl,fbe,damps);                      % LPF impulse response
fzc=zeros(1,fl+1); fzs=zeros(1,fl+1);       % initial state of filters=0
theta=zeros(1,N); theta(1)=-0.9;            % initial phase estimate
mu=.03; j=1; fc=f0;                         % algorithm stepsize mu
for k=1:length(rsc)
  cc=2*cos(2*pi*fc*t(k)+theta(j));          % cosine for demod
  ss=2*sin(2*pi*fc*t(k)+theta(j));          % sine for demod
  rc=rsc(k)*cc; rs=rsc(k)*ss;               % do the demods
  fzc=[fzc(2:fl+1),rc]; fzs=[fzs(2:fl+1),rs]; % states for LPFs
  x(k)=fliplr(h)*fzc';xder=fliplr(h)*fzs';  % LPFs give x and its derivative
  if mod(0.5*fl+M/2-k,M)==0                  % downsample to pick correct timing
    qx=quantalph(x(k),[-3,-1,1,3]);         % quantize to nearest symbol
    theta(j+1)=theta(j)-mu*(qx-x(k))*xder;  % algorithm update
    j=j+1;
  end
end
```

The same lowpass filter is used after demodulation with the cosine (to create x) and with the sine (to create its derivative xder). The filtering is done using the time domain method (the fourth method presented in waystofilt.m on page 133) because the demodulated signals are unavailable until the phase estimates are made. One subtlety in the decision directed phase tracking algorithm is that there are two time scales involved. The input, oscillators, and LPFs operate at the faster sampling rate T_s, while the algorithm update (10.17) operates at the slower symbol rate T. The correct relationship between these is maintained in the code by the mod function, which picks one out of each M T_s-rate sampled data points.

Typical output of plldd.m is shown in Figure 10.15. For initializations near the correct answer $\phi = -1.0$, the estimates converge to -1.0. Of course, there is the (by now familiar) π ambiguity. But there are also other values where the DD algorithm converges. What are these values?

As with any adaptive element, it helps to draw the error surface in order to understand its behavior. In this case, the error surface is the $J_{DD}(\theta)$ plotted as a function of the estimates θ. The following code approximates $J_{DD}(\theta)$ by averaging over N=1000 symbols drawn from the 4-PAM alphabet.

---

plldderrsys.m: error surface for decision directed phase tracking

```
N=1000;                          % average over N symbols,
m=pam(N,4,5);                    % use 4-PAM symbols
phi=-1.0;                        % unknown phase offset phi
theta=-2:.01:6;                  % grid for phase estimates theta
for k=1:length(theta)           % for each possible theta
  x=m*cos(phi-theta(k));        % find x with this theta
  qx=quantalph(x,[-3,-1,1,3]);  % q(x) for this theta
  jtheta(k)=(qx'-x)*(qx'-x)'/N; % cost for this theta
end
plot(theta,jtheta)              % plot J(theta) vs theta
```

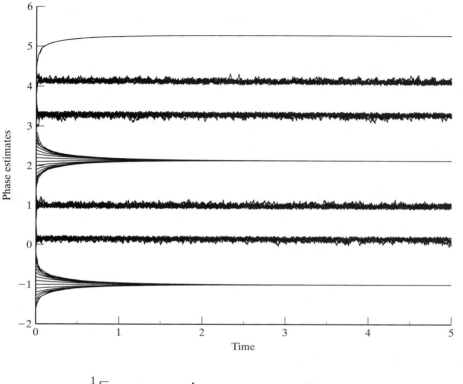

FIGURE 10.15 The decision directed tracking algorithm is adapted to locate a phase offset of $\phi = -1.0$. Many of the different initializations converge to $\phi = -1.0 \pm n\pi$, but there are also other convergent values.

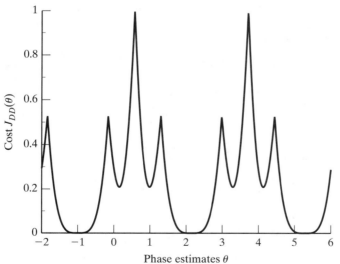

FIGURE 10.16 The error surface for the DD phase tracking algorithm (10.17) has several minima within each 2π repetition. The phase estimates will typically converge to the closest of these minima.

The output of `plldderrsys.m` is shown in Figure 10.16. First, the error surface is a periodic function of θ with period 2π, a property that it inherits from the cosine function. Within each period, there are six minima, two of which are broad and deep. One of these corresponds to the correct phase at $\phi = -1.0 \pm 2n\pi$. and the other (at $\phi = -1.0 + \pi \pm 2n\pi$) corresponds to the situation in which the cosine takes on a value of -1. This inverts each data symbol: ± 1 is mapped to ∓ 1, and ± 3 is mapped to ∓ 3. The other four occur near π multiples of $3\pi/8 - 1.0$ and $5\pi/8 - 1.0$, which correspond to values of the cosine that jumble the data sequence in various ways.

The implication of this error surface is clear: there are many values to which the decision directed method may converge. Only some of these correspond to desirable answers. Thus, the DD method is *local* in the same way that the steepest descent minimization of the function (6.8) (in Section 6.6) depended on the initial value of the estimate. If it is possible to start near the desired answer, then convergence can be assured. But if no good initialization is possible, then it may converge to one of the undesirable minima. This suggests that the decision directed method can perform acceptably in a tracking mode (when following a slowly varying phase), but would likely lose to the alternatives at start-up, when nothing is known about the correct value of the phase.

Problems

10.21. Use the code in `plldd.m` to "play with" the DD algorithm.
  (a) How large can the stepsize be made?
  (b) Is the LPF of the derivative really needed?
  (c) How crucial is it to the algorithm to pick the correct timing? Examine this question by choosing incorrect `j` at which to evaluate `x`.
  (d) What happens when the assumed frequency `fc` is not the same as `f0`?

10.22. The direct calculation of $\frac{dx(kT_s)}{d\theta}$ as a filtered version of (10.15) is only one way to calculate the derivative. Replace this using a numerical approximation (such as the forward or backward Euler, or the trapezoidal rule). Compare the performance of your algorithm to `plldd.m`.

10.23. Consider the DD phase tracking algorithm when the message alphabet is binary ± 1.
  (a) Modify `plldd.m` to simulate this case.
  (b) Modify `plldderrsys.m` to draw the error surface. Is the DD algorithm better (or worse) suited to the binary case than the 4-PAM case?

10.24. Consider the DD phase tracking algorithm when the message alphabet is 6-PAM.
  (a) Modify `plldd.m` to simulate this case.
  (b) Modify `plldderrsys.m` to draw the error surface. Is the DD algorithm better (or worse) suited to 6-PAM than to 4-PAM?

10.25. What happens when the number of inputs used to calculate the error surface is too small? Try $N = 100, 10, 1$. Can N be too large?

10.26. Investigate how the error surface depends on the input signal.
  (a) Draw the error surface for the DD phase tracking algorithm when the inputs are binary ± 1.
  (b) Draw the error surface for the DD phase tracking algorithm when the inputs are drawn from the 4-PAM constellation, for the case in which the symbol -3 never occurs.

10.6 Frequency Tracking

The problems inherent in even a tiny difference in the frequency of the carrier at the transmitter and the assumed frequency at the receiver are shown in (5.5) and illustrated graphically in Figure 9.19 on page 172. Since no two independent oscillators are ever exactly aligned, it is important to find ways of estimating the frequency from the received signal. The direct method of Section 10.6.1 derives an algorithm based on a performance function that uses a square difference in the time domain. Unfortunately, this does not work well, and its failure can be traced to the shape of the error surface.

Section 10.6.2 begins with the observation (familiar from Figures 10.9 and 10.12) that the estimates, we find that of phase made by the phase tracking algorithms over time lie on a line whose slope is proportional to the difference in frequency between the modulating and the demodulating oscillators. This slope contains valuable information that can be exploited to indirectly estimate the frequency.

10.6.1 Direct Frequency Estimation

Perhaps the simplest setting in which to begin frequency estimation is to assume that the received signal is $r(t) = \cos(2\pi f_0 t)$ where f_0 is unknown. By analogy with the squared difference method of phase estimation in Section 10.2, a reasonable strategy is to try to choose $\hat{f}$ so as to minimize

$$J(\hat{f}) = \frac{1}{2}\text{LPF}\{(r(t) - \cos(2\pi \hat{f} t))^2\}. \tag{10.18}$$

Following a gradient strategy for updating the estimates, we find that $\hat{f}$ leads to the algorithm

$$\hat{f}[k+1] = \hat{f}[k] - \mu \left. \frac{dJ(\hat{f})}{d\hat{f}} \right|_{\hat{f}=\hat{f}[k]} \tag{10.19}$$

$$= \hat{f}[k] - \mu\text{LPF}\{2\pi k T_s (r(kT_s) - \cos(2\pi k T_s \hat{f}[k])) \sin(2\pi k T_s \hat{f}[k])\}.$$

How well does this algorithm work? First, observe that the update is multiplied by $2\pi k T_s$. (This arises from application of the chain rule when taking the derivative of $\sin(2\pi k T_s \hat{f}[k])$ with respect to $\hat{f}[k]$.) This factor increases continuously, and acts like a stepsize that grows over time. Perhaps the easiest way to make any adaptive element fail is to use a stepsize that is too large; the form of this update ensures that eventually the "stepsize" will be too large.

Putting on our best engineering hat, let us just remove this offending term, and go ahead and simulate the method.[6] At first glance it might seem that the method works well. Figure 10.17 shows 20 different starting values. All 20 appear to converge nicely within one second to the unknown frequency value at f0=100. But then something strange happens: one by one, the estimates *diverge*. In the figure, one peels off at about 6 seconds, and one at about 17 seconds. Simulations can never prove conclusively that an algorithm is good for a given task, but if even simplified and idealized simulations function poorly, it is a safe bet that the algorithm is somehow flawed. What is the flaw in this case?

---

[6] The code is available in the program `pllfreqest.m` on the CD.

FIGURE 10.17 The fre-
quency estimation
algorithm (10.19)
appears to function
well at first. But over
time, the estimates
diverge from the
desired answer.

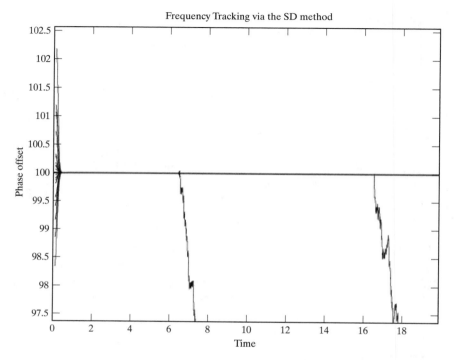

Recall that error surfaces are often a good way of picturing the behavior of gradient descent algorithms. Expanding the square and using the standard identities (A.4) and (A.9), $J(\hat{f})$ can be rewritten

$$J(\hat{f}) = \text{LPF}\left\{1 + \frac{1}{2}\cos(4\pi f_0 t) + \frac{1}{2}\cos(4\pi \hat{f} t) - \cos(2\pi(f_0 - \hat{f})t) - \cos(2\pi(f_0 + \hat{f})t)\right\}$$

$$= 1 - \text{LPF}\{\cos(2\pi(f_0 - \hat{f})t)\}, \tag{10.20}$$

assuming that the cutoff frequency of the lowpass filter is less than f_0 and that $\hat{f} \approx f_0$. At the point where $\hat{f} = f_0$, $J(\hat{f}) = 0$. For any other value of $\hat{f}$ other than f_0, however, as time t progresses, the cosine term undulates up and down with an average value of zero. Hence $J(\hat{f})$ averages 1 for any $\hat{f} \neq f_0$! This pathological situation is shown in Figure 10.18.

When $\hat{f}$ is far from f_0, this analysis does not hold because the LPF no longer removes the first two cosine terms in (10.20). Somewhat paradoxically, the algorithm behaves well until the answer is nearly correct. Once $\hat{f} \approx f_0$, the error surface flattens, and the estimates wander around. There is a slight possibility that it might accidently fall into the exact correct answer, but simulations suggest that such luck is rare. Oh well, whatever, never mind...

10.6.2 Indirect Frequency Estimation

Because the direct method of the previous section is unreliable, this section pursues an alternative strategy based on the observation that the phase estimates of the PLL

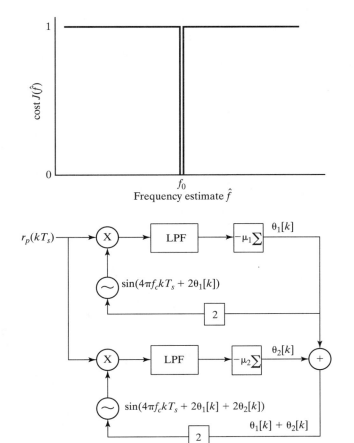

FIGURE 10.18 The error surface corresponding to the frequency estimation performance function (10.18) is flat everywhere except for a deep crevice at the correct answer $\hat{f} = f_0$.

FIGURE 10.19 A pair of PLLs can efficiently estimate the frequency offset at the receiver.

"converge" to a line that has a slope proportional to the difference between the actual frequency of the carrier and the frequency that is assumed at the receiver.[7] (Recall Figures 10.9 and 10.12.) The indirect method cascades two PLLs: the first finds this line (and hence indirectly specifies the frequency), the second converges to a constant appropriate for the phase offset.

The scheme is pictured in Figure 10.19. Suppose that the received signal has been preprocessed to form $r_p(t) = \cos(4\pi f_0 t + 2\phi)$. This is applied to the inputs of two PLLs.[8] The top PLL functions exactly as expected from previous sections: if the frequency of its oscillator is $2f_c$, then the phase estimates $2\theta_1$ converge to a ramp with slope $2\pi(f_0 - f_c)$, that is,

$$\theta_1(t) \rightarrow 2\pi(f_0 - f_c)t + b,$$

---

[7] In fact, this convergence can be substantiated analytically. See the document *Analysis of the Phase Locked Loop* on the CD.

[8] Or two SD phase tracking algorithms or two Costas loops, though in the latter case the squaring preprocessing is unnecessary.

where b is the y-intercept of the ramp. The θ_1 values are then added to θ_2, the phase estimate in the lower PLL. The output of the bottom oscillator is

$$\sin(4\pi f t + 2\theta_1(t) + 2\theta_2(t)) = \sin(4\pi f_c t + 4\pi(f_0 - f_c)t + 2b + 2\theta_2(t))$$

$$\rightarrow \sin(4\pi f_0 t + 2b + 2\theta_2(t)).$$

Effectively, the top loop has synthesized a signal that has the "correct" frequency for the bottom loop. Accordingly, $\theta_2(t) \rightarrow \phi - b$. Since a sinusoid with frequency $2\pi f_c t$ and "phase" $\theta_1(t) + \theta_2(t)$ is indistinguishable from a sinusoid with frequency $2\pi f_0 t$ and phase $\theta_2(t)$, these values can be used to generate a sinusoid that is aligned with $r_p(t)$ in both frequency and phase. This signal can then be used to demodulate the received signal.

Some MATLAB code to implement this dual PLL scheme is provided by dualplls.m.

---

dualplls.m: estimation of carrier via dual loop structure

```
Ts=1/10000; time=5; t=0:Ts:time-Ts;        % time vector
f0=1000; phoff=-2;                          % carrier freq. and phase
rp=cos(4*pi*f0*t+2*phoff);                  % preprocessed carrier
mu1=.01; mu2=.003;                          % algorithm stepsizes
fc=1001;                                    % assumed freq. at receiver
lent=length(t); th1=zeros(1,lent);          % initialize estimates
th2=zeros(1,lent); carest=zeros(1,lent);
for k=1:lent-1
  th1(k+1)=th1(k)-mu1*rp(k)*sin(4*pi*fc*t(k)+2*th1(k));            % top PLL
  th2(k+1)=th2(k)-mu2*rp(k)*sin(4*pi*fc*t(k)+2*th1(k)+2*th2(k));   % bottom PLL
  carest(k)=cos(4*pi*fc*t(k)+2*th1(k)+2*th2(k));                   % estimate
end
```

The output of this program is shown in Figure 10.20. The upper graph shows that θ_1, the phase estimate of the top PLL, converges to a ramp. The middle plot shows that θ_2, the phase estimate of the bottom PLL, converges to a constant. Thus the procedure is working. The bottom graph shows the error between the preprocessed signal rp and a synthesized carrier carest. The parameters fc, th1, and th2 can then be used to synthesize a cosine wave for the demodulation.

It is clear from the top plot of Figure 10.20 that θ_1 converges to a line. What line does it converge to? Looking carefully at the data generated by dualplls.m, the line can be calculated explicitly. The two points at $(2, -11.36)$ and $(4, -23.93)$ fit a line with slope $m = -6.28$ and an intercept $b = 1.21$. Thus,

$$2\pi(f_0 - f_c) = -6.28,$$

or $f_0 - f_c = -1$. Indeed, this was the value used in the simulation. Reading the final converged value of θ_2 from the simulation shown in middle plot gives -0.0627. $b - 0.627$ is 1.147, which is almost exactly π away from -2, the value used in phoff.

The dual PLL is certainly not the only way to proceed. A common approach is to use a higher order filter inside a single PLL. If this filter is chosen wisely, then even the single PLL can track modest phase and frequency changes. This is discussed at greater length in the document *Analysis of the Phase Locked Loop* which appears on the CD.

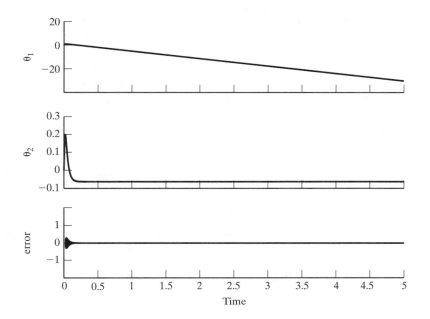

FIGURE 10.20 The output of MATLAB program `dualplls.m` shows the output of the first PLL converging to a line, which allows the second PLL to converge to a constant. The bottom figure shows that this estimator can be used to construct a sinusoid that is very close to the (preprocessed) carrier.

Problems

10.27. Use the preceding code to "play" with the frequency estimator.
  (a) How far can `f0` be from `fc` before the estimates deteriorate?
  (b) What is the effect of the two stepsizes `mu`? Should one be larger than other? Which one?
  (c) How does the method fare when the input is noisy?
  (d) What happens when the input is modulated by pulse-shaped data and not a simple sinusoid?

10.28. Build a frequency estimator using two SD phase tracking algorithms, rather than two PLLs. How does the performance change? Which do you think is preferable?

10.29. Build a frequency estimator that incorporates the preprocessing of the received signal from Figure 10.3 (as coded in `pllpreprocess.m`).

10.30. Build a frequency estimator using two Costas loops, rather than two PLLs. How does the performance change? Which do you think is preferable?

10.31. Investigate (via simulation) how the PLL functions when there is white noise (using `randn`) added to the received signal. Do the phase estimates become worse as the noise increases? Make a plot of the standard deviation of the noise versus the average value of the phase estimates (after convergence). Make a plot of the standard deviation of the noise versus the jitter in the phase estimates.

10.32. Repeat Problem 10.31 for the dual SD algorithm.

10.33. Repeat Problem 10.31 for the dual Costas loop algorithm.

10.34. Repeat Problem 10.31 for the dual DD algorithm.

10.35. Investigate (via simulation) how the PLL functions when there is intersymbol interference caused by a nonunity channel. Pick a channel (for instance `chan=[1, .5, .3, .1];`) and incorporate this into the simulation of the received signal. Using this received signal, are the phase estimates worse when the channel is present? Are they biased? Are they more noisy?

10.36. Repeat Problem 10.35 for the dual Costas loop.

10.37. Repeat Problem 10.35 for the Costas loop algorithm.

10.38. Repeat Problem 10.35 for the DD algorithm.

10.7 For Further Reading

The phase tracking algorithms of this chapter are only a few of the many possibilities. For example, the most common of the frequency estimation methods is probably the "second-order PLL" (rather than the dual PLL of Section 10.6.2) which replaces the LPF of Figure 10.7 with a higher order infinite impulse response filter. This is discussed in the article *Analysis of the PLL* on the CD.

- J.P. Costas, "Synchronous Communications," *Proceedings of the IRE*, pp. 1713–1718, Dec. 1956.

- L. E. Franks, "Carrier and Bit Synchronization in Data Communication—A Tutorial Review," *IEEE Transactions on Communications*, vol. COM-28, no. 8, pp. 1107–1120, August 1980.

11 Pulse Shaping and Receive Filtering

See first that the design is wise and just: that ascertained, pursue it resolutely; do not for one repulse forego the purpose that you resolved to effect.

— William Shakespeare

When the message is digital, it must be converted into an analog signal in order to be transmitted. This conversion is done by the "transmit" or "pulse-shaping" filter, which changes each symbol in the digital message into a suitable analog pulse. After transmission, the "receive" filter assists in recapturing the digital values from the received pulses. This chapter focuses on the design and specification of these filters.

The symbols in the digital input sequence $w(kT)$ are chosen from a finite set of values. For instance, they might be binary ± 1, or they may take values from a larger set such as the four-level alphabet ± 1, ± 3. As suggested in Figure 11.1, the sequence $w(kT)$ is indexed by the integer k, and the data rate is one symbol every T seconds. Similarly, the output $m(kT)$ assumes values from the same alphabet as $w(kT)$ and at the same rate. Thus the message is fully specified at times kT for all integers k. But what happens between these times, between kT and $(k + 1)T$? The analog modulation of Chapter 5 operates continuously, and some values must be used to fill in the digital input between the samples. This is the job of the pulse shaping filter: to turn a discrete-time sequence into an analog signal.

Each symbol $w(kT)$ of the message initiates an analog pulse that is scaled by the value of the signal. The pulse progresses through the communications system, and if all goes well, the output (after the decision) should be the same as the input, although perhaps with some delay. If the analog pulse is wider than the time between adjacent symbols, the outputs from adjacent symbols may overlap, a problem called *intersymbol interference*, which is abbreviated ISI. A series of examples in Section 11.2 shows how this happens, and the eye diagram is used in Section 11.3 to help visualize the impact of ISI.

What kinds of pulses minimize the ISI? One possibility is to choose a shape that is one at time kT and zero at mT for all $m \neq k$. Then the analog waveform at time kT contains only the value from the desired input symbol, and no interference from other nearby input symbols. These are called *Nyquist pulses* in Section 11.4. Yes, this is the same fellow who brought us the Nyquist sampling theorem and the Nyquist frequency.

Besides choosing the pulse shape, it is also necessary to choose a receive filter that helps decode the pulses. The received signal can be thought of as containing two parts: one part is due to the transmitted signal and the other part is due to the noise. The ratio

FIGURE 11.1 System schematic of a baseband communication system.

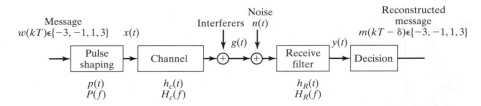

of the powers of these two parts is a kind of signal-to-noise ratio that can be maximized by choice of the pulse shape. This is discussed in Section 11.5. The chapter concludes in Section 11.6 by considering pulse shaping and receive filters that do both: provide a Nyquist pulse and maximize the signal-to-noise ratio.

The transmit and receive filter designs rely on the assumption that all other parts of the system are working well. For instance, the modulation and demodulation blocks have been removed from Figure 11.1, and the assumption is that they are perfect: the receiver knows the correct frequency and phase of the carrier. Similarly, the downsampling block has been removed, and the assumption is that this is implemented so that the decision device is a fully synchronized sampler and quantizer. Chapter 12 examines methods of satisfying these synchronization needs, but for now, they are assumed to be met. In addition, the channel is assumed benign.

11.1 Spectrum of the Pulse: Spectrum of the Signal

Probably the major reason that the design of the pulse shape is important is because the shape of the spectrum of the pulse shape dictates the spectrum of the whole transmission. To see this, suppose that the discrete-time message sequence $w(kT)$ is turned into the analog pulse train

$$w_a(t) = \sum_k w(kT)\delta(t - kT) = \begin{cases} w(kT) & t = kT \\ 0 & t \neq kT \end{cases} \qquad (11.1)$$

as it enters the pulse shaping filter. The response of the filter, with impulse response $p(t)$, is the convolution

$$x(t) = w_a(t) * p(t),$$

as suggested by Figure 11.1. Since the Fourier transform of a convolution is the product of the Fourier transforms (from (A.40)), it follows that

$$X(f) = W_a(f)P(f).$$

Though $W_a(f)$ is unknown, this shows that $X(f)$ can have no energy at frequencies where $P(f)$ vanishes. Whatever the spectrum of the message, the transmission is directly scaled by $P(f)$. In particular, the support of the spectrum $X(f)$ is no larger than the support of the spectrum $P(f)$.

As a concrete example, consider the pulse shape used in Chapter 9, which is the "blip" function shown in the top plot of Figure 11.2. The spectrum of this pulse shape can readily be calculated using `freqz`, and this is shown in the bottom plot of Figure 11.2. It is a kind of mild lowpass filter. The following code generates a sequence of N 4-PAM symbols, and then carries out the pulse shaping using the `filter` command.

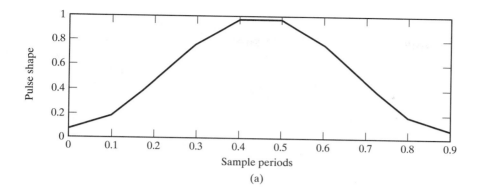

FIGURE 11.2 The `hamming` pulse shape and its magnitude spectrum.

---

pulsespec.m: spectrum of a pulse shape

```
N=1000; m=pam(N,4,5);              % 4-level signal of length N
M=10; mup=zeros(1,N*M); mup(1:M:end)=m;  % oversample by M
ps=hamming(M);                     % blip pulse shape of width M
x=filter(ps,1,mup);                % convolve pulse shape with data
```

The program `pulsespec.m` represents the "continuous-time" or analog signal by oversampling both the data sequence and the pulse shape by a factor of M. This technique was discussed in Section 6.3, where an "analog" sine wave `sine100hzsamp.m` was represented digitally at two sampling intervals, a slow symbol interval $T = MT_s$ and a faster rate (shorter interval) T_s representing the underlying analog signal. The pulse shape ps is a blip created by the `hamming` function, and this is also oversampled at the same rate. The convolution of the oversampled pulse shape and the oversampled data sequence is accomplished by the `filter` command. Typical output is shown in the top plot of Figure 11.3, which shows the "analog" signal over a time interval of about 25 symbols. Observe that the individual pulse shapes are clearly visible, one scaled blip for each symbol.

The spectrum of the output x is plotted in the bottom of Figure 11.3. As expected from the previous discussion, the spectrum $X(f)$ has the same contour as the spectrum of the individual pulse shape in Figure 11.2.

FIGURE 11.3 The top plot shows a segment of the output x of the pulse shaping filter. The bottom plots the magnitude spectrum of x, which has the same general contour as the spectrum of a single copy of the pulse. Compare with the bottom plot of Figure 11.2.

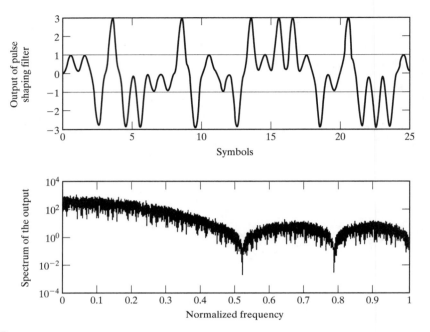

11.2 Intersymbol Interference

There are two situations when adjacent symbols may interfere with each other: when the pulse shape is wider than a single symbol interval T, and when there is a nonunity channel that "smears" nearby pulses, causing them to overlap. Both of these situations are called *intersymbol interference* (ISI). Only the first kind of ISI will be considered in this chapter; the second kind is postponed until Chapter 13. Before tackling the general setup, this section provides an instructive example.

Example 11.1   [ISI Caused by an Overly Wide Pulse Shape]

Suppose that the pulse shape in `pulsespec.m` is stretched so that its width is $3T$. This triple-wide Hamming pulse shape is shown in Figure 11.4, along with its spectrum. Observe that the spectrum has (roughly) one-third the null-to-null bandwidth of the single-symbol wide Hamming pulse. Since the width of the spectrum of the transmitted signal is dictated by the width of the spectrum of the pulse, this pulse shape is three times as parsimonious in its use of bandwidth. More FDM users can be active at the same time.

As might be expected, this boon has a price. Figure 11.5 shows the output of the pulse shaping filter over a time of about 25 symbols. There is no longer a clear separation of the pulse corresponding to one data point from the pulses of its neighbors. The transmission is correspondingly harder to properly decode. If the ISI caused by the overly wide pulse shape is too severe, symbol errors may occur.

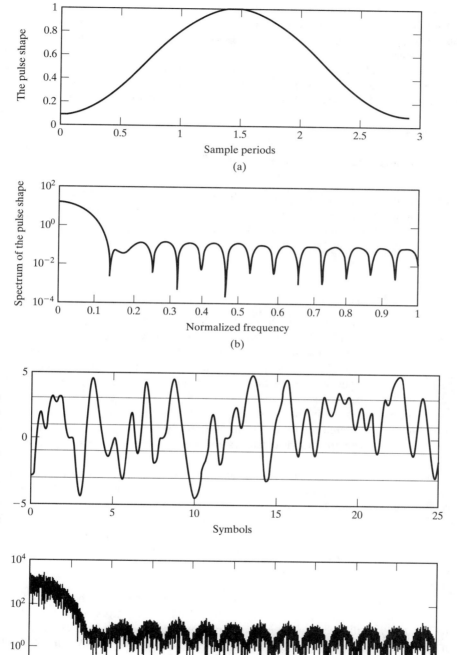

FIGURE 11.4 The triple-wide `hamming` pulse shape and its magnitude spectrum, which is drawn, using `freqz`.

(a)

(b)

FIGURE 11.5 The top plot shows a segment of the output `x` of the pulse shaping filter. With this $3T$-wide pulse shape, the pulses from adjacent symbols interfere with each other. The bottom shows the magnitude spectrum of the output, which has the same general contour as the spectrum of a single copy of the pulse, as in the bottom plot of Figure 11.4.

Thus, there is a tradeoff. Wider pulse shapes can occupy less bandwidth, which is always a good thing. On the other hand, a pulse shape like the Hamming blip does not need to be very many times wider before it becomes impossible to decipher the data because the ISI has become too severe. How much wider can it be without causing symbol errors? The next section provides a way of picturing ISI that answers this question. Subsequent sections discuss the practical issue of how such ISI can be prevented by a better choice of pulse shape. Yes, there are good pulse shapes that are wider than T.

Problems

11.1. Modify `pulsespec.m` to reproduce Figures 11.4 and 11.5 for the double-wide pulse shape.

11.2. Modify `pulsespec.m` to examine what happens when Hamming pulse shapes of width $4T$, $6T$, and $10T$ are used. What is the bandwidth of the resulting transmitted signals? Do you think it is possible to recover the message from the received signals? Explain.

11.3 Eye Diagrams

While the differences between the pulse shaped sequences in Figures 11.3 and 11.5 are apparent, it is difficult to see directly whether the distortions are serious; that is, whether they cause errors in the reconstructed data (i.e., the hard decisions) at the receiver. After all, if the reconstructed message is the same as the real message, then no harm has been done, even if the values of the received analog waveform are not identical. This section uses a visualization tool called *eye diagrams* that show how much smearing there is in the system, and whether symbol errors will occur. Eye diagrams were encountered briefly in Chapter 9 (refer back to Figure 9.8) when visualizing how the performance of the idealized system degraded when various impairments were added.

Imagine an oscilloscope that traces out the received signal, with the special feature that it is set to retrigger or restart the trace every nT seconds without erasing the screen. Thus the horizontal axis of an eye diagram is the time over which n symbols arrive, and the vertical axis is the value of the received waveform. In the ideal case, the trace begins with n pulses, each of which is a scaled copy of $p(t)$. Then the $n + 1$st to $2n$th pulses arrive, and overlay the first n, though each is scaled according to its symbol value. When there is noise, channel distortion, and timing jitter, the overlays will differ.

As the number of superimposed traces increases, the eye diagram becomes denser, and gives a picture of how the pulse shape, channel, and other factors combine to determine the reliability of the recovered message. Consider the $n = 2$ symbol eye diagram shown in Figure 11.6. In this figure, the message is taken from the 4-PAM alphabet $\pm 1 \pm 3$, and the Hamming pulse shape is used. The center of the "eye" gives the best times to sample, since the openings (i.e., the difference between the received pulse shape when the data value is -1 and the received pulse shape when the data value is 1, or between the received pulse shape when the data value is 1 and the received pulse shape when the data value is 3) are the largest. The width marked "sensitivity to timing

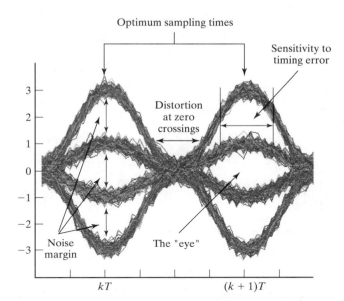

Optimum sampling times

Sensitivity to timing error

Distortion at zero crossings

Noise margin

The "eye"

kT

$(k + 1)T$

FIGURE 11.6 Interpreting eye diagrams: A T-wide Hamming blip is used to pulse shape a 4-PAM data sequence.

error" shows the range of time over which the samples quantize correctly. The noise margin is the smallest vertical distance between the bands, and is proportional to the amount of additive noise that can be resisted by the system without reporting erroneous values.

Thus, eye diagrams such as Figure 11.6 give a clear picture of how good (or how bad) a pulse shape may be. Sometimes the smearing in this figure is so great that the open segment in the center disappears. The eye is said to be *closed*, and this indicates that a simple quantizer (slicer) decision device will make mistakes in recovering the data stream. This is not good!

For example, reconsider the 4-PAM example of the previous section that used a triple-wide Hamming pulse shape. The eye diagram is shown in Figure 11.7. No noise was added when drawing this picture. In the top two plots there are clear regions about the symbol locations where the eye is open. Samples taken in these regions will be quantized correctly, though there are also regions where mistakes will occur. The other plots show the closed eye diagrams using $3T$-wide and $5T$-wide Hamming pulse shapes. Symbol errors will inevitably occur, even if all else in the system is ideal. All of the measures (noise margin, sensitivity to timing, and the distortion at zero crossings) become progressively worse, and ever smaller amounts of noise can cause decision errors.

The following code draws eye diagrams for the pulse shapes defined by the variable `ps`. As in the pulse shaping programs of the previous section, the N binary data points are oversampled by a factor of M and the convolution of the pulse shapes with the data uses the `filter` command. The `reshape(x,a,b)` command changes a vector x of size a*b into a matrix with a rows and b columns, which is used to segment x into b overlays, each a samples long. This works smoothly with the MATLAB `plot` function.

FIGURE 11.7 Eye diagrams for T, $2T$, $3T$, and $5T$-wide Hamming pulse shapes show how the sensitivity to noises and timing errors increases as the pulse shape widens. The closed eye in the bottom plot means that symbol errors are inevitable.

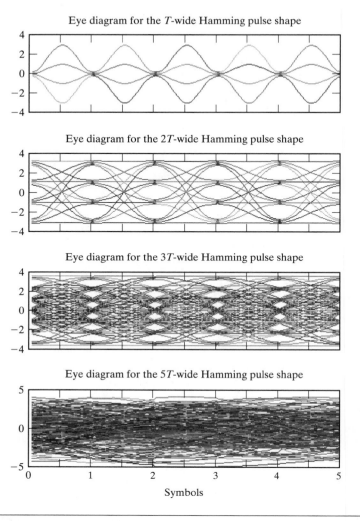

eyediag.m: plot eye diagrams for pulse shape ps

```
N=1000; m=pam(N,2,1);                    % random signal of length N
M=20; mup=zeros(1,N*M); mup(1:M:end)=m;  % oversampling by factor of M
ps=hamming(M);                           % hamming pulse of width M
x=filter(ps,1,mup);                      % convolve pulse shape with mup
neye=5; c=floor(length(x)/(neye*M));     % number of eyes to plot
xp=x(end-neye*M*c+1:end);                % dont plot transients at start
plot(reshape(xp,neye*M,c))               % overlay in groups of size neye
```

Typical output of eyediag.m is shown in Figure 11.8. The rectangular pulse shape in the top plot uses ps=ones(1,M), the Hamming pulse shape in the middle uses ps=hamming(M), and the bottom plot uses a truncated sinc pulse shape ps=SRRC(L,0,M) for L=10. The rectangular pulse is insensitive to timing errors, since sampling almost anywhere (except right at the transition boundaries) will return

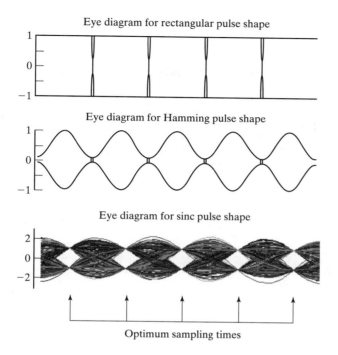

FIGURE 11.8 Eye diagrams for rectangular, Hamming, and sinc pulse shapes with binary data.

the correct values. The Hamming pulse shape has a wide eye, but may suffer from a loss of SNR if the samples are taken far from the center of the eye. Of the three, the sinc pulse is the most sensitive, since it must be sampled near the correct instants or erroneous values will result.

Problems

11.3. Modify `eyediag.m` so that the data sequence is drawn from the alphabet $\pm 1, \pm 3, \pm 5$. Draw the appropriate eye diagram for the rectangular, Hamming, and sinc pulse shapes.

11.4. Modify `eyediag.m` to add noise to the pulse shaped signal x. Use the MATLAB command `v*randn` for different values of v. Draw the appropriate eye diagrams. For each pulse shape, how large can v be and still have the eye remain open?

11.5. Combine the previous two problems. Modify `eyediag.m` as in Problem 11.3 so that the data sequence is drawn from the alphabet $\pm 1, \pm 3, \pm 5$. Add noise, and answer the same two questions as in Problem 11.4. Which alphabet is more susceptible to noise?

It is now easy to experiment with various pulse shapes. `pulseshape2.m` applies a sinc shaped pulse to a random binary sequence. Since the sinc pulse extends infinitely in time (both backward and forward), it cannot be represented exactly in the computer (or in a real communication system) and the parameter L specifies the duration of the sinc, in terms of the number of symbol periods.

FIGURE 11.9 A binary ±1 data sequence is pulse shaped using a sinc pulse.

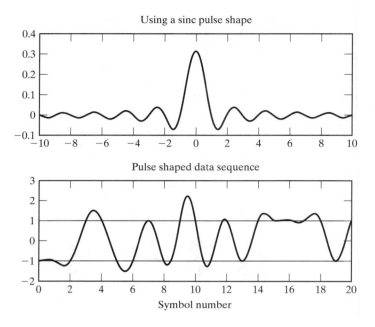

```
pulseshape2.m: pulse shape a (random) sequence

N=1000; m=pam(N,2,1);                        % 2-PAM signal of length N
M=10; mup=zeros(1,N*M); mup(1:M:end)=m;      % oversample by M
L=10; ps=SRRC(L,0,M);                        % sinc pulse shape 2L symbols wide
sc=sum(ps)/M; x=filter(ps/sc,1,mup);         % convolve pulse shape with data
```

Figure 11.9 plots the output of `pulseshape2.m`. The top figure shows the pulse shape while the bottom plot shows the "analog" pulse-shaped signal $x(t)$ over a duration of about 25 symbols. The function `SRRC.m` first appeared in the discussion of interpolation in Section 6.4 (and again in Problem 6.14), and is used here to generate the sinc pulse shape. The sinc function that `SRRC.m` produces is actually scaled, and this effect is removed by normalizing with the variable `sc`. Changing the second input argument from `beta=0` to other small positive numbers changes the shape of the curve, each with a "sinc-like" shape called a square root raised cosine. This will be discussed in greater detail in Sections 11.4 and 11.6. Typing `help srrc` in MATLAB gives useful information on using the function.

Observe that, though the signal oscillates above and below the ±1 lines, there is no intersymbol interference. When using the Hamming pulse as in Figure 11.3, each binary value was clearly delineated. With the sinc pulse of Figure 11.9, the analog waveform is more complicated. But at the correct sampling instances, it always returns to ±1 (the horizontal lines at ±1 are drawn to help focus the eye on the crossing times). Unlike the T-wide Hamming shape, the signal need not return near zero with each symbol.

Problems

11.6. In `pulseshape2.m`, examine the effect of using different oversampling rates M. Try `M=1, 5, 100`.

11.7. Change `pulseshape2.m` so that the data sequence is drawn from the alphabet $\pm 1, \pm 3, \pm 5$. Can you visually identify the correct values in the pulse shaped signal?

11.8. In `pulseshape2.m`, examine the effect of using sinc approximations of different lengths L. Try `L=1, 5, 100, 1000`.

11.9. In `pulseshape2.m`, examine the effect of adding noise to the received signal x. Try MATLAB commands `randn` and `rand`. How large can the noise be and still allow the data to be recognizable?

11.10. Using the code from Problem 11.7, examine the effects of adding noise in `pulseshape2.m`. Does the same amount of noise in the 6-level data have more or less effect than in the 2-level data?

11.11. Modify `pulseshape2.m` to include the effect of a nonunity channel. Try both a highpass channel and a bandpass channel. Which appears worse? What are reasonable criteria for "better" and "worse" in this context?

11.12. A MATLAB question: In `pulseshape2.m`, examine the effect of using the `filtfilt` command for the convolution instead of the `filter` command. Can you figure out why the results are different?

11.13. Another MATLAB question: In `pulseshape2.m`, examine the effect of using the `conv` command for the convolution instead of the `filter` command. Can you figure out how to make this work?

11.4 Nyquist Pulses

Consider a multilevel signal drawn from a finite alphabet with values $w(kT)$, where T is the sampling interval. Let $p(t)$ be the impulse response of the linear filter representing the pulse shape. The signal just after pulse shaping is

$$x(t) = w_a(t) * p(t),$$

where $w_a(t)$ is the pulse train signal (11.1).

The corresponding output of the received filter is

$$y(t) = w_a(t) * p(t) * h_c(t) * h_R(t),$$

as depicted in Figure 11.1, where $h_c(t)$ is the impulse response of the channel and $h_R(t)$ is the impulse response of the receive filter. Let $h_{equiv}(t) = p(t) * h_c(t) * h_R(t)$ be the overall equivalent impulse response. Then the equivalent overall frequency response (i.e., $\mathcal{F}\{h_{equiv}(t)\}$) is

$$H_{equiv}(f) = P(f)H_c(f)H_R(f). \qquad (11.2)$$

One approach would be to attempt to choose $H_R(f)$ so that $H_{equiv}(f)$ attained a desired value (such as a pure delay) for all f. This would be a specification of the impulse response $h_{equiv}(t)$ at all t, since the Fourier transform is invertible. But such a distortionless response is unnecessary, since it does not really matter what happens between samples, but only what happens at the sample instants. In other words, as long as the eye is open, the transmitted symbols are recoverable by sampling at the correct times. In general, if the pulse shape is zero at all integer multiples of kT but one, then it can have any shape in between without causing intersymbol interference.

The condition that one pulse does not interfere with other pulses at subsequent T-spaced sample instants is formalized by saying that $h_{NYQ}(t)$ is a *Nyquist pulse* if there is a τ such that

$$h_{NYQ}(kT + \tau) = \begin{cases} c & k = 0 \\ 0 & k \neq 0 \end{cases} \tag{11.3}$$

for all integers k, where c is some nonzero constant. The timing offset τ in (11.3) will need to be found by the receiver.

A rectangular pulse with time-width less than T certainly satisfies (11.3), as does any pulse shape that is less than T wide. But the bandwidth of the rectangular pulse (and other narrow pulse shapes such as the Hamming pulse shape) may be too wide. Narrow pulse shapes do not utilize the spectrum efficiently. But if just any wide shape is used (such as the multiple-T-wide Hamming pulses), then the eye may close. What is needed is a signal that is wide in time (and narrow in frequency) that also fulfills the Nyquist condition (11.3).

One possibility is the sinc pulse

$$h_{sinc}(t) = \frac{\sin(\pi f_0 t)}{\pi f_0 t},$$

with $f_0 = 1/T$. This has the narrowest possible spectrum, since it forms a rectangle in frequency (i.e., the frequency response of a lowpass filter). Assuming that the clocks at the transmitter and receiver are synchronized so that $\tau = 0$, the sinc pulse is Nyquist because $h_{sinc}(0) = 1$ and

$$h_{sinc}(kT) = \frac{\sin(\pi k)}{\pi k} = 0$$

for all integers $k \neq 0$. But there are several problems with the sinc pulse:

- It has infinite duration. In any real implementation, the pulse must be truncated.

- It is noncausal. In any real implementation, the truncated pulse must be delayed.

- The steep band edges of the rectangular frequency function $H_{sinc}(f)$ are difficult to approximate.

- The sinc function $\sin(t)/t$ decays slowly, at a rate proportional to $1/t$.

The slow decay (recall the plot of the sinc function in Figure 2.10 on page 32) means that samples that are far apart in time can interact with each other when there are even modest clock synchronization errors.

Fortunately, it is not necessary to choose between a pulse shape that is constrained to lie within a single symbol period T and the slowly decaying sinc. While the sinc has the smallest dispersion in frequency, there are other pulse shapes that are narrower in time and yet are only a little wider in frequency. Trading off time and frequency behaviors can be tricky. Desirable pulse shapes

(i) have appropriate zero crossings (i.e., they are Nyquist pulses),
(ii) have sloped band edges in the frequency domain, and
(iii) decay more rapidly in the time domain (compared with the sinc), while maintaining a narrow profile in the frequency domain.

One popular option is called the raised cosine-rolloff (or raised cosine) filter. It is defined by its Fourier transform

$$H_{RC}(f) = \begin{cases} 1 & |f| < f_1 \\ \frac{1}{2}\left(1 + \cos\left[\frac{\pi(|f|-f_1)}{2f_\Delta}\right]\right), & f_1 < |f| < B, \\ 0 & |f| > B \end{cases}$$

where

B is the absolute bandwidth,
f_0 is the 6 dB bandwidth, equal to $\frac{1}{2T}$, one half the symbol rate,
$f_\Delta = B - f_0$, and
$f_1 = f_0 - f_\Delta$.

The corresponding time domain function is

$$h_{RC}(t) = \mathcal{F}^{-1}\{H_{RC}(f)\} = 2f_0\left(\frac{\sin(2\pi f_0 t)}{2\pi f_0 t}\right)\left[\frac{\cos(2\pi f_\Delta t)}{1 - (4f_\Delta t)^2}\right]. \tag{11.4}$$

Define the *rolloff factor* $\beta = f_\Delta / f_0$. Figure 11.10 shows the magnitude spectrum $H_{RC}(f)$ of the raised cosine filter in the bottom and the associated time response $h_{RC}(t)$ on the top, for a variety of rolloff factors. With $T = \frac{1}{2f_0}$, $h_{RC}(kT)$ has a factor $\sin(\pi k)/\pi k$ which is zero for all integer $k \neq 0$. Hence the raised cosine is a Nyquist pulse. In fact, as $\beta \to 0$, $h_{RC}(t)$ becomes a sinc.

The raised cosine pulse $h_{RC}(t)$ with nonzero β has the following characteristics:

• zero crossings at desired times,

• band edges of $H_{RC}(f)$ that are less severe than with a sinc pulse,

• an envelope that falls off at approximately $1/|t|^3$ for large t (look at (11.4)). This is significantly faster than $1/|t|$. As the rolloff factor β increases from 0 to 1, the significant part of the impulse response gets shorter.

Thus, we have seen several examples of Nyquist pulses: rectangular, Hamming, sinc, and raised cosine with a variety of roll off factors. What is the general principle that distinguishes Nyquist pulses from all others? A necessary and sufficient condition for a signal $v(t)$ with Fourier transform $V(f)$ to be a Nyquist pulse is that the sum (over all n) of $V(f - nf_0)$ be constant. To see this, use the sifting property of an impulse (A.56) to factor $V(f)$ from the sum:

$$\sum_{n=-\infty}^{\infty} V(f - nf_0) = V(f) * \left[\sum_{n=-\infty}^{\infty} \delta(f - nf_0)\right].$$

FIGURE 11.10 Raised cosine pulse shape in the time and frequency domains.

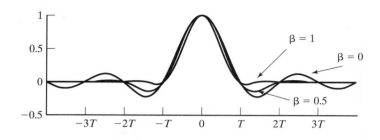

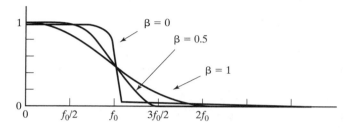

Given that convolution in the frequency domain is multiplication in the time domain (A.40), applying the definition of the Fourier transform, and using the transform pair (from (A.28) with $w(t) = 1$ and $W(f) = \delta(f)$)

$$\mathcal{F}\{\sum_{k=-\infty}^{\infty} \delta(t - kT)\} = \frac{1}{T} \sum_{n=-\infty}^{\infty} \delta(f - nf_0),$$

where $f_0 = 1/T$, this becomes

$$\sum_{n=-\infty}^{\infty} V(f - nf_0) = \int_{t=-\infty}^{\infty} [v(t)(T \sum_{k=-\infty}^{\infty} \delta(t - kT))]e^{-j2\pi ft} dt$$

$$= \sum_{k=-\infty}^{\infty} Tv(kT)e^{-j2\pi fkT}. \tag{11.5}$$

If $v(t)$ is a Nyquist pulse, the only nonzero term in the sum is $v(0)$, and

$$\sum_{n=-\infty}^{\infty} V(f - nf_0) = Tv(0).$$

Thus, the sum of the $V(f - nf_0)$ is a constant if $v(t)$ is a Nyquist pulse. Conversely, if the sum of the $V(f - nf_0)$ is a constant, then only the DC term in (11.5) can be nonzero, and so $v(t)$ is a Nyquist pulse.

Problems

11.14. Write a MATLAB routine that implements the raised cosine impulse response (11.4) with rolloff parameter β. Hint: If you have trouble with "divide by zero" errors, imitate the code in SRRC.m. Plot the output of your program for a variety of β. Hint 2: There is an easy way to use the function SRRC.m.

11.15. Use your code from the previous exercise, along with `pulseshape2.m` to apply raised cosine pulse shaping to a random binary sequence. Can you spot the appropriate times to sample "by eye?"

11.16. Use the code from the previous exercise and `eyediag.m` to draw eye diagrams for the raised cosine pulse with rolloff parameters $r = 0, 0.5, 0.9, 1.0, 5.0$. Compare these to the eye diagrams for rectangular and sinc functions. Consider
  (a) Sensitivity to timing errors
  (b) Peak distortion
  (c) Distortion of zero crossings
  (d) Noise margin

Intersymbol interference occurs when data values at one sample instant interfere with the data values at another sampling instant. Using Nyquist shapes such as the rectangle, sinc, and raised cosine pulses removes the interference, at least at the correct sampling instants, when the channel is ideal. The next sections parlay this discussion of isolated pulse shapes into usable designs for the pulse shaping and receive filters.

11.5 Matched Filtering

Communication systems must be robust to the presence of noises and other disturbances that arise in the channel and in the various stages of processing. Matched filtering is aimed at reducing the sensitivity to noise, which can be specified in terms of the power spectral density (this is reviewed in some detail in Appendix E).

Consider the filtering problem in which a message signal is added to a noise signal and then both are passed through a linear filter. This occurs, for instance, when the signal $g(t)$ of Figure 11.1 is the output of the pulse shaping filter (i.e., no interferers are present), the channel is the identity, and there is noise $n(t)$ present. Assume that the noise is "white"; that is, assume that its power spectral density $P_n(f)$ is equal to some constant η for all frequencies.

The output $y(t)$ of the linear filter with impulse response $h_R(t)$ can be described as the superposition of two components, one driven by $g(t)$ and the other by $n(t)$; that is,

$$y(t) = v(t) + w(t),$$

where

$$v(t) = h_R(t) * g(t) \text{ and } w(t) = h_R(t) * n(t).$$

This is shown in block diagram form in Figure 11.11. In both components, the processing and the output signal are the same. The bottom diagram separates out the component due to the signal ($v(kT)$, which contains the message filtered through the pulse shape and the receive filter), and the component due to the noise ($w(kT)$, which is the noise filtered through the receive filter). The goal of this section is to find the receive filter that maximizes the ratio of the power in the signal $v(kT)$ to the power in the noise $w(kT)$ at the sample instants.

Consider choosing $h_R(t)$ so as to maximize the power of the signal $v(t)$ at time $t = \tau$ compared with the power in $w(t)$ (i.e., to maximize $v^2(\tau)$ relative to the total power of the noise component $w(t)$). This choice of $h_R(t)$ tends to emphasize the signal $v(t)$ and

FIGURE 11.11 The two block diagrams result in the same output. The top shows the data flow in a normal implementation of pulse shaping and receive filtering. The bottom shows an equivalent that allows easy comparison between the parts of the output due to the signal (i.e., $v(kT)$) and the parts due to the noise (i.e., $w(kT)$).

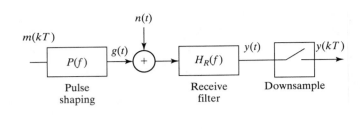

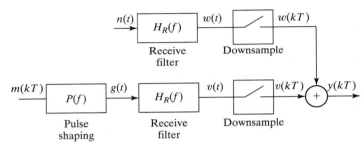

suppress the noise $w(t)$. The argument proceeds by finding the transfer function $H_R(f)$ that corresponds to this $h_R(t)$.

From (E.2), the total power in $w(t)$ is

$$P_w = \int_{-\infty}^{\infty} \mathcal{P}_w(f)df.$$

From the inverse Fourier transform,

$$v(\tau) = \int_{-\infty}^{\infty} V(f)e^{j2\pi f\tau}df,$$

where $V(f) = H_R(f)G(f)$. Thus,

$$v^2(\tau) = \left| \int_{-\infty}^{\infty} H_R(f)G(f)e^{j2\pi f\tau}df \right|^2.$$

Recall (E.3), which says that for $Y(f) = H_R(f)U(f)$, $\mathcal{P}_y(f) = |H_R(f)|^2\mathcal{P}_u(f)$. Thus,

$$\mathcal{P}_w(f) = |H_R(f)|^2\mathcal{P}_n(f) = \eta|H_R(f)|^2.$$

The quantity to be maximized can now be described by

$$\frac{v^2(\tau)}{P_w} = \frac{|\int_{-\infty}^{\infty} H_R(f)G(f)e^{j2\pi f\tau}df|^2}{\int_{-\infty}^{\infty} \eta|H_R(f)|^2df}. \qquad (11.6)$$

Schwarz's inequality (A.57) says that

$$\left| \int_{-\infty}^{\infty} a(x)b(x)dx \right|^2 \leq \left\{ \int_{-\infty}^{\infty} |a(x)|^2dx \right\} \left\{ \int_{-\infty}^{\infty} |b(x)|^2dx \right\},$$

and equality occurs only when $a(x) = kb^*(x)$. This converts (11.6) to

$$\frac{v^2(\tau)}{P_w} \leq \frac{\left(\int_{-\infty}^{\infty} |H_R(f)|^2 df\right) \left(\int_{-\infty}^{\infty} |G(f)e^{j2\pi f\tau}|^2 df\right)}{\eta \int_{-\infty}^{\infty} |H_R(f)|^2 df}, \qquad (11.7)$$

which is maximized with equality when

$$H_R(f) = k(G(f)e^{j2\pi f\tau})^*.$$

$H_R(f)$ must now be transformed to find the corresponding impulse response $h_R(t)$. Since $Y(f) = X(-f)$ when $y(t) = x(-t)$, (use the frequency scaling property of Fourier transforms with a scale factor of unity),

$$\mathcal{F}^{-1}\{W^*(-f)\} = w^*(t) \Rightarrow \mathcal{F}^{-1}\{W^*(f)\} = w^*(-t).$$

Applying the time shift property (A.38) yields

$$\mathcal{F}^{-1}\{W(f)e^{-j2\pi fT_d}\} = w(t - T_d).$$

Combining these two transform pairs yields

$$\mathcal{F}^{-1}\{(W(f)e^{j2\pi fT_d})^*\} = w^*(-(t - T_d)) = w^*(T_d - t).$$

Thus, when $g(t)$ is real,

$$\mathcal{F}^{-1}\{k(G(f)e^{j2\pi f\tau})^*\} = kg^*(\tau - t) = kg(\tau - t).$$

Observe that this filter has the following characteristics:

- This filter results in the maximum signal-to-noise ratio of $v^2(t)/P_w$ at the time instant $t = \tau$ for a noise signal with a flat power spectral density.

- Because the impulse response of this filter is a scaled time reversal of the pulse shape $p(t)$, it is said to be "matched" to the pulse shape, and is called a "matched filter."

- The shape of the magnitude spectrum of the matched filter $H_R(f)$ is the same as the magnitude spectrum $G(f)$.

- The shape of the magnitude spectrum of $G(f)$ is the same as the shape of the frequency response of the pulse shape $P(f)$ for a broadband $m(kT)$, as in Section 11.1.

- The matched filter for any filter with an even symmetric (about some t) time-limited impulse response is a delayed replica of that filter. The minimum delay is the upper limit of the time-limited range of the impulse response.

The following code allows hands-on exploration of this theoretical result. The pulse shape is defined by the variable ps (the default is the sinc function SRRC(L,0,M) for L=10). The receive filter is analogously defined by recfilt. As usual, the symbol alphabet is easily specified by the pam subroutine, and the system operates at an oversampling rate M. The noise is specified in n, and the ratio of the powers is output as powv/poww. Observe that, for any pulse shape, the ratio of the powers is maximized when the receive filter is the same as the pulse shape (the `fliplr` command carries out the time reversal). This holds no matter what the noise, no matter what the symbol alphabet, and no matter what the pulse shape.

---

matchfilt.m: test of SNR maximization

```
N=2^15; m=pam(N,2,1);                        % 2-PAM signal of length N
M=10; mup=zeros(1,N*M); mup(1:M:end)=m;      % oversample by M
L=10; ps=SRRC(L,0,M);                        % define pulse shape
ps=ps/sqrt(sum(ps.^2));                       % and normalize
n=0.5*randn(size(mup));                       % noise
g=filter(ps,1,mup);                           % convolve ps with data
recfilt=SRRC(L,0,M);                          % receive filter H sub R
recfilt=recfilt/sqrt(sum(recfilt.^2));        % normalize the pulse shape
v=filter(fliplr(recfilt),1,g);                % matched filter with data
w=filter(fliplr(recfilt),1,n);                % matched filter with noise
vdownsamp=v(1:M:end);                          % downsample to symbol rate
wdownsamp=w(1:M:end);                          % downsample to symbol rate
powv=pow(vdownsamp);                           % power in downsampled v
poww=pow(wdownsamp);                           % power in downsampled w
powv/poww                                      % ratio
```

In general, when the noise power spectral density is flat (i.e., $\mathcal{P}_n(f) = \eta$), the output of the matched filter may be realized by correlating the input to the matched filter with the pulse shape $p(t)$. To see this, recall that the output is described by the convolution

$$x(\alpha) = \int_{-\infty}^{\infty} s(\lambda)h(\alpha - \lambda)d\lambda$$

of the matched filter with the impulse response $h(t)$. Given the pulse shape $p(t)$ and the assumption that the noise has flat power spectral density, it follows that

$$h(t) = \begin{cases} p(\alpha - t), & 0 \leq t \leq T \\ 0, & \text{otherwise} \end{cases},$$

where α is the delay used in the matched filter. Because $h(t)$ is zero when t is negative and when $t > T$, $h(\alpha - \lambda)$ is zero for $\lambda > \alpha$ and $\lambda < \alpha - T$. Accordingly, the limits on the integration can be converted to

$$x(\alpha) = \int_{\lambda=-\alpha-T}^{\alpha} s(\lambda)p(\alpha - (\alpha - \lambda))d\lambda = \int_{\lambda=-\alpha-T}^{\alpha} s(\lambda)p(\lambda)d\lambda.$$

This is the cross-correlation of p with s as defined in (8.3).

When $\mathcal{P}_n(f)$ is not a constant, (11.6) becomes

$$\frac{v^2(\tau)}{P_w} = \frac{|\int_{-\infty}^{\infty} H(f)G(f)e^{j2\pi f\tau}df|^2}{\int_{-\infty}^{\infty} \mathcal{P}_n(f)|H(f)|^2df}.$$

To use the Schwarz inequality (A.57), associate a with $H\sqrt{\mathcal{P}_n}$ and b with $Ge^{j2\pi f\tau}/\sqrt{\mathcal{P}_n}$. Then (11.7) can be replaced by

$$\frac{v^2(\tau)}{P_w} \le \frac{\left(\int_{-\infty}^{\infty}|H(f)|^2\mathcal{P}_n(f)df\right)\left(\int_{-\infty}^{\infty}\frac{|G(f)e^{j2\pi f\tau}|^2}{\mathcal{P}_n(f)}df\right)}{\int_{-\infty}^{\infty}|H(f)|^2\mathcal{P}_n(f)df},$$

and equality occurs when $a(\cdot) = kb^*(\cdot)$; that is,

$$H(f) = \frac{kG^*(f)e^{-j2\pi f\tau}}{\mathcal{P}_n(f)}.$$

When the noise power spectral density $\mathcal{P}_n(f)$ is not flat, it shapes the matched filter. Recall that the power spectral density of the noise can be computed from its autocorrelation, as is shown in Appendix E.

Problems

11.17. Let the pulse shape be a T-wide Hamming blip. Use the code in `matchfilt.m` to find the ratio of the power in the downsampled v to that in the downsampled w when:
(a) the receive filter is a SRRC with `beta=0, 0.1, 0.5`,
(b) the receive filter is a rectangular pulse, and
(c) the receive filter is a $3T$-wide Hamming pulse.
When is the ratio largest?

11.18. Let the pulse shape be a SRRC with `beta=0.25`. Use the code in `match-filt.m` to find the ratio of the power in the downsampled v to that in the downsampled w when
(a) the receive filter is a SRRC with `beta=0, 0.1, 0.25, 0.5`,
(b) the receive filter is a rectangular pulse, and
(c) the receive filter is a T-wide Hamming pulse.
When is the ratio largest?

11.19. Let the symbol alphabet be 4-PAM.
(a) Repeat Problem 11.17.
(b) Repeat Problem 11.18.

11.20. Create a noise sequence that is uniformly distributed (using `rand`) with zero mean.
(a) Repeat Problem 11.17.
(b) Repeat Problem 11.18.

11.6 Matched Transmit and Receive Filters

While focusing separately on the pulse shaping and the receive filtering makes sense pedagogically, the two are intimately tied together in the communication system. This section notes that it is not really the pulse shape that should be Nyquist, but rather the convolution of the pulse shape with the receive filter.

FIGURE 11.12 Noisy
baseband communica-
tion system.

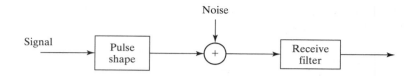

Recall the overall block diagram of the system in Figure 11.1, where it was assumed that the portion of the system from upconversion (to passband) to final downconversion (back to baseband) is done perfectly and that the channel is just the identity. Thus, the central portion of the system is effectively transparent (except for the intrusion of noise). This simplifies the system to the baseband model in Figure 11.12.

The task is to design an appropriate pair of filters: a pulse shape for the transmitter, and a receive filter that is matched to the pulse shape and the presumed noise description. It is not crucial that the transmitted signal itself have no intersymbol interference. Rather, the signal after the receive filter should have no ISI. Thus, it is not the pulse shape that should satisfy the Nyquist pulse condition, but the combination of the pulse shape and the receive filter.

The receive filter should simultaneously

(i) allow no intersymbol interference at the receiver, and
(ii) maximize the signal-to-noise ratio.

Hence, it is the convolution of the pulse shape and the receive filter that should be a Nyquist pulse, and the receive filter should be matched to the pulse shape. Considering candidate pulse shapes that are both symmetric and even about some time t, the associated matched filter (modulo the associated delay) is the same as the candidate pulse shape. What symmetric pulse shapes, when convolved with themselves, form a Nyquist pulse? Previous sections examined several Nyquist pulse shapes, the rectangle, the sinc, and the raised cosine. When convolved with themselves, do any of these shapes remain Nyquist?

For a rectangle pulse shape and its rectangular matched filter, the convolution is a triangle that is twice as wide as the original pulse shape. With precise timing, (so that the sample occurs at the peak in the middle), this triangular pulse shape is also a Nyquist pulse. This exact situation will be considered in detail in Section 12.2.

The convolution of a sinc function with itself is more easily viewed in the frequency domain as the point-by-point square of the transform. Since the transform of the sinc is a rectangle, its square is a rectangle as well. The inverse transform is consequently still a sinc, and is therefore a Nyquist pulse.

The raised cosine pulse fails. Its square in the frequency domain does not retain the odd symmetry around the band edges, and the convolution of the raised cosine with itself does not retain its original zero crossings. But the raised cosine was the preferred Nyquist pulse because it conserves bandwidth effectively and because its impulse response dies away quickly. One possibility is to define a new pulse shape that is the square root of the raised cosine (the square root is taken in the frequency domain, not the time domain). This is called the *square-root raised cosine filter* (SRRC). By definition, the square in frequency of the SRRC (which is the raised cosine) is a Nyquist pulse.

The time domain description of the SRRC pulse is found by taking the inverse Fourier transform of the square root of the spectrum of the raised cosine pulse. The answer is a bit complicated:

$$
v(t) = \begin{cases} \frac{1}{\sqrt{T}} \frac{\sin(\pi(1-\beta)t/T)+(4\beta t/T)\cos(\pi(1+\beta)t/T)}{(\pi t/T)(1-(4\beta t/T)^2)} & t \neq 0, \ t \neq \pm\frac{T}{4\beta} \\[2ex] \frac{1}{\sqrt{T}}(1 - \beta + (4\beta/\pi)) & t = 0 \\[2ex] \frac{\beta}{\sqrt{2T}}\left[\left(1+\frac{2}{\pi}\right)\sin\left(\frac{\pi}{4\beta}\right) + \left(1-\frac{2}{\pi}\right)\cos\left(\frac{\pi}{4\beta}\right)\right] & t = \pm\frac{T}{4\beta} \end{cases} \quad . \quad (11.8)
$$

Problem

11.21. Plot the SRRC pulse in the time domain and show that it is not a Nyquist pulse (because it does not cross zero at the desired times). The MATLAB routine SRRC.m will make this easier.

Though the SRRC is not itself a Nyquist pulse, the convolution in time of two SRRCs is a Nyquist pulse. The square root raised cosine is the most commonly used pulse in bandwidth-constrained communication systems.

Timing Recovery

All we have to decide is what to do with the time given us.
—Gandalf, in J. R. R. Tolkien's *Fellowship of the Ring*

When the signal arrives at the receiver, it is a complicated analog waveform that must be sampled in order to eventually recover the transmitted message. The timing offset experiments of Section 9.4.5 showed that one kind of "stuff" that can "happen" to the received signal is that the samples might inadvertently be taken at inopportune times. The "eye" becomes "closed" and the symbols are incorrectly decoded. Thus there needs to be a way to determine *when* to take the samples at the receiver. In accordance with the basic system architecture of Chapter 2, this chapter focuses on baseband methods of timing recovery (also called clock recovery). The problem is approached in a familiar way: find performance functions which have their maximum (or minimum) at the optimal point (i.e., at the correct sampling instants when the eye is open widest). These performance functions are then used to define adaptive elements that iteratively estimate the correct sampling times. As usual, all other aspects of the system are presumed to operate flawlessly: the up and down conversions are ideal, there are no interferers, and the channel is benign.

The discussion of timing recovery begins in Section 12.1 by showing how a sampled version of the received signal $x[k]$ can be written as a function of the timing parameter τ, which dictates when to take samples. Section 12.2 gives several examples that motivate several different possible performance functions, (functions of $x[k]$) which lead to "different" methods of timing recovery. The error between the received data values and the transmitted data (called the *source recovery error*) is an obvious candidate, but it can be measured only when the transmitted data are known or when there is an a priori known or agreed-upon header (or training sequence). An alternative is to use the *cluster variance*, which takes the square of the difference between the received data values and the nearest element of the source alphabet. This is analogous to the decision directed approach to carrier recovery (from Section 10.5), and an adaptive element based on the cluster variance is derived and studied in Section 12.3. A popular alternative is to measure the power of the T-spaced output of the matched filter. Maximizing this power (by choice of τ), also leads to a good answer, and an adaptive element based on output power maximization is detailed in Section 12.4.

In order to understand the various performance functions, the error surfaces are drawn. Interestingly, in many cases, the error surface for the cluster variance has minima wherever the error surface for the output power has maxima. In these cases, either method

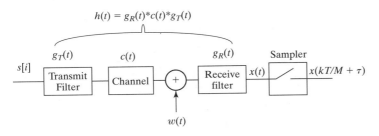

$h(t) = g_R(t) * c(t) * g_T(t)$

FIGURE 12.1 The transfer function h combines the effects of the transmitter pulse shaping g_T, the channel c, and the receive filter g_R.

can be used as the basis for timing recovery methods. On the other hand, there are also situations when the error surfaces have extremal points at different locations. In these cases, the error surface provides a simple way of examining which method is most fitting.

12.1 The Problem of Timing Recovery

The problem of timing recovery is to choose the instants at which to sample the incoming (analog) signal. This can be translated into the mathematical problem of finding a single parameter, the timing offset τ, which minimizes (or maximizes) some function (such as the source recovery error, the cluster variance, or the output power) of τ given the input. Clearly, the output of the sampler must also be a function of τ, since τ specifies when the samples are taken. The first step is to write out exactly how the values of the samples depend on τ. Suppose that the interval T between adjacent symbols is known exactly. Let $g_T(t)$ be the pulse shaping filter, $g_R(t)$ the receive filter, $c(t)$ the impulse response of the channel, $s[i]$ the data from the signal alphabet, and $w(t)$ the noise. Then the baseband waveform at the input to the sampler can be written explicitly as

$$x(t) = \sum_{i=-\infty}^{\infty} s[i]\delta(t - iT) * g_T(t) * c(t) * g_R(t) + w(t) * g_R(t),$$

Combining the three linear filters with

$$h(t) = g_T(t) * c(t) * g_R(t), \tag{12.1}$$

as shown in Figure 12.1, and sampling at interval $\frac{T}{M}$ (M is again the *oversampling factor*) yields the sampled output at time $\frac{kT}{M} + \tau$:

$$x\left(\frac{kT}{M} + \tau\right) = \sum_{i=-\infty}^{\infty} s[i]h(t - iT) + w(t) * g_R(t)\Bigg|_{t=\frac{kT}{M}+\tau}.$$

Assuming the noise has the same distribution no matter when it is sampled, the noise term $v[k] = w(t) * g_R(t)|_{t=\frac{kT}{M}+\tau}$ is independent of τ. Thus, the goal of the optimization is to find τ so as to maximize or minimize some simple function of the samples, such as

$$x[k] = x\left(\frac{kT}{M} + \tau\right) = \sum_{i=-\infty}^{\infty} s[i]h\left(\frac{kT}{M} + \tau - iT\right) + v[k]. \tag{12.2}$$

There are three ways that timing recovery algorithms can be implemented, and these are shown in Figure 12.2. In the first, an analog processor determines when the sampling

FIGURE 12.2 Three generic structures for timing recovery. In (a), an analog processor determines when the sampling instants will occur. In (b), a digital post processor is used to determine when to sample. In (c), the sampling instants are chosen by a free running clock, and digital post processing is used to recover the values of the received signal that would have occurred at the optimal sampling instants.

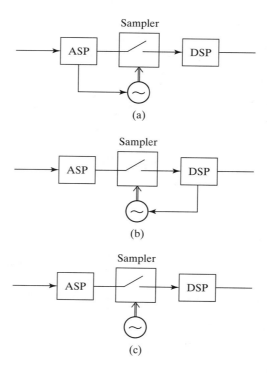

(a)

(b)

(c)

instants will occur. In the second, a digital post processor is used to determine when to sample. In the third, the sampling instants are chosen by a free running clock, and digital post processing (interpolation) is used to recover the values of the received signal that would have occurred at the optimal sampling instants. The adaptive elements of the next sections can be implemented in any of the three ways, though in digital radio systems the trend is to remove as much of the calculation from analog circuitry as possible.

12.2 An Example

This section works out in complete and gory detail what may be the simplest case of timing recovery. More realistic situations will be considered (by numerical methods) in later sections.

Consider a noise-free binary ± 1 baseband communication system in which the transmitter and receiver have agreed on the rate of data flow (one symbol every T seconds, with an oversampling factor of $M = 1$). The goal is to select the instants $kT + \tau$ at which to sample, that is, to find the offset τ. Suppose that the pulse shaping filter is chosen so that $h(t)$ is Nyquist; that is

$$h(kT) = \begin{cases} 1 & k = 1 \\ 0 & k \neq 1 \end{cases}.$$

The sampled output sequence is the amplitude modulated impulse train $s[i]$ convolved with a filter that is the concatenation of the pulse shaping, the channel, and the receive

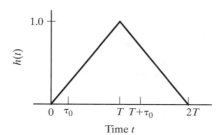

FIGURE 12.3 For the example of this section, the concatenation of the pulse shape, the channel, and the receive filtering ($h(t)$ of (12.1)) is assumed to be a symmetric triangle wave with unity amplitude and support $2T$.

filtering, and evaluated at the sampler closure times, as in (12.2). Thus,

$$x[k] = \sum_i s[i]h(t - iT)\Bigg|_{t=kT+\tau}.$$

To keep the computations tractable, suppose that $h(t)$ has the triangular shape shown in Figure 12.3. This might occur, for instance, if the pulse shaping filter and the receive filter are both rectangular pulses of width T and the channel is the identity. There are three cases to consider: $\tau = 0$, $\tau > 0$, and $\tau < 0$.

- With $\tau = 0$, which synchronizes the sampler to the transmitter pulse times,

$$h(t - iT)|_{t=kT+\tau} = h(kT + \tau - iT) = h((k - i)T + \tau) = h((k - i)T)$$

$$= \begin{cases} 1 & k - i = 1 \Rightarrow i = k - 1 \\ 0 & \text{otherwise} \end{cases}.$$

In this case, $x[k] = s[k - 1]$ and the system is a pure delay.

- With $\tau = \tau_0 > 0$, the only two nonzero points among the sampled impulse response are at $h(\tau_0)$ and $h(T + \tau_0)$, as illustrated in Figure 12.3. Therefore,

$$h(t - iT)|_{t=kT+\tau_0} = h((k - i)T + \tau_0)$$

$$= \begin{cases} 1 - \frac{\tau_0}{T} & k - i = 1 \\ \frac{\tau_0}{T} & k - i = 0 \\ 0 & \text{otherwise} \end{cases}.$$

To work out a numerical example, let $k = 6$. Then

$$x[6] = \sum_i s[i]h((6-i)T+\tau_0) = s[6]h(\tau_0)+s[5]h(T+\tau_0) = s[6]\frac{\tau_0}{T}+s[5]\left(1 - \frac{\tau_0}{T}\right).$$

Since the data are binary, there are four possibilities for the pair $(s[5], s[6])$:

$$\begin{aligned}
(s[5], s[6]) &= (+1, +1) \Rightarrow x[6] = \tfrac{\tau_0}{T} + 1 - \tfrac{\tau_0}{T} = 1 \\
(s[5], s[6]) &= (+1, -1) \Rightarrow x[6] = \tfrac{-\tau_0}{T} + 1 - \tfrac{\tau_0}{T} = 1 - \tfrac{2\tau_0}{T} \\
(s[5], s[6]) &= (-1, +1) \Rightarrow x[6] = \tfrac{\tau_0}{T} - 1 + \tfrac{\tau_0}{T} = -1 + \tfrac{2\tau_0}{T} \\
(s[5], s[6]) &= (-1, -1) \Rightarrow x[6] = \tfrac{-\tau_0}{T} - 1 + \tfrac{\tau_0}{T} = -1
\end{aligned} \qquad (12.3)$$

Note that two of the possibilities for $x[6]$ give correct values for $s[5]$, while two are incorrect.

- With $\tau = \tau_0 < 0$, the only two nonzero points among the sampled impulse response are at $h(2T + \tau_0)$ and $h(T + \tau_0)$. In this case,

$$h(t - iT)|_{t=kT+\tau_0} == \begin{cases} 1 - \frac{|\tau_0|}{T} & k - i = 1 \\ \frac{|\tau_0|}{T} & k - i = 2 \\ 0 & \text{otherwise} \end{cases}.$$

The next two examples look at two possible measures of the quality of τ: the cluster variance and the output power.

Example 12.1   Cluster Variance

The decision device $Q(x[k])$ quantizes its argument to the nearest member of the symbol alphabet. For binary data, this is the signum operator that maps any positive number to $+1$ and any negative number to -1. If $-T/2 < \tau_0 < T/2$, then $Q(x[k]) = s[k - 1]$ for all k, the eye is open, and the source recovery error can be written as $e[k] = s[k - 1] - x[k] = Q(x[k]) - x[k]$. Continuing the example, and assuming that all symbol pair choices are equally likely, the average squared error at time $k = 6$ is

$$\text{avg}\{e^2[6]\} = \left(\frac{1}{4}\right)\left\{(1 - 1)^2 + \left(1 - \left(1 - \frac{2|\tau_0|}{T}\right)\right)^2 + \left(-1 - \left(-1 + \frac{2|\tau_0|}{T}\right)\right)^2 + (-1 - (-1))^2\right\}$$

$$= \left(\frac{1}{4}\right)\left(\frac{4\tau_0^2}{T^2} + \frac{4\tau_0^2}{T^2}\right) = \frac{2\tau_0^2}{T^2}.$$

The same result occurs for any other k.

If τ_0 is outside the range of $(-T/2, T/2)$ then $Q(x[k])$ no longer equals $s[k - 1]$ (but it does equal $s[j]$ for some $j \neq k - 1$). The cluster variance

$$CV = \text{avg}\{e^2[k]\} = \text{avg}\{(Q(x[k]) - x[k])^2\} \tag{12.4}$$

is a useful measure, and this is plotted in Figure 12.4 as a function of τ. The periodic nature of the function is clear, and the problem of timing recovery can be viewed as a one-dimensional search for τ that minimizes the CV.

FIGURE 12.4 Cluster variance as a function of offset timing τ.

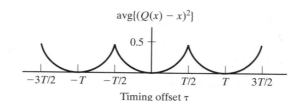

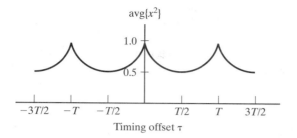

avg$\{x^2\}$

1.0

0.5

$-3T/2$ $-T$ $-T/2$ $T/2$ T $3T/2$

Timing offset τ

FIGURE 12.5 Average squared output as a function of timing offset τ.

Example 12.2   Output Power Maximization

Another measure of the quality of the timing parameter τ is given by the power (average energy) of the $x[k]$. Using the four formulas (12.3), and observing that analogous formulas also apply when $\tau_0 < 0$, the average energy can be calculated for any k by

$$\text{avg}\{x^2[k]\} = (1/4)[(1)^2 + (1 - (2|\tau|/T))^2 + (-1 + (2|\tau|/T))^2 + (-1)^2]$$
$$= (1/4)[2 + 2(1 - (2|\tau|/T))^2]$$
$$= 1 - (2|\tau|/T) + (2|\tau|^2/T^2),$$

assuming that the four symbol pairs are equally likely. The average of $x^2[k]$ is plotted in Figure 12.5 as a function of τ. Over $-T/2 < \tau < T/2$, this average is maximized with $\tau = 0$. Thus, the problem of timing recovery can also be viewed as a one-dimensional search for the τ that maximizes $\text{avg}\{x^2[k]\}$.

Thus, at least in the simple case of binary transmission with $h(t)$ a triangular pulse, the optimal timing offset (for the plots in Figures 12.4 and 12.5, at $\tau = nT$ for integer n) can be obtained either by minimizing the cluster variance or by maximizing the output power. In more general situations, the two measures may not be optimized at the same point. Which approach is best when:

- There is channel noise?
- The source alphabet is multilevel?
- More common pulse shapes are used?
- There is intersymbol interference?

The next two sections show how to design adaptive elements that carry out these minimizations and maximizations. The error surfaces corresponding to the performance functions will be used to gain insight into the behavior of the methods even in nonideal situations.

12.3 Decision-Directed Timing Recovery

If the combination of the pulse shape, channel, and matched filter has the Nyquist property, then the value of the waveform is exactly equal to the value of the data at the correct sampling times. Thus, there is an obvious choice for the performance function: find the sampling instants at which the difference between the received value and the transmitted values are smallest. This is called the *source recovery error* and can be used when the transmitted data are known—for instance, when there is a training sequence. But if the data are unavailable (which is the normal situation), then the source recovery error cannot be measured and hence cannot form the basis of a timing recovery algorithm.

The previous section suggested that a possible substitute is to use the cluster variance $\text{avg}\{(Q(x[k]) - x[k])^2\}$. Remember that the samples $x[k] = x(\frac{kT}{M} + \tau)$ are functions of τ because τ specifies when the samples are taken, as is evident from (12.2). Thus, the goal of the optimization is to find τ so as to minimize

$$J_{CV}(\tau) = \text{avg}\{(Q(x[k]) - x[k])^2\}. \tag{12.5}$$

Solving for τ directly is nontrivial, but $J_{CV}(\tau)$ can be used as the basis for an adaptive element

$$\tau[k+1] = \tau[k] - \bar{\mu} \left. \frac{dJ_{CV}(\tau)}{d\tau} \right|_{\tau=\tau[k]}. \tag{12.6}$$

Using the approximation (G.13), which swaps the order of the derivative and the average, yields

$$\frac{dJ_{CV}(\tau)}{d\tau} \approx \text{avg} \left\{ \frac{d\,(Q(x[k]) - x[k])^2}{d\tau} \right\} = -2\text{avg} \left\{ (Q(x[k]) - x[k]) \frac{dx[k]}{d\tau} \right\}. \tag{12.7}$$

The derivative of $x[k]$ can be approximated numerically. One way of doing this is to use

$$\frac{dx[k]}{d\tau} = \frac{dx(\frac{kT}{M} + \tau)}{d\tau} \approx \frac{x(\frac{kT}{M} + \tau + \delta) - x(\frac{kT}{M} + \tau - \delta)}{2\delta}, \tag{12.8}$$

which is valid for small δ. Substituting (12.8) and (12.7) into (12.6) and evaluating at $\tau = \tau[k]$ gives the algorithm

$$\tau[k+1] = \tau[k] + \mu\text{avg} \left\{ (Q(x[k]) - x[k]) \left[x \left(\frac{kT}{M} + \tau[k] + \delta \right) - x \left(\frac{kT}{M} + \tau[k] - \delta \right) \right] \right\},$$

where the stepsize $\mu = \frac{\bar{\mu}}{\delta}$. As usual, this algorithm acts as a lowpass filter to smooth or average the estimates of τ, and it is common to remove the explicit outer averaging operation from the update, which leads to

$$\tau[k+1] = \tau[k] + \mu(Q(x[k]) - x[k]) \left[x \left(\frac{kT}{M} + \tau[k] + \delta \right) - x \left(\frac{kT}{M} + \tau[k] - \delta \right) \right]. \tag{12.9}$$

If the $\tau[k]$ are too noisy, then the stepsize μ can be decreased (or the length of the average, if present, can be increased), although these will inevitably slow the convergence of the algorithm.

The algorithm (12.9) is easy to implement, though it requires samples of the waveform $x(t)$ at three different points: $x(\frac{kT}{M} + \tau[k] - \delta)$, $x(\frac{kT}{M} + \tau[k])$, and $x(\frac{kT}{M} + \tau[k] + \delta)$. One possibility is to straightforwardly sample three times. Since sampling is done by

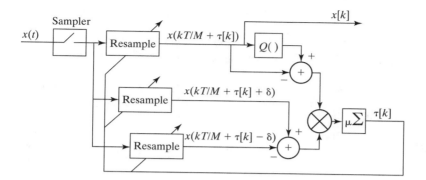

FIGURE 12.6 One implementation of the adaptive element (12.9) uses three digital interpolations (resamplers). After the $\tau[k]$ converge, the output $x[k]$ is a sampled version of the input $x(t)$, with the samples taken at times that minimize the cluster variance.

hardware, this is a hardware intensive solution. Alternatively, the values can be interpolated. Recall from the sampling theorem that a waveform can be reconstructed exactly at any point, as long as it is sampled faster than twice the highest frequency. This is useful since the values at $x(\frac{kT}{M} + \tau[k] - \delta)$ and at $x(\frac{kT}{M} + \tau[k] + \delta)$ can be interpolated from the nearby samples $x[k]$. Recall that interpolation was discussed in Section 6.4, and the MATLAB routine `interpsinc.m` on page 103 makes it easy to implement bandlimited interpolation and reconstruction. Of course, this requires extra calculations, and so is a more "software intensive" solution. This strategy is diagrammed in Figure 12.6.

The following code prepares the transmitted signal that will be used subsequently to simulate the timing recovery methods. The user specifies the signal constellation (default is 4-PAM), the number of data points n, and the oversampling factor m. The channel is allowed to be nonunity, and a square-root raised cosine pulse of width `2*1` and with rolloff `beta` is used as the default transmit (pulse shaping) filter. An initial timing offset is specified in `toffset`, and the code implements this delay with an offset in the SRRC function. The matched filter is implemented using the same SRRC (but without the time delay). Thus, the timing offset is not known at the receiver.

---

clockrecDD.m: (part 1) prepare transmitted signal

```
n=10000;                          % number of data points
m=2;                              % oversampling factor
constel=4;                        % 4-pam constellation
beta=0.5;                         % rolloff parameter for SRRC
l=50;                             % 1/2 length of pulse shape (in symbols)
chan=[1];                         % T/m "channel"
toffset=-0.3;                     % initial timing offset
pulshap=SRRC(l,beta,m,toffset);   % SRRC pulse shape with timing offset
s=pam(n,constel,5);               % random data sequence with var=5
sup=zeros(1,n*m);                 % upsample the data by placing...
sup(1:m:end)=s;                   % ... m-1 zeros between each data point
hh=conv(pulshap,chan);            % ... and pulse shape
r=conv(hh,sup);                   % ... to get received signal
matchfilt=SRRC(l,beta,m,0);       % matched filter = SRRC pulse shape
x=conv(r,matchfilt);              % convolve signal with matched filter
```

The goal of the timing recovery in `clockrecDD.m` is to find (the negative of) the value of `toffset` using only the received signal—that is, to have `tau` converge to `-toffset`. The adaptive element is implemented in `clockrecDD.m` using the iterative cluster variance algorithm (12.9). The algorithm is initialized with an offset estimate of `tau=0` and stepsize `mu`. The received signal is sampled at `m` times the symbol rate, and the `while` loop runs though the data, incrementing `i` once for each symbol (and incrementing `tnow` by `m` for each symbol). The offsets `tau` and `tau+m` are indistinguishable from the point of view of the algorithm. The update term contains the interpolated value `xs` as well as two other interpolated values to the left and right that are used to approximate the derivative term.

---

clockrecDD.m: (part 2) clock recovery minimizing cluster variance

```
tnow=l*m+1; tau=0; xs=zeros(1,n);           % initialize variables
tausave=zeros(1,n); tausave(1)=tau; i=0;
mu=0.01;                                     % algorithm stepsize
delta=0.1;                                   % time for derivative
while tnow<length(x)-2*l*m                   % run iteration
  i=i+1;
  xs(i)=interpsinc(x,tnow+tau,l);            % interpolated value at tnow+tau
  x_deltap=interpsinc(x,tnow+tau+delta,l);   % get value to the right
  x_deltam=interpsinc(x,tnow+tau-delta,l);   % get value to the left
  dx=x_deltap-x_deltam;                      % calculate numerical derivative
  qx=quantalph(xs(i),[-3,-1,1,3]);           % quantize xs to nearest 4-PAM symbol
  tau=tau+mu*dx*(qx-xs(i));                  % alg update: DD
  tnow=tnow+m; tausave(i)=tau;               % save for plotting
end
```

Typical output of the program is plotted in Figure 12.7, which shows the 4-PAM constellation diagram, along with the trajectory of the offset estimation as it converges towards the negative of the "unknown" value -0.3. Observe that, initially, the values are widely dispersed about the required 4-PAM values, but as the algorithm nears its convergent point, the estimated values of the symbols converge nicely.

As usual, a good way to conceptualize the action of the adaptive element is to draw the error surface—in this case, to plot $J_{CV}(\tau)$ of (12.5) as a function of the timing offset τ. In the examples of Section 12.2, the error surface was drawn by exhaustively writing down all the possible input sequences, and evaluating the performance function explicitly in terms of the offset τ. In the binary setup with an identity channel, where the pulse shape is only $2T$ long, and with $M = 1$ oversampling, there were only four cases to consider. But when the pulse shape and channel are long and the constellation has many elements, the number of cases grows rapidly. Since this can get out of hand, an "experimental" method can be used to approximate the error surface. For each timing offset, the code in `clockrecDD-cost.m` chooses n random input sequences, evaluates the performance function, and averages.

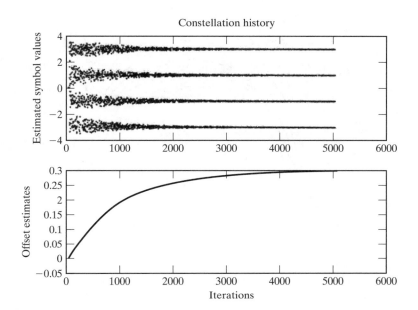

FIGURE 12.7 Output of the program `clock-recDD.m` shows the symbol estimates in the top plot and the trajectory of the offset estimation in the bottom.

---

clockrecDDcost.m: error surfaces for cluster variance performance function

```
l=10;                              % 1/2 duration of pulse shape in symbols
beta=0.75;                         % rolloff for pulse shape
m=20;                              % evaluate at m different points
ps=srrc(l,beta,m);                 % make srrc pulse shape
psrc=conv(ps,ps);                  % convolve 2 srrc's to get rc
psrc=psrc(l*m+1:3*l*m+1);          % truncate to same length as ps
cost=zeros(1,m); n=20000;          % calculate perf via "experimental" method
x=zeros(1,n);
for i=1:m                          % for each offset
  pt=psrc(i:m:end);                % rc is shifted i/m of a symbol
  for k=1:n                        % do it n times
    rd=pam(length(pt),4,5);        % random 4-PAM vector
    x(k)=sum(rd.*pt);              % received data point w/ISI
  end
  err=quantalph(x,[-3,-1,1,3])-x'; % quantize to nearest 4-PAM
  cost(i)=sum(err.^2)/length(err); % DD performance function
end
```

The output of `clockrecDDcost.m` is shown in Figure 12.8. The error surface is plotted for the SRRC with five different rolloff factors. For all β, the correct answer at $\tau = 0$ is a minimum. For small values of β, this is the only minimum and the error surface is unimodal over each period. In these cases, no matter where τ is initialized, it should converge to the correct answer. As β is increased, however, the error surface flattens across its top and gains two extra minima. These represent erroneous values of τ to which the adaptive element may converge. Thus, the error surface can warn the system designer to expect certain kinds of failure modes in certain situations (such as certain pulse shapes).

FIGURE 12.8 The performance function (12.5) is plotted as a function of the timing offset τ for five different pulse shapes characterized by different rolloff factors β. The correct answer is at the global minimum at $\tau = 0$.

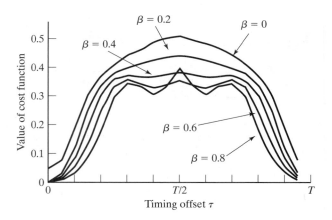

Problems

12.1. Use `clockrecDD.m` to "play with" the clock recovery algorithm.
  (a) How does `mu` affect the convergence rate? What range of stepsizes works?
  (b) How does the signal constellation of the input affect the convergent value of `tau`? (Try 2-PAM and 6-PAM. Remember to quantize properly in the algorithm update.)

12.2. Implement a rectangular pulse shape. Does this work better or worse than the SRRC?

12.3. Add noise to the signal (add a zero mean noise to the received signal using the MATLAB `randn` function). How does this affect the convergence of the timing offset parameter `tau`. Does it change the final converged value?

12.4. Modify `clockrecDD.m` by setting `toffset=-0.8`. This starts the iteration in a closed eye situation. How many iterations does it take to open the eye? What is the convergent value?

12.5. Modify `clockrecDD.m` by changing the channel. How does this affect the convergence speed of the algorithm? Do different channels change the convergent value? Can you think of a way to predict (given a channel) what the convergent value will be?

12.6. Modify the algorithm (12.9) so that it minimizes the source recovery error ($(s[k - d] - x[k])^2$, where d is some (integer) delay. You will need to assume that the message $s[k]$ is known at the receiver. Implement the algorithm by modifying the code in `clockrecDDcost.m`. Compare the new algorithm with the old in terms of convergence speed and final convergent value.

12.7. Using the source recovery error algorithm of Problem 12.6, examine the effect of different pulse shapes. Draw the error surfaces (mimic the code in `clockrecDDcost.m`). What happens when you have the wrong d? The right d?

12.8. Investigate how the error surface depends on the input signal.
  (a) Draw the error surface for the DD timing recovery algorithm when the inputs are binary ± 1.

(b) Draw the error surface when the inputs are drawn from the 4-PAM constellation, for the special case in which the symbol -3 never occurs.

12.4 Timing Recovery via Output Power Maximization

Any timing recovery algorithm must choose the instants at which to sample the received signal. The previous section showed that this can be translated into the mathematical problem of finding a single parameter, the timing offset τ, which minimizes the cluster variance. The extended example of Section 12.2 suggests that maximizing the average of the received power (i.e., maximizing avg$\{x^2[k]\}$) leads to the same solutions as minimizing the cluster variance. Accordingly, this section builds an element that adapts τ so as to find the sampling instants at which the power (in the sampled version of the received signal) is maximized.

To be precise, the goal of the optimization is to find τ so as to maximize

$$J_{OP}(\tau) = \text{avg}\{x^2[k]\} = \text{avg}\left\{x^2\left(\frac{kT}{M} + \tau\right)\right\}, \qquad (12.10)$$

which can be optimized using an adaptive element

$$\tau[k+1] = \tau[k] + \bar{\mu} \left.\frac{dJ_{OP}(\tau)}{d\tau}\right|_{\tau=\tau[k]}. \qquad (12.11)$$

The updates proceed in the same direction as the gradient (rather than minus the gradient) because the goal is to maximize, to find the τ that leads to the largest value of $J_{OP}(\tau)$ (rather than the smallest). The derivative of $J_{OP}(\tau)$ can be approximated using (G.13) to swap the differentiation and averaging operations

$$\frac{dJ_{OP}(\tau)}{d\tau} \approx \text{avg}\left\{\frac{dx^2[k]}{d\tau}\right\} = 2\text{avg}\left\{x[k]\frac{dx[k]}{d\tau}\right\}. \qquad (12.12)$$

The derivative of $x[k]$ can be approximated numerically. One way of doing this is to use (12.8), which is valid for small δ. Substituting (12.8) and (12.12) into (12.11) and evaluating at $\tau = \tau[k]$ gives the algorithm

$$\tau[k+1] = \tau[k] + \mu\text{avg}\left\{x[k]\left[x\left(\frac{kT}{M} + \tau[k] + \delta\right) - x\left(\frac{kT}{M} + \tau[k] - \delta\right)\right]\right\},$$

where the stepsize $\mu = \frac{\bar{\mu}}{\delta}$. As usual, the small stepsize algorithm acts as a lowpass filter to smooth the estimates of τ, and it is common to remove the explicit outer averaging operation, leading to

$$\tau[k+1] = \tau[k] + \mu x[k]\left[x\left(\frac{kT}{M} + \tau[k] + \delta\right) - x\left(\frac{kT}{M} + \tau[k] - \delta\right)\right]. \qquad (12.13)$$

If $\tau[k]$ is noisy, then μ can be decreased (or the length of the average, if present, can be increased), although these will inevitably slow the convergence of the algorithm.

Using the algorithm (12.13) is similar to implementing the cluster variance scheme (12.9), and a "software intensive" solution is diagrammed in Figure 12.9. This uses interpolation (resampling) to reconstruct the values of $x(t)$ at $x(\frac{kT}{M} + \tau[k] - \delta)$ and at

FIGURE 12.9 One implementation of the adaptive element (12.13) uses three digital interpolations (resamplers). After the $\tau[k]$ converge, the output $x[k]$ is a sampled version of the input $x(t)$, with the samples taken at times that maximize the power of the output.

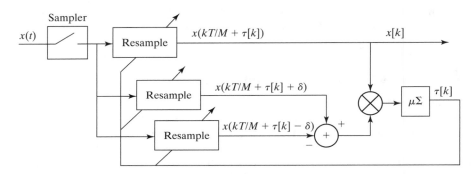

$x(\frac{kT}{M} + \tau[k] + \delta)$ from nearby samples $x[k]$. As suggested by Figure 12.2, the same idea can be implemented in analog, hybrid, or digital form.

The following program implements the timing recovery algorithm using the recursive output power maximization algorithm (12.13). The user specifies the transmitted signal, channel, and pulse shaping exactly as in part 1 of clockrecDD.m. An initial timing offset toffset is specified, and the algorithm in clockrecOP.m tries to find (the negative of) this value using only the received signal.

---

clockrecOP.m: clock recovery maximizing output power

```
tnow=1*m+1; tau=0; xs=zeros(1,n);            % initialize variables
tausave=zeros(1,n); tausave(1)=tau; i=0;
mu=0.05;                                      % algorithm stepsize
delta=0.1;                                    % time for derivative
while tnow<length(x)-1*m                      % run iteration
  i=i+1;
  xs(i)=interpsinc(x,tnow+tau,1);             % interpolated value at tnow+tau
  x_deltap=interpsinc(x,tnow+tau+delta,1);    % get value to the right
  x_deltam=interpsinc(x,tnow+tau-delta,1);    % get value to the left
  dx=x_deltap-x_deltam;                       % calculate numerical derivative
  tau=tau+mu*dx*xs(i);                        % alg update (energy)
  tnow=tnow+m; tausave(i)=tau;                % save for plotting
end
```

Typical output of the program is plotted in Figure 12.10. For this plot, the message was drawn from a 2-PAM binary signal, which is recovered nicely by the algorithm, as shown in the top plot. The bottom plot shows the trajectory of the offset estimation as it converges to the "unknown" value at -toffset.

The error surface for the output power maximization algorithm can be drawn using the same "experimental" method as was used in clockrecDDcost.m. Replacing the line that calculates the performance function with

```
cost(i)=sum(x.^2)/length(x);        % cost (energy)
```

calculates the error surface for the output power algorithm (12.13). Figure 12.11 shows this, along with three variants:

1. the average value of the absolute value of the output of the sampler avg$\{|x[k]|\}$,
2. the average of the fourth power of the output of the sampler avg$\{x^4[k]\}$, and
3. the average of the dispersion avg$\{(x^2[k] - 1)^2\}$.

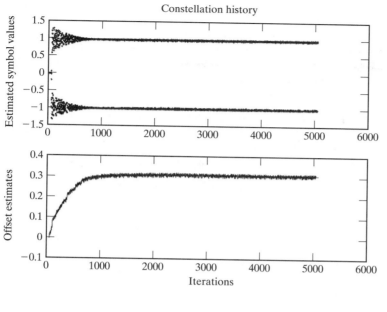

FIGURE 12.10 Output of the program `clockrecOP.m` shows the estimates of the symbols in the top plot and the trajectory of the offset estimates in the bottom.

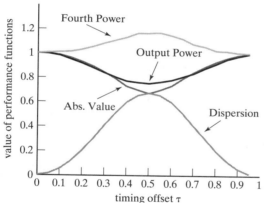

FIGURE 12.11 Four performance functions that can be used for timing recovery, plotted as a function of the timing offset τ. In this figure, the optimal answer is at $\tau = 0$. Some of the performance functions must be minimized and some must be maximized.

Clearly, some of these require maximization (the output power and the absolute value), while others require minimization (the fourth power and the dispersion). While they all behave more or less analogously in this easy setting (the figure shows the 2-PAM case with a SRRC pulse shape with `beta=0.5`), the maxima (or minima) may occur at different values of τ in more extreme settings.

Problems

12.9. Use the code in `clockrecOP.m` to "play with" the output power clock recovery algorithm. How does `mu` affect the convergence rate? What range of stepsizes works? How does the signal constellation of the input affect the convergent value of `tau` (try 4-PAM and 8-PAM)?

12.10. Implement a rectangular pulse shape. Does this work better or worse than the SRRC?

12.11. Add noise to the signal (add a zero mean noise to the received signal using the MATLAB `randn` function). How does this affect the convergence of the timing offset parameter `tau`. Does it change the final converged value?

12.12. Modify `clockrecOP.m` by setting `toffset=-1`. This starts the iteration in a closed eye situation. How many iterations does it take to open the eye? What is the convergent value? Try other values of `toffset`. Can you predict what the final convergent value will be? Try `toffset=-2.3`. Now let the oversampling factor be $m = 4$ and answer the same questions.

12.13. Redo Figure 12.11 using a sinc pulse shape. What happens to the Output Power performance function?

12.14. Redo Figure 12.11 using a T-wide Hamming pulse shape. Which of the four performance functions need to be minimized and which need to be maximized?

12.5 Two Examples

This section presents two examples in which timing recovery plays a significant role. The first looks at the behavior of the algorithms in the nonideal setting. When there is channel ISI, the answer to which the algorithms converge is not the same as in the ISI-free setting. This happens because the ISI of the channel causes an effective delay in the energy that the algorithm measures. The second example shows how the timing recovery algorithms can be used to estimate (slow) changes in the optimal sampling time. When these changes occur linearly, they are effectively a change in the underlying period, and the timing recovery algorithms can be used to estimate the offset of the period in the same way that the phase estimates of the PLL in Section 10.6 can be used to find a (small) frequency offset in the carrier.

Example 12.3

Modify the simulation in clockrecDD.m by changing the channel:

```
chan=[1 0.7 0 0 .5];                    % T/m "channel"
```

With an oversampling of `m=2`, 2-PAM constellation, and `beta=0.5`, the output of the output power maximization algorithm clockrecOP.m is shown in Figure 12.12. With these parameters, the iteration begins in a closed eye situation. Because of the channel, no single timing parameter can hope to achieve a perfect ±1 outcome. Nonetheless, by finding a good compromise position (in this case, converging to an offset of about 0.6), the hard decisions are correct once the eye has opened (which first occurs around iteration 500).

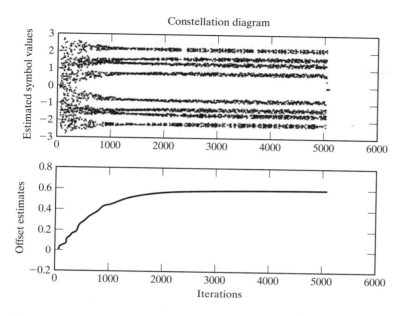

FIGURE 12.12 Output of the program `clockrecOP.m` as modified for Example 12.3 shows the constellation history in the top plot and the trajectory of the offset estimation in the bottom.

Example 12.3 shows that the presence of ISI changes the convergent value of the timing recovery algorithm. Why is this?

Suppose first that the channel was a pure delay. (For instance, set `chan=[0 1]` in Example 12.3). Then the timing algorithm will change the estimates `tau` (in this case, by one) to maximize the output power to account for the added delay. When the channel is more complicated, the timing recovery again moves the estimates to that position which maximizes the output power, but the actual value attained is a weighted version of all the taps. For example, with `chan=[1 1]`, the energy is maximized halfway between the two taps and the answer is offset by 0.5. Similarly, with `chan=[3 1]`, the energy is located a quarter of the way between the taps and the answer is offset by 0.25. In general, the offset is (roughly) proportional to the size of the taps and their delay.

To see the general situation, consider the received analog signal due to a single symbol triggering the pulse shape filter and passing through a channel with ISI. An adjustment in the baud-timing setting at the receiver will sample at slightly different points on the received analog signal. A change in τ is effectively equivalent to a change in the channel ISI. This will be dealt with in Chapter 13 when designing equalizers.

Example 12.4

With the signal generated as in `clockrecDD.m` on page 233, the following code resamples (using sinc interpolation) the received signal to simulate a change in the underlying period by a factor of `fac`.

FIGURE 12.13 Output
of the program
`clockrecpe-`
`riod.m` as modified
for Example 12.4
shows the constel-
lation history in the
top plot and the tra-
jectory of the offset
estimation in the bot-
tom. The slope of the
estimates is propor-
tional to the difference
between the nominal
and the actual clock
period.

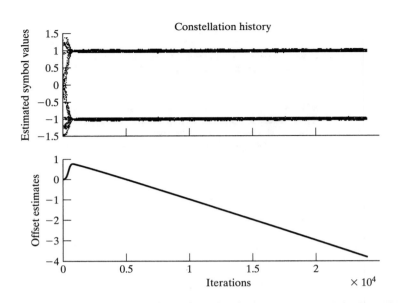

Example 12.4   Continued

clockrecperiod.m: resample to change the period

```
fac=1.0001; z=zeros(size(x));     % percent change in period
t=1+1:fac:length(x)-2*l;          % vector of new times
for i=1:length(t)                 % resample x at new rate
  z(i)=interpsinc(x,t(i),l);      % to create received signal
end                               % with period offset
x=z;                              % relabel signal
```

If this code is followed by one of the timing recovery schemes, then the timing
parameter τ follows the changing period. For instance, in Figure 12.13, the timing
estimation converges rapidly to a "line" with slope that is proportional to the difference
in period between the assumed value of the period at the receiver and the actual value
used at the transmitter.

Thus, the standard timing recovery algorithms can handle the case in which the clock
periods at the transmitter and receiver are somewhat different. More accurate estimates
could be made using two timing recovery algorithms analogous to the dual-carrier recovery structure of Section 10.6.2 or by mimicking the second-order filter structure of the
PLL in the article *Analysis of the Phase Locked Loop*, which can be found on the CD.
There are also other common timing recovery algorithms such as the early–late method,
the method of Mueller and Müller, and band-edge timing algorithms.

Problems

12.15. Modify `clockrecOP.m` to implement one of the alternative performance functions of Figure 12.11: $\text{avg}\{|x[k]|\}$, $\text{avg}\{x^2[k]\}$, or $\text{avg}\{(x^2[k] - 1)^2\}$.

12.16. Modify `clockrecOP.m` by changing the channel as in Example 12.3. Use different values of `beta` in the SRRC pulse shape routine. How does this affect the convergence speed of the algorithm? Do different pulse shapes change the convergent value?

12.17. Investigate how the error surface depends on the input signal.
 (a) Draw the error surface for the output energy maximization timing recovery algorithm when the inputs are binary ± 1.
 (b) Draw the error surface when the inputs are drawn from the 4-PAM constellation, for the case in which the symbol -3 never occurs.

12.18. Imitate Example 12.3 using a channel of your own choosing. Do you expect that the eye will always be able to open?

12.19. Instead of the ISI channel used in Example 12.3, include a white noise channel. How does this change the timing estimates?

12.20. Explore the limits of the period tracking in Example 12.4. How large can `fac` be made and still have the estimates converge to a line? What happens to the cluster variance when the estimates cannot keep up? Does it help to increase the size of the stepsize `mu`?

12.6 For Further Reading

A comprehensive collection of timing and carrier recovery schemes can be found in the following two texts:

- H. Meyr, M. Moeneclaey, and S. A. Fechtel, *Digital Communication Receivers,* Wiley, 1998.

- J. A. C. Bingham, *The Theory and Practice of Modem Design*, Wiley Interscience, 1988.

13 Linear Equalization

The revolution in data communications technology can be dated from the invention of automatic and adaptive channel equalization in the late 1960s.
—Gitlin, Hayes, and Weinstein, *Data Communication Principles*, 1992

When all is well in the receiver, there is no interaction between successive symbols; each symbol arrives and is decoded independently of all others. But when symbols interact, when the waveform of one symbol corrupts the value of a nearby symbol, then the received signal becomes distorted. It is difficult to decipher the message from such a received signal. This impairment is called "intersymbol interference" and was discussed in Chapter 11 in terms of non-Nyquist pulse shapes overlapping in time. This chapter considers another source of interference between symbols that is caused by multipath reflections (or frequency-selective dispersion) in the channel.

When there is no intersymbol interference (from a multipath channel, from imperfect pulse shaping, or from imperfect timing), the impulse response of the system from the source to the recovered message has a single nonzero term. The amplitude of this single "spike" depends on the transmission losses, and the delay is determined by the transmission time. When there is intersymbol interference caused by a multipath channel, this single spike is "scattered," duplicated once for each path in the channel. The number of nonzero terms in the impulse response increases. The channel can be modeled as a finite-impulse-response, linear filter C, and the *delay spread* is the total time interval during which reflections with significant energy arrive. The idea of the equalizer is to build (another) filter in the receiver that counteracts the effect of the channel. In essence, the equalizer must "unscatter" the impulse response. This can be stated as the goal of designing the equalizer E so that the impulse response of the combined channel and equalizer CE has a single spike. This can be cast as an optimization problem, and can be solved using techniques familiar from Chapters 6, 10, and 12.

The transmission path may also be corrupted by additive interferences such as those caused by other users. These noise components are usually presumed to be uncorrelated with the source sequence and they may be broadband or narrowband, in band or out of band relative to the bandlimited spectrum of the source signal. Like the multipath channel interference, they cannot be known to the system designer in advance. The second job of the equalizer is to reject such additive narrowband interferers by designing appropriate linear notch filters "on-the-fly" based on the received signal. At the same time, it is important that the equalizer not unduly enhance the broadband noise.

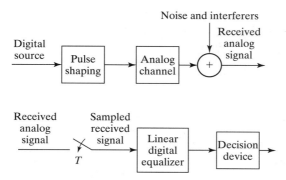

FIGURE 13.1 The base-band linear (digital) equalizer is intended to (automatically) cancel unwanted effects of the channel and to cancel certain kinds of additive interferences.

The signal path of a baseband digital communication system is shown in Figure 13.1, which emphasizes the role of the equalizer in trying to counteract the effects of the multipath channel and the additive interference. As in previous chapters, all of the inner parts of the system are assumed to operate precisely: thus, the upconversion and down-conversion, the timing recovery, and the carrier synchronization (all those parts of the receiver that are *not* shown in Figure 13.1) are assumed to be flawless and unchanging. Modelling the channel as a time-invariant FIR filter, the next section focuses on the task of selecting the coefficients in the block labelled "linear digital equalizer," with the goal of removing the intersymbol interference and attenuating the additive interferences. These coefficients are to be chosen based on the sampled received signal sequence and (possibly) knowledge of a prearranged "training sequence." While the channel may actually be time varying, the variations are often much slower than the data rate, and the channel can be viewed as (effectively) time invariant over small time scales.

This chapter suggests several different ways that the coefficients of the equalizer can be chosen. The first procedure, in Section 13.2.1, minimizes the square of the symbol recovery error[1] over a block of data, which can be done using a matrix pseudoinversion. Minimizing the (square of the) error between the received data values and the transmitted values can also be achieved using an adaptive element, as detailed in Section 13.3. When there is no training sequence, other performance functions are appropriate, and these lead to equalizers such as the decision-directed approach in Section 13.4 and the dispersion minimization method in Section 13.5. The adaptive methods considered here are only modestly complex to implement, and they can potentially track time variations in the channel model, assuming the changes are sufficiently slow.

13.1 Multipath Interference

The villains of this chapter are multipath and other additive interferers. Both should be familiar from Section 4.1.

The distortion caused by an analog wireless channel can be thought of as a combination of scaled and delayed reflections of the original transmitted signal. These reflections occur when there are different paths from the transmitting antenna to the receiving antenna. Between two microwave towers, for instance, the paths may include one along the line-of-sight, reflections from nearby hills, and bounces from a field or lake between the towers.

---

[1] This is the error between the equalizer output and the transmitted symbol, and is known whenever there is a training sequence.

For indoor digital TV reception, there are many (local) time-varying reflectors, including people in the receiving room, and nearby vehicles. The strength of the reflections depends on the physical properties of the reflecting objects, while the delay of the reflections is primarily determined by the length of the transmission path. Let $u(t)$ be the transmitted signal. If N delays are represented by Δ_1, Δ_2, ..., Δ_N, and the strength of the reflections is α_1, α_2, ..., α_N, then the received signal is

$$y(t) = \alpha_1 u(t - \Delta_1) + \alpha_2 u(t - \Delta_2) + \ldots + \alpha_N u(t - \Delta_N) + \eta(t), \qquad (13.1)$$

where $\eta(t)$ represents additive interferences. This model of the transmission channel has the form of a finite impulse response filter, and the total length of time $\Delta_N - \Delta_1$ over which the impulse response is nonzero is called the *delay spread* of the physical medium.

This transmission channel is typically modelled digitally assuming a fixed sampling period T_s. Thus, (13.1) is approximated by

$$y(kT_s) = a_1 u(kT_s) + a_2 u((k-1)T_s) + \ldots + a_n u((k-n)T_s) + \eta(kT_s). \qquad (13.2)$$

In order for the model (13.2) to closely represent the system (13.1), the total time over which the impulse response is nonzero (the time nT_s) must be at least as large as the maximum delay Δ_N. Since the delay is not a function of the symbol period T_s, smaller T_s require more terms in the filter (i.e., larger n).

For example, consider a sampling interval of $T_s \approx 40$ nanoseconds (i.e., a transmission rate of 25 MHz). A delay spread of approximately 4 microseconds would correspond to one hundred taps in the model (13.2). Thus, at any time instant, the received signal would be a combination of (up to) one hundred data values. If T_s were increased to 0.4 microsecond (i.e., 2.5 MHz), only 10 terms would be needed, and there would be interference with only the 10 nearest data values. If T_s were larger than 4 microseconds (i.e., 0.25 MHz), only one term would be needed in the discrete-time impulse response. In this case, adjacent sampled symbols would not interfere. Such finite duration impulse response models as (13.2) can also be used to represent the frequency-selective dynamics that occur in the wired local end-loop in telephony, and other (approximately) linear, finite-delay-spread channels.

The design objective of the equalizer is to undo the effects of the channel and to remove the interference. Conceptually, the equalizer attempts to build a system that is a "delayed inverse" of (13.2), removing the intersymbol interference while simultaneously rejecting additive interferers uncorrelated with the source. If the interference $\eta(kT_s)$ is unstructured (for instance white noise) then there is little that a linear equalizer can do to remove it. But when the interference is highly structured (such as narrowband interference from another user), then the linear filter can often notch out the offending frequencies.

As shown in Example 12.3 of Section 12.5, the solution for the optimal sampling times found by the clock recovery algorithms depend on the ISI in the channel. Consequently, the digital model (such as (13.2)) formed by sampling an analog transmission path (such as (13.1)) depends on when the samples are taken within each period T_s. To see how this can happen in a simple case, consider a two-path transmission channel

$$\delta(t) + 0.6\delta(t - \Delta),$$

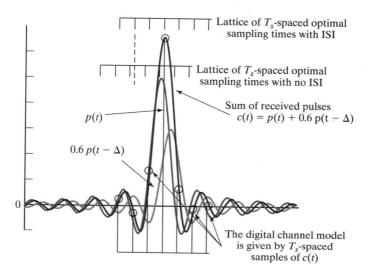

Lattice of T_s-spaced optimal sampling times with ISI

Lattice of T_s-spaced optimal sampling times with no ISI

Sum of received pulses
$c(t) = p(t) + 0.6\, p(t - \Delta)$

$p(t)$

$0.6\, p(t - \Delta)$

0

The digital channel model is given by T_s-spaced samples of $c(t)$

FIGURE 13.2 The optimum sampling times (as found by the energy maximization algorithm) differ when there is ISI in the transmission path, and change the effective digital model of the channel.

where Δ is some fraction of T_s. For each transmitted symbol, the received signal will contain two copies of the pulse shape $p(t)$, the first undelayed and the second delayed by Δ and attenuated by a factor of 0.6. Thus, the receiver sees

$$c(t) = p(t) + 0.6p(t - \Delta).$$

This is shown in Figure 13.2 for $\Delta = 0.7T_s$. The clock recovery algorithms cannot separate the individual copies of the pulse shapes. Rather, they react to the complete received shape, which is their sum. The power maximization will locate the sampling times at the peak of this curve, and the lattice of sampling times will be different from what would be expected without ISI. The effective (digital) channel model is thus a sampled version of $c(t)$. This is depicted in Figure 13.2 by the small circles that occur at T_s spaced intervals.

   In general, an accurate digital model for a channel depends on many things: the underlying analog channel, the pulse shaping used, and the timing of the sampling process. At first glance, this seems like it might make designing an equalizer for such a channel almost impossible. But there is good news. No matter what timing instants are chosen, no matter what pulse shape is used, and no matter what the underlying analog channel may be (as long as it is linear), there is a FIR linear representation of the form (13.2) that closely models its behavior. The details may change, but it is always a sampling of the smooth curve (like $c(t)$ in Figure 13.2) that defines the digital model of the channel. As long as the digital model of this channel does not have deep nulls (i.e., a frequency response that practically zeroes out some important band of frequencies), there is a good chance that the equalizer can undo the effects of the channel.

13.2 Trained Least-Squares Linear Equalization

When there is a training sequence available (for instance, in the known frame information that is used in synchronization), this can also be used to help build or "train" an equalizer.

The basic strategy is to find a suitable function of the unknown equalizer parameters that can be used to define an optimization problem. Then, applying the techniques of Chapters 6, 10, and 12, the optimization problem can be solved in a variety of ways.

13.2.1 A Matrix Description

The linear equalization problem is depicted in Figure 13.3. A prearranged training sequence $s[k]$ is assumed known at the receiver. The goal is to find an FIR filter (called the *equalizer*) so that the output of the equalizer is approximately equal to the known source, though possibly delayed in time. Thus, the goal is to choose the impulse response f_i so that $y[k] \approx s[k - \delta]$ for some specific δ.

The input–output behavior of the FIR linear equalizer can be described as the convolution

$$y[k] = \sum_{j=0}^{n} f_j r[k - j], \tag{13.3}$$

where the lower index on j can be no lower than zero (or else the equalizer is noncausal; that is, it can illogically respond to an input before the input is applied). This convolution is illustrated in Figure 13.4 as a "direct form FIR" or "tapped delay line."

The summation in (13.3) can also be written (e.g., for $k = n + 1$) as the inner product of two vectors

$$y[n + 1] = [r[n + 1], \ r[n], \dots, \ r[1]] \begin{bmatrix} f_0 \\ f_1 \\ \vdots \\ f_n \end{bmatrix}. \tag{13.4}$$

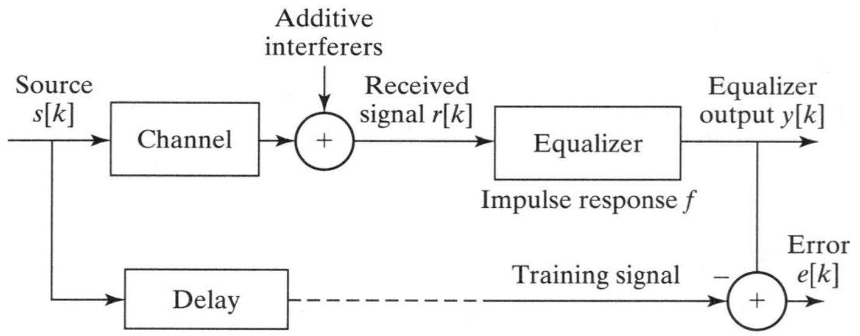

FIGURE 13.3 The problem of linear equalization is to find a linear system f that undoes the effects of the channel while minimizing the effects of the interferences.

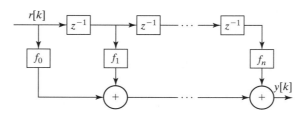

FIGURE 13.4 Direct form FIR as tapped delay line.

Note that $y[n + 1]$ is the earliest output that can be formed given no knowledge of $r[i]$ for $i < 1$. Incrementing the time index in (13.4) gives

$$y[n + 2] = [r[n + 2], \ r[n + 1], \ldots, \ r[2]] \begin{bmatrix} f_0 \\ f_1 \\ \vdots \\ f_n \end{bmatrix}$$

and

$$y[n + 3] = [r[n + 3], \ r[n + 2], \ldots, \ r[3]] \begin{bmatrix} f_0 \\ f_1 \\ \vdots \\ f_n \end{bmatrix}.$$

Observe that each of these uses the same equalizer parameter vector. Concatenating $p - n$ of these measurements into one matrix equation over the available data set for $i = 1$ to p gives

$$\begin{bmatrix} y[n + 1] \\ y[n + 2] \\ y[n + 3] \\ \vdots \\ y[p] \end{bmatrix} = \begin{bmatrix} r[n + 1] & r[n] & \cdots & r[1] \\ r[n + 2] & r[n + 1] & \cdots & r[2] \\ r[n + 3] & r[n + 2] & \cdots & r[3] \\ \vdots & \vdots & & \vdots \\ r[p] & r[p - 1] & \cdots & r[p - n] \end{bmatrix} \begin{bmatrix} f_0 \\ f_1 \\ \vdots \\ f_n \end{bmatrix}, \qquad (13.5)$$

or, with the appropriate matrix definitions,

$$Y = RF. \qquad (13.6)$$

Note that R has a special structure, that the entries along each diagonal are the same. R is known as a *Toeplitz* matrix and the `toeplitz` command in MATLAB makes it easy to build matrices with this structure.

13.2.2 Source Recovery Error

The delayed source recovery error is

$$e[k] = s[k - \delta] - y[k] \qquad (13.7)$$

for a particular δ. This section shows how the source recovery error can be used to define a performance function that depends on the unknown parameters f_i. Calculating the parameters that minimize this performance function provides a good solution to the equalization problem.

Define

$$S = \begin{bmatrix} s[n + 1 - \delta] \\ s[n + 2 - \delta] \\ s[n + 3 - \delta] \\ \vdots \\ s[p - \delta] \end{bmatrix} \qquad (13.8)$$

and

$$E = \begin{bmatrix} e[n+1] \\ e[n+2] \\ e[n+3] \\ \vdots \\ e[p] \end{bmatrix}.$$
(13.9)

Using (13.6), write

$$E = S - Y = S - RF.$$
(13.10)

As a measure of the performance of the f_i in F, consider

$$J_{LS} = \sum_{i=n+1}^{p} e^2[i].$$
(13.11)

J_{LS} is nonnegative since it is a sum of squares. Minimizing such a summed squared delayed source recovery error is a common objective in equalizer design, since the f_i that minimize J_{LS} cause the output of the equalizer to become close to the values of the (delayed) source.

Given (13.9) and (13.10), J_{LS} in (13.11) can be written as

$$J_{LS} = E^T E = (S - RF)^T (S - RF)$$
$$= S^T S - (RF)^T S - S^T RF + (RF)^T RF.$$
(13.12)

Because J_{LS} is a scalar, $(RF)^T S$ and $S^T RF$ are also scalars. Since the transpose of a scalar is equal to itself, $(RF)^T S = S^T RF$, and (13.12) can be rewritten as

$$J_{LS} = S^T S - 2S^T RF + (RF)^T RF.$$
(13.13)

The issue is now one of choosing the $n+1$ entries of F to make J_{LS} as small as possible.

13.2.3 The Least-Squares Solution

Define the matrix

$$\Psi = [F - (R^T R)^{-1} R^T S]^T (R^T R)[F - (R^T R)^{-1} R^T S]$$
$$= F^T (R^T R)F - S^T RF - F^T R^T S + S^T R(R^T R)^{-1} R^T S.$$

The purpose of this definition is to rewrite (13.13) in terms of Ψ:

$$J_{LS} = \Psi + S^T S - S^T R(R^T R)^{-1} R^T S$$
$$= \Psi + S^T [I - R(R^T R)^{-1} R^T] S.$$
(13.14)

Since $S^T [I - R(R^T R)^{-1} R^T] S$ is not a function of F, the minimum of J_{LS} occurs at the F that minimizes Ψ. This occurs when

$$F^\dagger = (R^T R)^{-1} R^T S,$$
(13.15)

assuming that $(R^T R)^{-1}$ exists.[2] The corresponding minimum achievable by J_{LS} at $F = F^\dagger$ is the summed squared delayed source recovery error. This is the remaining term in (13.14); that is,

$$J_{LS}^{\min} = S^T[I - R(R^T R)^{-1} R^T]S. \qquad (13.16)$$

The formulas for the optimum F in (13.15) and the associated minimum achievable J_{LS} in (13.16) are for a specific δ. To complete the design task, it is also necessary to find the optimal delay δ. The most straightforward approach is to set up a series of $S = RF$ calculations, one for each possible δ, to compute the associated values of $J_{LS}^{\min}$, and pick the delay associated with the smallest one.

This procedure is straightforward to implement in MATLAB, and the program LSe-qualizer.m allows you to play with the various parameters to get a feel for their effect. Much of this program will be familiar from openclosed.m. The first three lines define a channel, create a binary source, and then transmit the source through the channel using the filter command. At the receiver, the data are put through a quantizer, and then the error is calculated for a range of delays. The new part is in the middle.

LSequalizer.m find a LS equalizer f for the channel b

```
b=[0.5 1 -0.6];                      % define channel
m=1000; s=sign(randn(1,m));          % binary source of length m
r=filter(b,1,s);                     % output of channel
n=3;                                 % length of equalizer - 1
delta=3;                             % use delay <=n
p=length(r)-delta;
R=toeplitz(r(n+1:p),r(n+1:-1:1));    % build matrix R
S=s(n+1-delta:p-delta)';             % and vector S
f=inv(R'*R)*R'*S                     % calculate equalizer f
Jmin=S'*S-S'*R*inv(R'*R)*R'*S        % Jmin for this f and delta
y=filter(f,1,r);                     % equalizer is a filter
dec=sign(y);                         % quantize and find errors
err=0.5*sum(abs(dec(delta+1:end)-s(1:end-delta)))
```

The variable n defines the length of the equalizer, and delta defines the delay that will be used in constructing the vector S defined in (13.8) (observe that delta must be positive and less than or equal to n). The Toeplitz matrix R is defined in (13.5) and (13.6), and the equalizer coefficients f are computed as in (13.15). The value of minimum achievable performance is Jmin, which is calculated as in (13.16). To demonstrate the effect of the equalizer, the received signal r is filtered by the equalizer coefficients, and the output is then quantized. If the equalizer has done its job (i.e., if the eye is open), then there should be some shift sh at which no errors occur.

---

[2] A matrix is invertible as long as it has no eigenvalues equal to zero. Since $R^T R$ is a quadratic form it has no negative eigenvalues. Thus, all eigenvalues must be positive in order for it to be invertible.

For example, using the default channel b=[0.5 1 -0.6], and length 4 equalizer (n=3), four values of the delay delta give

delay delta	Jmin	equalizer f
0	832	{0.33, 0.027, 0.070, 0.01}
1	134	{0.66, 0.36, 0.16, 0.08}
2	30	{−0.28, 0.65, 0.30, 0.14}
3	45	{0.1, −0.27, 0.64, 0.3}

$$(13.17)$$

The best equalizer is the one corresponding to a delay of 2, since this Jmin is the smallest. In this case, however, any of the last three open the eye. Observe that the number of errors (as reported in err) is zero when the eye is open.

Problems

13.1. Plot the frequency response (using freqz) of the channel b in LSequal-izer.m. Plot the frequency response of each of the four equalizers found by the program. For each channel/equalizer pair, form the product of the magnitude of the frequency responses. How close are these products to unity?

13.2. Add (uncorrelated, normally distributed) noise into the simulation using the command r=filter(b,1,s)+sd*randn(size(s)).
   (a) For the equalizer with delay 2, what is the largest sd you can add and still have no errors?
   (b) Make a plot of Jmin as a function of sd.
   (c) Now try the equalizer with delay 1. What is the largest sd you can add, and still have no errors?
   (d) Which is a better equalizer?

13.3. Use LSequalizer.m to find an equalizer that can open the eye for the channel b=[1 1 -0.8 -.3 1 1].
   (a) What equalizer length n is needed?
   (b) What delays delta give zero error at the output of the quantizer?
   (c) What is the corresponding Jmin?
   (d) Plot the frequency response of this channel.
   (e) Plot the frequency response of your equalizer.
   (f) Calculate and plot the product of the two.

13.4. Modify LSequalizer.m to generate a source sequence from the alphabet ±1, ±3. For the default channel [0.5 1 -0.6], find an equalizer that opens the eye.
   (a) What equalizer length n is needed?
   (b) What delays delta give zero error at the output of the quantizer?
   (c) What is the corresponding Jmin?
   (d) Is this a fundamentally easier or more difficult task than when equalizing a binary source?
   (e) Plot the frequency response of the channel and of the equalizer.

There is a way to convert the exhaustive search over all the delays δ in the previous approach into a single matrix operation. Construct the $(p - \alpha) \times (\alpha + 1)$ matrix of training data

$$\bar{S} = \begin{bmatrix} s[\alpha + 1] & s[\alpha] & \cdots & s[1] \\ s[\alpha + 2] & s[\alpha + 1] & \cdots & s[2] \\ \vdots & \vdots & & \vdots \\ s[p] & s[p - 1] & \cdots & s[p - \alpha] \end{bmatrix}, \tag{13.18}$$

where α specifies the number of delays δ that will be searched, from $\delta = 0$ to $\delta = \alpha$. The $(p - \alpha) \times (n + 1)$ matrix of received data is

$$\bar{R} = \begin{bmatrix} r[\alpha + 1] & r[\alpha] & \cdots & r[\alpha - n + 1] \\ r[\alpha + 2] & r[\alpha + 1] & \cdots & r[\alpha - n + 2] \\ \vdots & \vdots & & \vdots \\ r[p] & r[p - 1] & \cdots & r[p - n] \end{bmatrix}, \tag{13.19}$$

where each column corresponds to one of the possible delays. Note that $\alpha > n$ is required to keep the lowest index of $r[\cdot]$ positive. In the $(n + 1) \times (\alpha + 1)$ matrix

$$\bar{F} = \begin{bmatrix} f_{00} & f_{01} & \cdots & f_{0\alpha} \\ f_{10} & f_{11} & \cdots & f_{1\alpha} \\ \vdots & \vdots & & \vdots \\ f_{n0} & f_{n1} & \cdots & f_{n\alpha} \end{bmatrix},$$

each column is a set of equalizer parameters, one corresponding to each of the possible delays. The strategy is to use $\bar{S}$ and $\bar{R}$ to find $\bar{F}$. The column of $\bar{F}$ that results in the smallest value of the cost J_{LS} is then the optimal receiver at the optimal delay.

The jth column of $\bar{F}$, corresponds to the equalizer parameter vector choice for $\delta = j - 1$. The product of $\bar{R}$ with this jth column from $\bar{F}$ is intended to approximate the jth column of $\bar{S}$. The least-squares solution of $\bar{S} \approx \bar{R}\bar{F}$ is

$$\bar{F}^\dagger = (\bar{R}^T \bar{R})^{-1} \bar{R}^T \bar{S}, \tag{13.20}$$

where the number of columns of $\bar{R}$ (i.e., $n + 1$) must be less than or equal to the number of rows of $\bar{R}$ (i.e., $p - \alpha$) for $(\bar{R}^T \bar{R})^{-1}$ to exist. Consequently, $p - \alpha \geq n + 1$ implies that $p > n + \alpha$. If so, the minimum value associated with a particular column of $\bar{F}^\dagger$ (e.g., $\bar{F}_\ell^\dagger$) is, from (13.16),

$$J_{LS}^{\min, \ell} = \bar{S}_\ell^T [I - \bar{R}(\bar{R}^T \bar{R})^{-1} \bar{R}^T] \bar{S}_\ell, \tag{13.21}$$

where $\bar{S}_\ell$ is the ℓth column of $\bar{S}$. Thus, the set of these $J_{LS}^{\min, \ell}$ are all along the diagonal of

$$\Phi = \bar{S}^T [I - \bar{R}(\bar{R}^T \bar{R})^{-1} \bar{R}^T] \bar{S}. \tag{13.22}$$

Hence, the minimum value on the diagonal of Φ (e.g., at the (j, j)th entry) corresponds to the optimum δ.

Example 13.1   A Low-Order Example

Consider the low-order example with $n = 1$ (so F has two parameters), $\alpha = 2$ (so $\alpha > n$), and $p = 5$ (so $p > n + \alpha$). Thus,

$$\bar{S} = \begin{bmatrix} s[3] & s[2] & s[1] \\ s[4] & s[3] & s[2] \\ s[5] & s[4] & s[3] \end{bmatrix},$$

$$\bar{R} = \begin{bmatrix} r[3] & r[2] \\ r[4] & r[3] \\ r[5] & r[4] \end{bmatrix},$$

and

$$\bar{F} = \begin{bmatrix} f_{00} & f_{01} & f_{02} \\ f_{10} & f_{11} & f_{12} \end{bmatrix}.$$

For the example, assume that the true channel is

$$r[k] = ar[k-1] + bs[k-1].$$

A two-tap equalizer $F = [f_0 \ \ f_1]^T$ can provide perfect equalization for $\delta = 1$ with $f_0 = 1/b$, $f_1 = -a/b$, since

$$y[k] = f_0 r[k] + f_1 r[k-1] = \frac{1}{b}[r[k] - ar[k-1]]$$

$$= \frac{1}{b}[ar[k-1] + bs[k-1] - ar[k-1]] = s[k-1].$$

Consider

$$\{s[1], \ s[2], \ s[3], \ [s4], \ s[5]\} = \{1, \ -1, \ -1, \ 1, \ -1\},$$

which results in

$$\bar{S} = \begin{bmatrix} -1 & -1 & 1 \\ 1 & -1 & -1 \\ -1 & 1 & -1 \end{bmatrix}.$$

With $a = 0.6$, $b = 1$, and $r[1] = 0.8$,

$$r[2] = ar[1] + bs[1] = 0.48 + 1 = 1.48,$$

$$r[3] = 0.888 - 1 = -0.112,$$

$$r[4] = -1.0672,$$

and

$$r[5] = 0.3597.$$

Example 13.1   A Low-Order Example (Continued)

The effect of channel noise will be simulated by rounding these values for r in composing

$$\bar{R} = \begin{bmatrix} -0.1 & 1.5 \\ -1.1 & -0.1 \\ 0.4 & -1.1 \end{bmatrix}.$$

Thus, from (13.22),

$$\Phi = \begin{bmatrix} 1.2848 & 0.0425 & 0.4778 \\ 0.0425 & 0.0014 & 0.0158 \\ 0.4778 & 0.0158 & 0.1777 \end{bmatrix},$$

and from (13.20),

$$\bar{F}^\dagger = \begin{bmatrix} -1.1184 & 0.9548 & 0.7411 \\ -0.2988 & -0.5884 & 0.8806 \end{bmatrix}.$$

Since the second diagonal term in Φ is the smallest diagonal term, $\delta = 1$ is the optimum setting (as expected) and the second column of $\bar{F}^\dagger$ is the minimum summed squared delayed recovery error solution (i.e., $f_0 = 0.9548$ ($\approx 1/b = 1$) and $f_1 = -0.5884$ ($\approx -a/b = -0.6$)).

With a "better" received signal measurement, for instance,

$$\bar{R} = \begin{bmatrix} -0.11 & 1.48 \\ -1.07 & -0.11 \\ 0.36 & -1.07 \end{bmatrix},$$

the diagonal of Φ is [1.3572, 0.0000, 0.1657] and the optimum delay is again $\delta = 1$, and the optimum equalizer settings are 0.9960 and −0.6009, which is a better fit to the ideal noise-free answer. Infinite precision in $\bar{R}$ (measured without channel noise or other interferers) produces a perfect fit to the "true" f_0 and f_1 and a zeroed delayed source recovery error.

13.2.4 Summary of Least-squares Equalizer Design

The steps of the linear FIR equalizer design strategy are as follows:

1. Select the order n for the FIR equalizer in (13.3).
2. Select maximum of candidate delays α ($> n$) used in (13.18) and (13.19).
3. Utilize set of p training signal samples $\{s[1], \ s[2], \ldots, s[p]\}$ with $p > n + \alpha$.
4. Obtain corresponding set of p received signal samples $\{r[1], \ r[2], \ldots, r[p]\}$.
5. Compose $\bar{S}$ in (13.18).
6. Compose $\bar{R}$ in (13.19).

7. Check if $\bar{R}^T \bar{R}$ has poor conditioning induced by any (near) zero eigenvalues. MATLAB will return a warning (or an error) if the matrix is too close to singular.[3]
8. Compute $\bar{F}^\dagger$ from (13.20).
9. Compute Φ by substituting $\bar{F}^\dagger$ into (13.22), rewritten as

$$\Phi = \bar{S}^T [\bar{S} - \bar{R}\bar{F}^\dagger].$$

10. Find the minimum value on the diagonal of Φ. This index is $\delta + 1$. The associated diagonal element of Φ is the minimum achievable summed squared delayed source recovery error $\sum_{i=\alpha+1}^{p} e^2[i]$ over the available data record.
11. Extract the $(\delta + 1)$th column of the previously computed $\bar{F}^\dagger$. This is the impulse response of the optimum equalizer.
12. Test the design. Test it on synthetic data, and then on measured data (if available). If inadequate, repeat design, perhaps increasing n or twiddling some other designer-selected quantity.

This procedure, along with three others that will be discussed in the ensuing sections, is available on the CD in the program `dae.m`. Combining the various approaches makes it easier to compare their behaviors in the examples of Section 13.6.

13.2.5 Complex Signals and Parameters

The preceding development assumes that the source signal and channel, and therefore the received signal, equalizer, and equalizer output, are all real valued. However, the source signal and channel may be modeled as complex valued when using modulations such as QAM of Section 5.3. This is explored in some detail in the document *A Digital Quadrature Amplitude Modulation Radio*, which can be found on the CD. The same basic strategy for equalizer design can also be used in the complex case.

Consider a complex delayed source recovery error

$$e[k] = e_R[k] + je_I[k],$$

where $j = \sqrt{-1}$. Consider its square,

$$e^2[k] = e_R^2[k] + 2je_R[k]e_I[k] - e_I^2[k],$$

which is typically complex valued, and potentially real valued and negative when $e_R \approx 0$. Thus, a sum of e^2 is no longer a suitable measure of performance since $|e|$ might be nonzero but its squared average might be zero.

Instead, consider the product of a complex e with its complex conjugate $e^* = e_R - je_I$; that is,

$$e[k]e^*[k] = e_R^2[k] - je_R[k]e_I[k] + je_R[k]e_I[k] - j^2e_I^2[k]$$
$$= e_R^2[k] + e_I^2[k].$$

---

[3] The condition number (= maximum eigenvalue / minimum eigenvalue) of $\bar{R}^T \bar{R}$ should be checked. If the condition number is extremely large, start over with different $\{s[\cdot]\}$. If all choices of $\{s[\cdot]\}$ result in poorly conditioned $\bar{R}^T \bar{R}$, then most likely the channel has deep nulls that prohibit the successful application of a T-spaced linear equalizer.

In vector form, the summed squared error of interest is $E^H E$ (rather than the $E^T E$ of (13.12)), where the superscript H denotes the operations of both transposition and complex conjugation. Thus, (13.15) becomes

$$F^\dagger = (R^H R)^{-1} R^H S.$$

Note that in implementing this refinement in the MATLAB code, the symbol pair `.'` implements a transpose, while `'` alone implements a conjugate transpose.

13.2.6 Fractionally Spaced Equalization

The preceding development assumes that the sampled input to the equalizer is symbol spaced with the sampling interval equal to the symbol interval of T seconds. Thus, the unit delay in realizing the tapped-delay-line equalizer is T seconds. Sometimes, the input to the equalizer is oversampled such that the sample interval is shorter than the symbol interval, and the resulting equalizer is said to be fractionally spaced. The same kinds of algorithms and solutions can be used to calculate the coefficients in fractionally spaced equalizers as are used for T-spaced equalizers. Of course, details of the construction of the matrices corresponding to $\bar{S}$ and $\bar{R}$ will necessarily differ due to the structural differences. The more rapid sampling allows greater latitude in the ordering of the blocks in the receiver. This is discussed at length in *Equalization* on the CD.

13.3 An Adaptive Approach to Trained Equalization

The block oriented design of the previous section requires substantial computation even when the system delay is known since it requires calculating the inverse of an $(n + 1) \times (n + 1)$ matrix, where n is the largest delay in the FIR linear equalizer. This section considers using an adaptive element to minimize the average of the squared error,

$$J_{LMS} = \frac{1}{2}\text{avg}\{e^2[k]\}.$$

Observe that J_{LMS} is a function of all the equalizer coefficients f_i, since

$$e[k] = s[k - \delta] - y[k] = s[k - \delta] - \sum_{j=0}^{n} f_j r[k - j], \tag{13.23}$$

which combines (13.7) with (13.3), and where $r[k]$ is the received signal at baseband after sampling. An algorithm for the minimization of J_{LMS} with respect to the ith equalizer coefficient f_i is

$$f_i[k + 1] = f_i[k] - \mu \left. \frac{dJ_{LMS}}{df_i} \right|_{f_i = f_i[k]}. \tag{13.24}$$

To create an algorithm that can easily be implemented, it is necessary to evaluate this derivative with respect to the parameter of interest. This is

$$\frac{dJ_{LMS}}{df_i} = \frac{d\text{avg}\{\frac{1}{2}e^2[k]\}}{df_i} \approx \text{avg}\left\{\frac{\frac{1}{2}de^2[k]}{df_i}\right\} = \text{avg}\left\{e[k]\frac{de[k]}{df_i}\right\}, \tag{13.25}$$

where the approximation follows from (G.13) and the final equality from the chain rule (A.59). Using (13.23), the derivative of the source recovery error $e[k]$ with respect to the ith equalizer parameter f_i is

$$\frac{de[k]}{df_i} = \frac{ds[k-\delta]}{df_i} - \sum_{j=0}^{n} \frac{df_j r[k-j]}{df_i} = -r[k-i], \qquad (13.26)$$

since $\frac{ds[k-\delta]}{df_i} = 0$ and $\frac{df_j r[k-j]}{df_i} = 0$ for all $i \neq j$. Substituting (13.26) into (13.25) and then into (13.24), the update for the adaptive element is

$$f_i[k+1] = f_i[k] + \mu \text{avg}\{e[k]r[k-i]\}.$$

Typically, the averaging operation is suppressed since the iteration with small stepsize μ itself has a lowpass (averaging) behavior. The result is commonly called the least mean squares (LMS) algorithm for direct linear equalizer impulse response coefficient adaptation:

$$f_i[k+1] = f_i[k] + \mu e[k]r[k-i]. \qquad (13.27)$$

This adaptive equalization scheme is illustrated in Figure 13.5.

When all goes well, the recursive algorithm (13.27) converges to the vicinity of the block least-squares answer for the particular δ used in forming the delayed recovery error. As long as μ is nonzero, if the underlying composition of the received signal changes so that the error increases and the desired equalizer changes, then the f_i react accordingly. It is this tracking ability that earns it the label adaptive.[4]

The following MATLAB code implements an adaptive equalizer design. The beginning and ending of the program are familiar from `openclosed.m` and `LSequal-izer.m`. The heart of the recursion lies in the `for` loop. For each new data point, a vector is built containing the new value and the past n values of the received signal. This is multiplied by `f` to make a prediction of the next source symbol, and the error is the difference between the prediction and the reality. (This is the calculation of $e[k]$ from (13.23).) The equalizer coefficients `f` are then updated as in (13.27).

FIGURE 13.5 Trained adaptive linear equalizer.

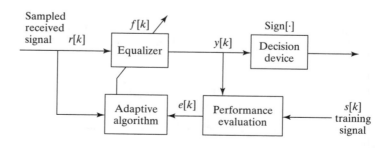

---

[4] To provide tracking capability, the matrix solution of Section 13.2.1 could be recomputed for successive data blocks, but this requires significantly more computation.

LMSequalizer.m: find a LMS equalizer f for the channel b

```
b=[0.5 1 -0.6];                    % define channel
m=1000; s=sign(randn(1,m));        % binary source of length m
r=filter(b,1,s);                   % output of channel
n=4; f=zeros(n,1);                 % initialize equalizer at 0
mu=.1; delta=2;                    % stepsize and delay delta
for i=n+1:m                        % iterate
  rr=r(i:-1:i-n+1)';               % vector of received signal
  e=s(i-delta)-f'*rr;              % calculate error
  f=f+mu*e*rr;                     % update equalizer coefficients
end
y=filter(f,1,r);                   % equalizer is a filter
dec=sign(y);                       % quantization
for sh=0:n;                        % error at different delays
err(sh+1)=0.5*sum(abs(dec(sh+1:end)-s(1:end-sh)));
end
```

As with the matrix approach, the default channel b=[0.5 1 -0.6] can be equalized easily with a short equalizer (one with a small n). The convergent values of the f are very close to the final values of the matrix approach; that is, for a given channel, the value of f given by LMSequalizer.m is very close to the value found using LSequalizer.m. A design consideration in the adaptive approach to equalization involves the selection of the stepsize. Smaller stepsizes μ mean that the trajectory of the estimates is smoother (tends to reject noises better) but it also results in a slower convergence and slower tracking when the underlying solution is time varying. Similarly, if the explicit averaging operation is retained, longer averages imply smoother estimates but slower convergence. Similar tradeoffs appear in the block approach in the choice of block size: larger blocks average the noise better, but give no details about changes in the underlying solution within the time span covered by a block.

This trained adaptive approach, along with several others, is implemented in the program dae.m, which is available on the CD. Simulated examples of LMS with training and other adaptive equalization methods are presented in Section 13.6.

Problems

13.5. Verify that, by proper choice of n and delta, the convergent values of f in LMSequalizer.m are close to the values shown in (13.17).

13.6. What happens in LMSequalizer.m when the stepsize parameter mu is too large? What happens when it is too small?

13.7. Add (uncorrelated, normally distributed) noise into the simulation using the command r=filter(b,1,s)+sd*randn(size(s)).
   (a) For the equalizer with delay 2, what is the largest sd you can add, and still have no errors? How does this compare with the result from Problem 13.2. Hint: It may be necessary to simulate for more than the default m data points.
   (b) Now try the equalizer with delay 1. What is the largest sd you can add, and still have no errors?

(c) Which is a better equalizer?

13.8. Use LMSequalizer.m to find an equalizer that can open the eye for the channel b=[1 1 -0.8 -.3 1 1].

(a) What equalizer length n is needed?

(b) What delays delta give zero error in the output of the quantizer?

(c) How does the answer compare with the design in Problem 13.3?

13.9. Modify LMSequalizer.m to generate a source sequence from the alphabet ±1, ±3. For the default channel [0.5 1 -0.6], find an equalizer that opens the eye.

(a) What equalizer length n is needed?

(b) What delays delta give zero error in the output of the quantizer?

(c) Is this a fundamentally easier or more difficult task than when equalizing a binary source?

(d) How does the answer compare with the design in Problem 13.4?

13.4 Decision-Directed Linear Equalization

During the training period, the communication system does not transmit any message data. Commonly, a block of training data is followed by a block of message data. The fraction of time devoted to training should be small, but can be up to 20% in practice. If it were possible to adapt the equalizer parameters without using the training data, then the message bearing (and revenue generating) capacity of the channel would be enhanced.

Consider the situation in which some procedure has produced an equalizer setting that opens the eye of the channel. Thus, all decisions are perfect, but the equalizer parameters may not yet be at their optimal values. In such a case, the output of the decision device is an exact replica of the delayed source (i.e., it is as good as a training signal). For a binary ±1 source and decision device that is a sign operator, the delayed source recovery error can be computed as $\text{sign}\{y[k]\} - y[k]$, where $y[k]$ is the equalizer output and $\text{sign}\{y[k]\}$ equals $s[k - \delta]$. Thus, the trained adaptive equalizer of Figure 13.5 can be replaced by the decision-directed equalizer shown in Figure 13.6. This converts (13.27) to decision-directed LMS, which has the update

$$f_i[k + 1] = f_i[k] + \mu(\text{sign}(y[k]) - y[k])r[k - i]. \qquad (13.28)$$

When the signal $s[4]$ is multilevel instead of binary, the sign function in (13.28) can be replaced with a quantizer.

FIGURE 13.6 Decision-directed adaptive linear equalizer.

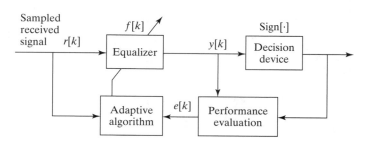

Problem

13.10. Show that the decision-directed LMS algorithm (13.28) can be derived as an adaptive element with performance function $(1/2)\text{avg}\{(\text{sign}\{y[k]\} - y[k])^2\}$. Hint: Suppose that the derivative of the sign function $\frac{d\text{sign}\{x\}}{dx}$ is zero everywhere.

Observe that the source signal $s[k]$ does not appear in (13.28). Thus, no training signal is required for its implementation and the decision-directed LMS equalizer adaptation law of (13.28) is called a "blind" equalizer. Given its genesis, one should expect decision-directed LMS to exhibit poor behavior when the assumption regarding perfect decisions is violated. The basic rule of thumb is that 5% (or so) decision errors can be tolerated before decision-directed LMS fails to converge properly.

The MATLAB program `DDequalizer.m` has a familiar structure. The only code changed from `LMSequalizer.m` is the calculation of the error term, which implements $e[k] = \text{sign}\{y[k]\} - y[k]$ rather than the LMS error (13.23), and the initialization of the equalizer. Because the equalizer must begin with an open eye, `f=0` is a poor choice. The initialization that follows starts all taps at zero except for one in the middle that begins at unity. This is called the "center-spike" initialization. If the channel eye is open, then the combination of the channel and equalizer will also have an open eye when initialized with the center spike. The exercises ask you to explore the issue of finding good initial values for the equalizer parameters.

DDequalizer.m: find a DD equalizer f for the channel b

```
b=[0.5 1 -0.6];              % define channel
m=1000; s=sign(randn(1,m));  % binary source of length m
r=filter(b,1,s);             % output of channel
n=4; f=[0 1 0 0]';           % initialize equalizer
mu=.1;                       % stepsize
for i=n+1:m                  % iterate
   rr=r(i:-1:i-n+1)';        % vector of received signal
   e=sign(f'*rr)-f'*rr;      % calculate error
   f=f+mu*e*rr;              % update equalizer coefficients
end
y=filter(f,1,r);            % equalizer is a filter
dec=sign(y);                % quantization
for sh=0:n;                 % error at different delays
  err(sh+1)=0.5*sum(abs(dec(sh+1:end)-s(1:end-sh)));
end
```

Problems

13.11. Try the initialization `f=[0 0 0 0]'` in `DDequalizer.m`. With this initialization, can the algorithm open the eye? Try increasing `m`. Try changing the stepsize `mu`. What other initializations will work?

13.12. What happens in `DDequalizer.m` when the stepsize parameter `mu` is too large? What happens when it is too small?

13.13. Add (uncorrelated, normally distributed) noise into the simulation using the command `r=filter(b,1,s)+sd*randn(size(s))`. What is the largest `sd`

you can add, and still have no errors? Does the initial value for f influence this number? Try at least three initializations.

13.14. Use DDequalizer.m to find an equalizer that can open the eye for the channel b=[1 1 -0.8 -.3 1 1].
   (a) What equalizer length n is needed?
   (b) What initializations for f did you use?
   (c) How does the converged answer compare with the design in Problem 13.3 and Problem 13.8?

13.15. Modify DDequalizer.m to generate a source sequence from the alphabet ±1, ±3. For the default channel [0.5 1 -0.6], find an equalizer that opens the eye.
   (a) What equalizer length n is needed?
   (b) What initializations for f did you use?
   (c) Is this a fundamentally easier or more difficult task than when equalizing a binary source?
   (d) How does the answer compare with the design in Problem 13.4 and Problem 13.9?

Section 13.6 provides the opportunity to view the simulated behavior of the decision-directed equalizer, and to compare its performance with the other methods.

13.5 Dispersion-Minimizing Linear Equalization

This section considers an alternative performance function that leads to another kind of blind equalizer. Observe that for a binary ± 1 source, the square of the source is known, even when the particular values of the source are not. Thus $s^2[k] = 1$ for all k. This suggests creating a performance function that penalizes the deviation from this known squared value $\gamma = 1$. In particular, consider

$$J_{DM} = \frac{1}{4}\text{avg}\{(\gamma - y^2[k])^2\},$$

which measures the *dispersion* of the equalizer output about its desired squared value γ.

The associated adaptive element for updating the equalizer coefficients is

$$f_i[k + 1] = f_i[k] - \mu \left. \frac{dJ_{DM}}{df_i} \right|_{f_i = f_i[k]}.$$

Mimicking the derivation in (13.24) through (13.27) yields the dispersion-minimizing algorithm[5] (DMA) for blindly adapting the coefficients of a linear equalizer. The algorithm is

$$f_i[k + 1] = f_i[k] + \mu\text{avg}\{(1 - y^2[k])y[k]r[k - i]\}.$$

Suppressing the averaging operation, this becomes

$$f_i[k + 1] = f_i[k] + \mu(1 - y^2[k])y[k]r[k - i], \qquad (13.29)$$

which is shown in the block diagram of Figure 13.7.

---

[5] This is also known as the constant modulus algorithm and as the Godard algorithm.

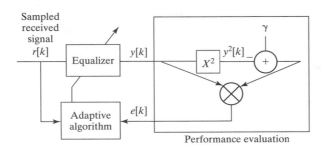

FIGURE 13.7 Dispersion-minimizing adaptive linear equalizer.

When the source alphabet is ± 1, then $\gamma = 1$. When the source is multilevel, it is still useful to minimize the dispersion, but the constant should change to $\gamma = \frac{\text{avg}\{s^4\}}{\text{avg}\{s^2\}}$.

While DMA typically may converge to the desired answer from a worse initialization than decision-directed LMS, it is not as robust as trained LMS. For a particular delay δ, the (average) squared recovery error surface being descended (approximately) along the gradient by trained LMS is unimodal (i.e., it has only one minimum). Therefore, no matter where the search is initialized, it finds the desired sole minimum, associated with the δ used in computing the source recovery error. The dispersion performance function is multimodal with separate minima corresponding to different achieved delays and polarities. To see this in the simplest case, observe that an answer in which all $+1$'s are swapped with all -1's has the same value at the optimal point. Thus, the convergent delay and polarity achieved depend on the initialization used. A typical initialization for DMA is a single nonzero spike located near the center of the equalizer. The multimodal nature of DMA can be observed in the examples in the next section.

A simple MATLAB program that implements the DMA algorithm is given in `DMAequalizer.m`. The first few lines define the channel, create the binary source, and pass the input through the channel. The last few lines implement the equalizer and calculate the error between the output of the equalizer and the source as a way of measuring the performance of the equalizer. These parts of the code are familiar from `LSequalizer.m`. The new part of the code is in the center, which defines the length n of the equalizer, the stepsize mu of the algorithm, and the initialization of the equalizer (which defaults to a "center spike" initialization). The coefficients of the equalizer are updated as in (13.29).

DMAequalizer.m: find a DMA equalizer f for the channel b

```
b=[0.5 1 -0.6];                  % define channel
m=1000; s=sign(randn(1,m));      % binary source of length m
r=filter(b,1,s);                 % output of channel
n=4; f=[0 1 0 0]';               % center spike initialization
mu=.01;                          % algorithm stepsize
for i=n+1:m                      % iterate
  rr=r(i:-1:i-n+1)';             % vector of received signal
  e=(f'*rr)*(1-(f'*rr)^2);       % calculate error
  f=f+mu*e*rr;                   % update equalizer coefficients
end
y=filter(f,1,r);                 % equalizer is a filter
dec=sign(y);                     % quantization
for sh=0:n;                      % error at different delays
  err(sh+1)=0.5*sum(abs(dec(sh+1:end)-s(1:end-sh)));
end
```

Running `DMAequalizer.m` results in an equalizer that is numerically similar to the equalizers of the previous two sections. Initializing with the "spike" at different locations results in equalizers with different effective delays. The following exercises are intended to encourage you to explore the DMA equalizer method.

Problems

13.16. Try the initialization `f=[0 0 0 0]'` in `DMAequalizer.m`. With this initialization, can the algorithm open the eye? Try increasing `m`. Try changing the stepsize `mu`. What other nonzero initializations will work?

13.17. What happens in `DMAequalizer.m` when the stepsize parameter `mu` is too large? What happens when it is too small?

13.18. Add (uncorrelated, normally distributed) noise into the simulation using the command `r=filter(b,1,s)+sd*randn(size(s))`. What is the largest `sd` you can add, and still have no errors? Does the initial value for `f` influence this number? Try at least three initializations.

13.19. Use `DMAequalizer.m` to find an equalizer that can open the eye for the channel `b=[1 1 -0.8 -.3 1 1]`.
  (a) What equalizer length `n` is needed?
  (b) What initializations for `f` did you use?
  (c) How does the converged answer compare with the designs in Problems 13.3, 13.8, and 13.14?

13.20. Modify `DMAequalizer.m` to generate a source sequence from the alphabet $\pm 1, \pm 3$. For the default channel `[0.5 1 -0.6]`, find an equalizer that opens the eye.
  (a) What equalizer length `n` is needed?
  (b) What is an appropriate value of γ?
  (c) What initializations for `f` did you use?
  (d) Is this a fundamentally easier or more difficult task than when equalizing a binary source?
  (e) How does the answer compare with the designs in Problems 13.4, 13.9, and 13.15?

13.21. Consider a DMA-like performance function $J = \frac{1}{2}\text{avg}\{|1 - y^2[k]|\}$. Show that the resulting gradient algorithm is

$$f_i[k+1] = f_i[k] + \mu\text{avg}\{\text{sign}(1 - y^2[k])y[k]r[k-i]\}.$$

Hint: Assume that the derivative of the absolute value is the sign function. Implement the algorithm and compare its performance with the DMA of (13.29) in terms of
  (a) speed of convergence,
  (b) number of errors in a noisy environment (recall Problem 13.18), and
  (c) ease of initialization.

13.22. Consider a DMA-like performance function $J = \text{avg}\{|1 - |y[k]||\}$. What is the resulting gradient algorithm? Implement your algorithm and compare its performance with the DMA of (13.29) in terms of

   (a) speed of convergence of the equalizer coefficients f,

   (b) number of errors in a noisy environment (recall Problem 13.18), and

   (c) ease of initialization.

13.6 Examples and Observations

This section uses the MATLAB program dae.m which is available on the CD. The program demonstrates some of the properties of the least squares solution to the equalization problem and its adaptive cousins: LMS, decision-directed LMS, and DMA.[6]

The default settings in dae.m are used to perform the equalizer designs for three channels. The source alphabet is a binary ± 1 signal. Each channel has a FIR impulse response, and its output is summed with a sinusoidal interference and some uniform white noise before reaching the receiver. The user is prompted for

1. choice of channels (0, 1, or 2),
2. maximum delay of the equalizer,
3. number of samples of training data,
4. gain of the sinusoidal interferer,
5. frequency of the sinusoidal interferer (in radians), and
6. magnitude of the white noise.

The program returns plots of the

1. received signal,
2. optimal equalizer output,
3. impulse response of the optimal equalizer and the channel,
4. recovery error at the output of the decision device,
5. zeros of the channel and the combined channel–equalizer pair, and
6. magnitude and phase frequency responses of the channel, equalizer, and the combined channel–equalizer pair.

For the default channels and values, these plots are shown in Figures 13.8–13.13. The program also prints the condition number of $\bar{R}^T \bar{R}$, the minimum average squared recovery error (i.e., the minimum value achieved by the performance function by the optimal equalizer for the optimum delay δ_{opt}), the optimal value of the delay δ_{opt}, and the percentage of decision device output errors in matching the delayed source. These values were as follows:

- Channel 0

  – condition number: 130.2631

  – minimum value of performance function: 0.0534

  – optimum delay: 16

  – percentage of errors: 0

---

[6] Throughout these simulations, other aspects of the system are assumed optimal; thus, the downconversion is numerically perfect and the synchronization algorithms are assumed to have attained their convergent values.

FIGURE 13.8 Trained least-squares equalizer for Channel 0: Time responses.

FIGURE 13.9 Trained least-squares equalizer for Channel 0: Singularities and frequency responses. The large circles show the locations of the zeros of the channel in the upper left plot and the locations of the zeros of the combined channel–equalizer pair in the lower left. The *** represents the frequency response of the channel, — is the frequency response of the equalizer, and the solid line is the frequency response of the combined channel–equalizer pair.

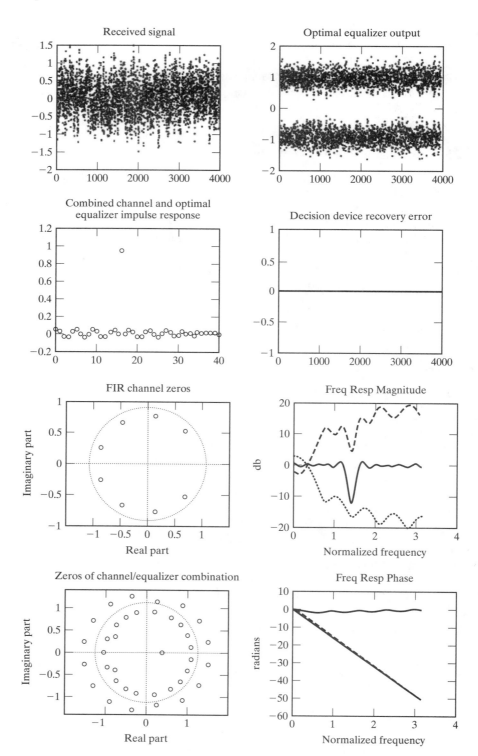

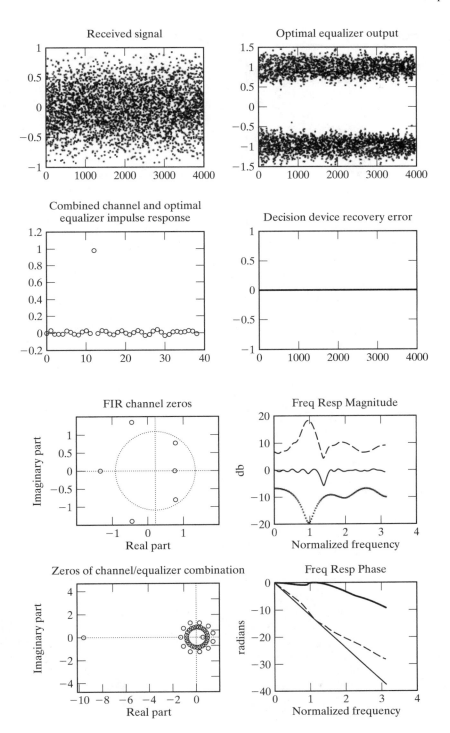

FIGURE 13.10 Trained least-squares equalizer for Channel 1: Time responses.

FIGURE 13.11 Trained least-squares equalizer for Channel 1: Singularities and frequency responses. The large circles show the locations of the zeros of the channel in the upper left plot and the locations of the zeros of the combined channel–equalizer pair in the lower left. The *** represents the frequency response of the channel, — is the frequency response of the equalizer, and the solid line is the frequency response of the combined channel–equalizer pair.

FIGURE 13.12 Trained least-squares equalizer for Channel 2: Time responses.

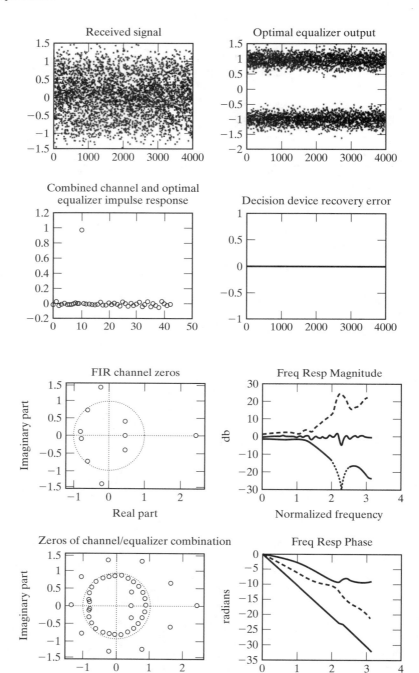

FIGURE 13.13 Trained least-squares equalizer for Channel 2: Singularities and frequency responses. The large circles show the locations of the zeros of the channel in the upper left plot and the locations of the zeros of the combined channel–equalizer pair in the lower left. The *** represents the frequency response of the channel, — is the frequency response of the equalizer, and the solid line is the frequency response of the combined channel–equalizer pair.

- Channel 1

  - condition number: 14.795
  - minimum value of performance function: 0.0307
  - optimum delay: 12
  - percentage of errors: 0

- Channel 2

  - condition number: 164.1081
  - minimum value of performance function: 0.0300
  - optimum delay: 10
  - percentage of errors: 0

To see what these figures mean, consider the eight plots contained in Figures 13.8 and 13.9. The first plot is the received signal, which contains the transmitted signal corrupted by the sinusoidal interferer and the white noise. After the equalizer design, this received signal is passed through the equalizer, and the output is shown in the plot entitled "optimal equalizer output." The equalizer transforms the data in the received signal into two horizontal stripes. Passing this through a simple sign device recovers the transmitted signal.[7] The width of these stripes is related to the cluster variance. The difference between the sign of the output of the equalizer, and the transmitted data is shown in the plot labelled "decision device recovery error." This is zero, indicating that the equalizer has done its job. The plot entitled "combined channel and optimal equalizer impulse response" shows the convolution of the impulse response of the channel with the impulse response of the equalizer. If the design were perfect and there was no interference present, then one tap of this combination would be unity and all the rest would be zero. In this case, the actual design is close to this ideal.

The plots in Figure 13.9 show the same situation, but in the frequency domain. The zeros of the channel are depicted in the plot in the upper left. This constellation of zeros corresponds to the darkest of the frequency responses drawn in the second plot. The primarily lowpass character of the channel can be intuited directly from the zero plot with the technique of Section F.2. The T-spaced equalizer, accordingly, has a primarily highpass character, as can be seen from the dashed frequency response in the upper right plot of Figure 13.9. Combining these two gives the response in the middle. This middle response (plotted with the solid line) is mostly flat, except for a large dip at 1.4 radians. This is exactly the frequency of the sinusoidal interferer, and this demonstrates the second major use of the equalizer; it is capable of removing uncorrelated interferences. Observe that the equalizer design is given no knowledge of the frequency of the interference, nor even that any interference exists. Nonetheless, it automatically compensates for the narrow band interference by building a notch at the offending frequency. The plot labelled "channel-optimum equalizer combination zeros" shows the zeros of the convolution of the impulse response of the channel and the impulse response of the optimal equalizer. Were the ring of zeros at a uniform distance from the unit circle, then the magnitude of the frequency response would be nearly flat. But observe that one pair of zeros (at ± 1.4 radians) is considerably closer to the circle than all the others. Since the magnitude of

---

[7] Without the equalizer, the sign function would be applied directly to the received signal, and the result would bear little relationship to the transmitted signal.

the frequency response is the product of the distances from the zeros to the unit circle, this distance becomes small where the zero comes close. This causes the notch.[8]

The eight plots for each of the other channels are displayed in similar fashion in Figures 13.10 to 13.13.

Figures 13.14–13.16 demonstrate equalizer design using the various recursive methods of Sections 13.3 to 13.5 on the same problem. After running the least-square design in dae.m, the script asks if you wish to simulate a recursive solution. If yes, then you can choose

- which algorithm to run: trained LMS, decision-directed LMS, or blind DMA,
- the stepsize, and
- the initialization: a scale factor specifies the size of the ball about the optimum equalizer within which the initial value for the equalizer is randomly chosen

As is apparent from Figures 13.14–13.16, all three adaptive schemes are successful with the recommended "default" values, which were used in equalizing channel 0. All three exhibit, in the upper left plots of Figures 13.14-13.16, decaying averaged squared parameter error relative to their respective trained least-squares equalizer for the data block. This means that all are converging to the vicinity of the trained least-squares equalizer about which dae.m initializes the algorithms. The collapse of the squared prediction error is apparent from the upper right plot in each of the same figures. An initially closed eye appears for a short while in each of the lower left plots of equalizer output history in the same figures. The match of the magnitudes of the frequency responses of the trained (block) least-squares equalizer (plotted with the solid line) and the last adaptive equalizer setting (plotted with asterisks) from the data block stream is quite striking in the lower right plots in the same figures.

As expected,

- With modest noise, as in the cases here outside the frequency band occupied by the single narrowband interferer, the magnitude of the frequency response of the trained least-squares solution exhibits peaks (valleys) where the channel response has valleys (peaks) so that the combined response is nearly flat. The phase of the trained least-squares equalizer adds with the channel phase so that their combination approximates a linear phase curve. Refer to plots in the right columns of Figures 13.9, 13.11, and 13.13.

- With modest channel noise and interferers, as the length of the equalizer increases, the zeros of the combined channel and equalizer form rings. The rings are denser the nearer the channel zeros are to the unit circle.

There are many ways that the program dae.m can be used to investigate and learn about equalization. Try to choose the various parameters to observe that

1. Increasing the power of the channel noise suppresses the frequency response of the least-squares equalizer, with those frequency bands most suppressed being those in which the channel has a null (and the equalizer—without channel noise—would have a peak).

---

[8] If this kind of argument relating the zeros of the transfer function to the frequency response of the system seems unfamiliar, see Appendix F.

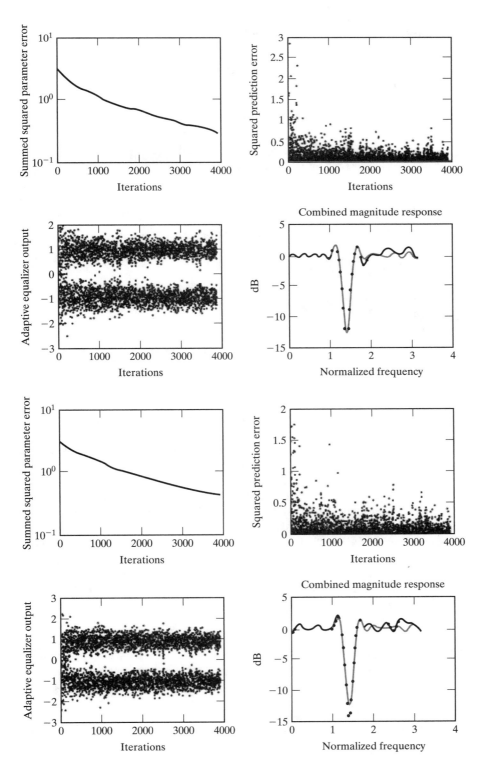

FIGURE 13.14 Trained LMS equalizer for Channel 0. The *** represents the achieved frequency response of the equalizer while the solid line represents the frequency response of the desired (optimal) mean square error solution.

FIGURE 13.15 Decision-directed LMS equalizer for Channel 0. The *** represents the achieved frequency response of the equalizer while the solid line represents the frequency response of the desired (optimal) mean square error solution.

FIGURE 13.16 Blind DMA equalizer for Channel 0. The *** represents the achieved frequency response of the equalizer while the solid line represents the frequency response of the desired (optimal) mean square error solution.

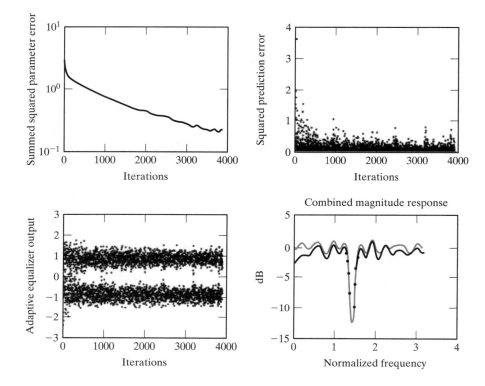

2. Increasing the gain of a narrowband interferer results in a deepening of a notch in the trained least squares equalizer at the frequency of the interferer.
3. DMA is considered slower than trained LMS. Do you find that DMA takes longer to converge? Can you think of why it might be slower?
4. DMA typically accommodates larger initialization error than decision-directed LMS. Can you find cases where, with the same initialization, DMA converges to an error-free solution but the decision directed LMS does not? Do you think there are cases in which the opposite holds?
5. It is necessary to specify the delay δ for the trained LMS, whereas the blind methods do not require the parameter δ. Rather, the selection of an appropriate delay is implicit in the initialization of the equalizer coefficients. Can you find a case in which, with the delay poorly specified, DMA outperforms trained LMS from the same initialization?

13.7 For Further Reading

A comprehensive survey of trained adaptive equalization can be found in

- S. U. H. Qureshi, "Adaptive equalization," *Proceedings of the IEEE*, pp. 1349–1387, 1985.

An overview of the analytical tools that can be used to analyze LMS-style adaptive algorithms can be found in

- W. A. Sethares, "The LMS Family," in *Efficient System Identification and Signal Processing Algorithms*, Ed. N. Kalouptsidis and S. Theodoridis, Prentice Hall, 1993.

A copy of this paper can also be found on the accompanying CD.
One of our favorite discussions of adaptive methods is

- C. R. Johnson Jr., *Lectures on Adaptive Parameter Estimation,* Prentice-Hall, 1988.

This whole book can be found in .pdf form on the CD.
An extensive discussion of equalization can also be found in *Equalization* on the CD.

14 *Coding*

> Before Shannon it was commonly believed that the only way of achieving arbitrarily small probability of error on a communications channel was to reduce the transmission rate to zero. Today we are wiser. Information theory characterizes a channel by a single parameter; the channel capacity. Shannon demonstrated that it is possible to transmit information at any rate below capacity with an arbitrarily small probability of error.
>
> —from A. R. Calderbank, "The Art of Signaling: Fifty Years of Coding Theory," *IEEE Transactions on Information Theory,* p. 2561, October 1998.

The underlying purpose of any communication system is to transmit information. But what exactly is information? How is it measured? Are there limits to the amount of data that can be sent over a channel, even when all the parts of the system are operating at their best? This chapter addresses these fundamental questions using the ideas of Claude Shannon (1916–2001), who defined a measure of information in terms of bits. The number of bits per second that can be transmitted over the channel (taking into account its bandwidth, the power of the signal, and the noise) is called the *bit rate*, and can be used to define the *capacity* of the channel.

Unfortunately, Shannon's results do not give a recipe for how to construct a system that achieves the optimal bit rate. Earlier chapters have highlighted several problems that can arise in communication systems (including synchronization errors such as phase offsets, and clock jitter, frequency offsets, and intersymbol interference). This chapter assumes that all of these are perfectly mitigated. Thus, in Figure 14.1, the inner parts of the communication system are assumed to be ideal, except for the presence of channel noise. Even so, most systems still fall far short of the optimal performance promised by Shannon.

There are two problems. First, most messages that people want to send are redundant, and the redundancy squanders the capacity of the channel. A solution is to preprocess the message so as to remove the redundancies. This is called *source coding*, and is discussed in Section 14.5. For instance, as demonstrated in Section 14.2, any natural language (such as English), whether spoken or written, is repetitive. Information theory (as Shannon's approach is called) quantifies the repetitiveness, and gives a way to judge the efficiency of a source code by comparing the information content of the message to the number of bits required by the code.

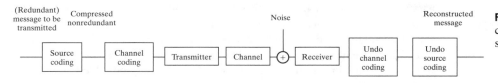

FIGURE 14.1 Digital communication system.

The second problem is that messages must be resistant to noise. If a message arrives at the receiver in garbled form, then the system has failed. A solution is to preprocess the message by adding extra bits, which can be used to determine if an error has occurred, and to correct errors when they do occur. For example, one simple system would transmit each bit three times. Whenever a single bit error occurs in transmission, then the decoder at the receiver can figure out by a simple voting rule that the error has occurred and what the bit should have been. Schemes for finding and removing errors are called *error-correcting codes* or *channel codes*, and are discussed in Section 14.6.

At first glance, this appears paradoxical; source coding is used to remove redundancy, while channel coding is used to add redundancy. But it is not really self-defeating or contradictory because the redundancy that is removed by source coding does not have a structure or pattern that a computer algorithm at the receiver can exploit to detect or correct errors. The redundancy that is added in channel coding is highly structured, and can be exploited by computer programs implementing the appropriate decoding routines. Thus Figure 14.1 begins with a message, and uses a source code to remove the redundancy. This is then coded again by the channel encoder to add structured redundancy, and the resulting signal provides the input to the transmitter of the previous chapters. One of the triumphs of modern digital communications systems is that, by clever choice of source and channel codes, it is possible to get close to the Shannon limits and to utilize all the capacity of a channel.

14.1 What Is Information?

Like many common English words, *information* has many meanings. The American Heritage Dictionary catalogs six:

1. Knowledge derived from study, experience, or instruction.
2. Knowledge of a specific event or situation; intelligence.
3. A collection of facts or data.
4. The act of informing or the condition of being informed; communication of knowledge.
5. Computer Science. A nonaccidental signal or character used as an input to a computer or communication system.
6. A numerical measure of the uncertainty of an experimental outcome.

It would clearly be impossible to capture all of these senses in a technical definition that would be useful in transmission systems. The final definition is closest to our needs, though it does not specify exactly how the numerical measure should be calculated. Shannon does. Shannon's insight was that there is a simple relationship between the amount of information conveyed in a message and the probability of the message being sent. This does not apply directly to "messages" such as sentences, images, or .wav files, but to the symbols of the alphabet that are transmitted.

For instance, suppose that a fair coin has heads H on one side and tails T on the other. The two outcomes are equally uncertain, and receiving either H or T removes the same amount of uncertainty (conveys the same amount of information). But suppose the coin is biased. The extreme case is occurs when the probability of H is 1. Then, when H is received, *no* information is conveyed, because H is the only possible choice! Now suppose that the probability of sending H is 0.9 while the probability of sending T is 0.1. Then, if H is received, it removes a little uncertainty, but not much. H is expected, since it usually occurs. But if T is received, it is somewhat unusual, and hence conveys a lot of information. In general, events that occur with high probability give little information, while events of low probability give considerable information.

To make this relationship between the probability of events and information more plain, imagine a game in which you must guess a word chosen at random from the dictionary. You are given the starting letter as a hint. If the hint is that the first letter is "t," then this does not narrow down the possibilities very much, since so many words start with "t." But if the hint is that the first letter is "x," then there are far fewer choices. The likely letter (the highly probable "t") conveys little information, while the unlikely letter (the improbable "x") conveys a lot more information by narrowing down the choices.

Here's another everyday example. Someone living in Ithaca (New York) would be completely unsurprised that the weather forecast called for rain, and such a prediction would convey little real information since it rains frequently. On the other hand, to someone living in Reno (Nevada), a forecast of rain would be very surprising, and would convey that very unusual meteorological events were at hand. In short, it would convey considerable information. Again, the amount of information conveyed is inversely proportional to the probabilities of the events.

To transform this informal argument into a mathematical statement, consider a set of N possible events x_i, for $i = 1, 2, \ldots, N$. Each event represents one possible outcome of an experiment, such as the flipping of a coin or the transmission of a symbol across a communication channel. Let $p(x_i)$ be the probability that the ith event occurs, and suppose that some event must occur.[1] This means that $\sum_{i=1}^{N} p(x_i) = 1$. The goal is find a function $I(x_i)$ that represents the amount of information conveyed by each outcome.

The following three qualitative conditions all relate the probabilities of events with the amount of information they convey:

$$
\begin{aligned}
(i) & \quad p(x_i) = p(x_j) & \Rightarrow & \quad I(x_i) = I(x_j); \\
(ii) & \quad p(x_i) < p(x_j) & \Rightarrow & \quad I(x_i) > I(x_j); \\
(iii) & \quad p(x_i) = 1 & \Rightarrow & \quad I(x_i) = 0.
\end{aligned}
\tag{14.1}
$$

Thus, receipt of the symbol x_i should

1. give the same information as receipt of x_j if they are equally likely,
2. give more information if x_i is less likely than x_j, and
3. convey no information if it is known a priori that x_i is the only alternative.

What kinds of functions $I(x_i)$ fulfill these requirements? There are many. For instance, $I(x_i) = \frac{1}{p(x_i)} - 1$ and $I(x_i) = \frac{1 - p^2(x_i)}{p(x_i)}$ both fulfill (i)–(iii).

To narrow down the possibilities, consider what happens when a series of experiments are conducted, or equivalently, when a series of symbols are transmitted. Intuitively, it

---

[1] When flipping the coin, it cannot roll into the corner and stand on its edge; each flip results in either H or T.

seems reasonable that if x_i occurs at one trial and x_j occurs at the next, then the total information in the two trials should be the sum of the information conveyed by receipt of x_i and the information conveyed by receipt of x_j; that is, $I(x_i) + I(x_j)$. This assumes that the two trials are independent of each other, that the second trial is not influenced by the outcome of first (and vice versa).

Formally, two events are defined to be *independent* if the probability that both occur is equal to the product of the individual probabilities—that is, if

$$p(x_i \text{ and } x_j) = p(x_i)p(x_j), \tag{14.2}$$

where $p(x_i \text{ and } x_j)$ means that x_i occurred in the first trial and x_j occurred in the second. This additivity requirement for the amount of information conveyed by the occurrence of independent events is formally stated in terms of the information function as

$$(iv) \quad I(x_i \text{ and } x_j) = I(x_i) + I(x_j)$$

when the events x_i and x_j are independent.

Combining the additivity in (iv) with the three conditions (i)–(iii), there is one (and only one) possibility for $I(x_i)$:

$$I(x_i) = \log\left(\frac{1}{p(x_i)}\right) = -\log(p(x_i)). \tag{14.3}$$

It is easy to see that (i)–(iii) are fulfilled, and (iv) follows from the properties of the log (recall that $\log(ab) = \log(a) + \log(b)$). Therefore,

$$I(x_i \text{ and } x_j) = \log\left(\frac{1}{p(x_i \text{ and } x_j)}\right) = \log\left(\frac{1}{p(x_i)p(x_j)}\right)$$
$$= \log\left(\frac{1}{p(x_i)}\right) + \log\left(\frac{1}{p(x_j)}\right)$$
$$= I(x_i) + I(x_j).$$

The base of the logarithm can be any (positive) number. The most common choice is base 2, in which case the measurement of information is called *bits*. Unless otherwise stated explicitly, all logs in this chapter are assumed to be base 2.

Example 14.1

Suppose there are $N = 3$ symbols in the alphabet, which are transmitted with probabilities $p(x_1) = 1/2$, $p(x_2) = 1/4$, and $p(x_3) = 1/4$. Then the information conveyed by receiving x_1 is 1 bit, since

$$I(x_1) = \log\left(\frac{1}{p(x_1)}\right) = \log(2) = 1.$$

Similarly, the information conveyed by receiving either x_2 or x_3 is $I(x_2) = I(x_3) = \log(4) = 2$ bits.

Example 14.2

Suppose that a length m binary sequence is transmitted, with all symbols equally probable. Thus $N = 2^m$, x_i is the binary representation of the ith symbol for $i = 1, 2, \ldots, N$, and $p(x_i) = 2^{-m}$. The information contained in the receipt of any given symbol is

$$I(x_i) = \log\left(\frac{1}{p(x_i)}\right) = \log(2^m) = m \text{ bits.}$$

Problems

14.1. Consider a standard six-sided die. Identify N, x_i, and $p(x_i)$. How many bits of information are conveyed if a 3 is rolled. Now roll two dice, and suppose the total is 12. How many bits of information does this represent?

14.2. Consider transmitting a signal with values chosen from the six-level alphabet $\pm 1, \pm 3, \pm 5$.
   (a) Suppose that all six symbols are equally likely. Identify N, x_i and $p(x_i)$, and calculate the information $I(x_i)$ associated with each i.
   (b) Suppose instead that the symbols ± 1 occur with probability $1/4$ each, ± 3 occur with probability $1/8$ each, and 5 occurs with probability $1/4$. What percentage of the time is -5 transmitted? What is the information conveyed by each of the symbols?

14.3. The 8-bit binary ASCII representation of any letter (or any character of the keyboard) can be found using the MATLAB command `dec2bin(text)` where `text` is any string. Using ASCII, how much information is contained in the letter "a," assuming that all the letters are equally probable?

14.4. Consider a decimal representation of $\pi = 3.1415926\ldots$ Calculate the information (number of bits) required to transmit successive digits of π, assuming that the digits are independent. Identify N, x_i, and $p(x_i)$. How much information is contained in the first million digits of π?

There is an alternative definition of information (in common usage in the mathematical logic and computer science communities) which defines information in terms of the complexity of representation, rather than in terms of the reduction in uncertainty. Informally speaking, this alternative defines the complexity (or information content) of a message by the length of the shortest computer program that can replicate the message. For many kinds of data, such as a sequence of random numbers, the two measures agree because the shortest program that can represent the sequence is just a listing of the sequence. But in other cases, they can differ dramatically. Consider transmitting the first million digits of the number π. Shannon's definition gives a large information content (as in

Problem 14.4), while the complete sequence can, in principle, be transmitted with a very short computer program.

14.2 Redundancy

All the examples in the previous section presume that there is no relationship between successive symbols. (This was the independence assumption in (14.2).) This section shows by example that real messages often have significant correlation between symbols, which is a kind of redundancy. Consider the following sentence from Shannon's paper *A Mathematical Theory of Communication*:

```
It is clear, however, that by sending the information in a
redundant form the probability of errors can be reduced.
```

This sentence contains 20 words and 115 characters, including the commas, period, and spaces. It can be "coded" into the 8-bit binary ASCII character set recognized by computers as the "text" format, which translates the character string (that is readable by humans) into a binary string containing 920 ($= 8 * 115$) bits.

Suppose that Shannon's sentence is transmitted, but that errors occur so that 1% of the bits are flipped from one to zero (or from zero to one). Then about 3.5% of the letters have errors:

```
It is clea2, however, that by sendhng the information in a
redundaNt form the probability of errors can be reduced.
```

The message is comprehensible, although it appears to have been typed poorly. With 2% bit error, about 7% of the letters have errors:

```
It is clear, howaver, thad by sending the information in a
redundan4 form phe prkbability of errors cAf be reduced.
```

Still the underlying meaning is decipherable. A dedicated reader can often decipher text with up to about 3% bit error (10% symbol error). Thus, the message has been conveyed, despite the presence of the errors. The reader, with an extensive familiarity with English words, sentences, and syntax, is able to recognize the presence of the errors and to correct them.

As the bit error rate grows to 10%, about one third of the letters have errors, and many words have become incomprehensible. Because "space" is represented as an ASCII character just like all the other symbols, errors can transform spaces into letters or letters into spaces, thus blurring the true boundaries between the words.

```
Wt is ahear, h/wav3p, dhat by sending phc )hformatIon if a
rEdundaft fnre thd prkba@)hity ob erropc can be reduaed.
```

With 20% bit error, about half of the letters have errors and the message is completely illegible:

```
I4 "s C'd'rq h+Ae&d"( '(At by s'jdafd th$ hfFoPmati/. )f a
p(d5jdan' fLbe thd 'r''ab!DITy o& dr'kp1 aa& bE rd@u!ed.
```

These examples were all generated using the following MATLAB program `redun-dant.m` which takes the text `textm`, translates it into a binary string, and then causes `per` percent of the bits to be flipped. The program then gathers statistics on the resulting numbers of bit errors and symbol errors (how many letters were changed).

---

redundant.m: redundancy of written english in bits and letters

```
textm='It is clear, however, that by sending the information in a . . .
     redundant form the probability of errors can be reduced.'
ascm=dec2bin(textm);                    % 8-bit ascii (binary) equivalent of text
binm=reshape(ascm',1,8*length(textm)); % turn into one long binary string
per=.01;                                % probability of bit error
for i=1:8*length(textm)
   r=rand;                              % swap 0 and 1 with probability per
   if (r>1-per) & binm(i)=='0', binm(i)='1'; end
   if (r>1-per) & binm(i)=='1', binm(i)='0'; end
end
ascr=reshape(binm',8,length(textm))'; % back into ascii binary
textr=setstr(bin2dec(ascr)')          % back into text
biterror=sum(sum(abs(ascr-ascm)))     % total number of bit errors
symerror=sum(sign(abs(textm-textr)))  % total number of symbol errors
letterror=symerror/length(textm)      % number of "letter" errors
```

Problem

14.5. Read in a large text file using the following MATLAB code. (Use one of your own or use one of the included text files.)[2] Make a plot of the symbol error rate as a function of the bit error rate by running `redundant.m` for a variety of values of `per`. Examine the resulting text. At what value of `per` does the text become unreadable? What is the corresponding symbol error rate?

---

readtext.m: read in a text document and translate to character string

```
[fid,messagei] = fopen('OZ.txt','r'); % file must be in text format
fdata=fread(fid)';                     % read text as a vector
text=setstr(fdata);                    % change to a character string
```

Thus, for English text encoded as ASCII characters, a significant number of errors can occur (about 10% of the letters can be arbitrarily changed), without altering the meaning of the sentence. While these kinds of errors can be corrected by a human reader, the redundancy is not in a form that is easily exploited by a computer. Even imagining that the computer could look up words in a dictionary, the person knows from context that "It is clear" is a more likely phrase than "It is clean" when correcting Shannon's sentence with 1% errors. The person can figure out from context that "cAf" (from the phrase with 2% bit errors) must have had two errors by using the long term correlation of the sentence (i.e., its meaning). Computers do not deal readily with meaning.[3]

---

[2] *Through the Looking Glass* by Lewis Carroll (carroll.txt) and *Wonderful Wizard of Oz* by Frank Baum (OZ.txt) are available on the CD.

[3] A more optimistic rendering of this sentence: "Computers do not yet deal readily with meaning."

In the previous section, the information contained in a message was defined to depend on two factors: the number of symbols and their probability of occurrence. But this assumes that the symbols do not interact—that the letters are independent. How good an assumption is this for English text? It is a poor assumption. As the preceding examples suggest, normal English is highly correlated.

It is easy to catalog the frequency of occurrence of the letters. The letter "e" is the most common. In Frank Baum's *Wizard of Oz*, for instance, "e" appears 20,345 times and "t" appears 14,811 times, but the letters "q" and "x" appear only 131 and 139 times, respectively. ("z" might be a bit more common in Baum's book than normal because of the title). The percentage of occurrence for each letter in the *Wizard of Oz* is as follows:

$$
\begin{array}{llllllll}
a & 6.47 & h & 5.75 & o & 6.49 & v & 0.59 \\
b & 1.09 & i & 4.63 & p & 1.01 & w & 2.49 \\
c & 1.77 & j & 0.08 & q & 0.07 & x & 0.07 \\
d & 4.19 & k & 0.90 & r & 4.71 & y & 1.91 \\
e & 10.29 & l & 3.42 & s & 4.51 & z & 0.13 \\
f & 1.61 & m & 1.78 & t & 7.49 & & \\
g & 1.60 & n & 4.90 & u & 2.05 & &
\end{array}
\qquad (14.4)
$$

"Space" is the most frequent character, occurring 20% of the time. It was easier to use the following MATLAB code in conjunction with `readtext.m`, than to count the letters by hand.

---

freqtext.m: frequency of occurrence of letters in text

```
little=length(find(text=='t'));    % how many times t occurs
big=length(find(text=='T'));       % how many times T occurs
freq=(little+big)/length(text)     % percentage
```

If English letters were truly independent, then it should be possible to generate "English-like" text using this table of probabilities. Here is a sample:

```
Od m shous t ad schthewe be amalllingod ongoutorend youne he
Any bupecape tsooa w beves p le t ke teml ley une weg rloknd
```

which does not look anything like English. How can the nonindependence of the text be modeled? One way is to consider the probabilities of successive pairs of letters instead of the probabilities of individual letters. For instance, the pair "th" is quite frequent, occurring 11,014 times in the *Wizard of Oz*, while "sh" occurs 861 times. Unlikely pairs such as "wd" occur in only five places[4] and "pk" not at all. For example, suppose that "He" was chosen first. The next pair would be "e" followed by something, with the probability of the something dictated by the entries in the table. Following this procedure results in output like this:

```
Her gethe womfor if you the to had the sed th and the wention
At th youg the yout by and a pow eve cank i as saing paill
```

Observe that most of the two-letter combinations are actual words, as well as many three-letter words. Longer sets of symbols tend to wander improbably. While, in principle, it

---

[4] In the words "crowd" and "sawdust."

would be possible to continue gathering probabilities of all three-letter combinations, then four, etc., the table begins to get rather large (a matrix with 26^n elements would be needed to store all the *n*-letter probabilities). Shannon[5] suggests another way:

> ...one opens a book at random and selects a letter on the page. This letter is recorded. The book is then opened to another, page and one reads until this letter is encountered. The succeeding letter is then recorded. Turning to another page, this second letter is searched for, and the succeeding letter recorded, etc.

Of course, Shannon did not have access to MATLAB when he was writing in 1948. If he had, he might have written a program like textsim.m, which allows specification of any text (with default being *The Wizard of Oz*) and any number of terms for the probabilities. For instance, with m=1, the letters are chosen completely independently; with m=2, the letters are chosen from successive pairs; and with m=3, they are chosen from successive triplets. Thus, the probabilities of clusters of letters are defined implicitly by the choice of the source text.

---

textsim.m: use (large) text to simulate transition probabilities

```
m=1;                                % # terms for transition
linelength=60;                      % approx # letters in each line
load OZ.mat                         % file for input
n=text(1:m); nline=n; nlet='x';     % initialize variables
for i=1:100                         % length of output in lines
  j=1;
  while j<linelength | nlet~=' '    % scan through file
    k=findstr(text,n);              % find all occurrences of seed
    ind=round((length(k)-1)*rand)+1; % pick one
    nlet=text(k(ind)+m);
    if abs(nlet)==13                % pretend carriage returns
      nlet=' ';                     % are spaces
    end
    nline=[nline, nlet];            % add next letter
    n=[n(2:m),nlet];                % new seed
    j=j+1;
  end
  nline=[nline setstr(13)]          % format output/ add CRs
  nline='';                         % initialize next line
end
```

Typical output of textsim.m depends heavily on the number of terms m used for the transition probabilities. With m=1 or m=2, the results appear much as above. When m=3,

```
Be end claime armed yes a bigged wenty for me fearabbag girl
Humagine ther mightmarkling the many the scarecrow pass and
I havely and lovery wine end at then only we pure never
```

---

[5] "A Mathematical Theory of Communication," *The Bell System Technical Journal*, Vol 27, 1948.

many words appear, and many combinations of letters that might be words, but aren't quite. "Humagine" is suggestive, though it is not clear exactly what "mightmarkling" might mean. When m=4,

```
Water of everythinkies friends of the scarecrow no head we
She time toto be well as some although to they would her been
Them became the small directions and have a thing woodman
```

the vast majority of words are actual English, though the occasional conjunction of words (such as "everythinkies") is not uncommon. The output also begins to strongly reflect the text used to derive the probabilities. Since many four-letter combinations occur only once, there is no choice for the method to continue spelling a longer word; this is why the "scarecrow" and the "woodman" figure prominently. For m=5 and above, the "random" output is recognizably English, and strongly dependent on the text used:

```
Four trouble and to taken until the bread hastened from its
Back to you over the emerald city and her in toward the will
Trodden and being she could soon and talk to travely lady i
```

Problems

14.6. Run the program `textsim.m` using the input file `carroll.mat`, which contains the text to Lewis Carroll's *Through the Looking Glass*, with m=1, 2, ..., 8. At what point does the output repeat large phrases from the input text?

14.7. Run the program `textsim.m` using the input file `foreign.mat`, which contains a book that is not in English. Looking at the output for various m, can you tell what language the input is? What is the smallest m (if any) at which it becomes obvious?

The following two problems may not appeal to everyone:

14.8. The program `textsim.m` operates at the level of letters and the probabilities of transition between successive sets of m-length letter sequences. Write an analogous program that operates at the level of words and the probabilities of transition between successive sets of m-length word sequences. Does your program generate plausible sounding phrases or sentences?

14.9. There is nothing about the technique of `textsim.m` that is inherently limited to dealing with text sequences. Consider a piece of (notated) music as a sequence of symbols, labelled so that each "C" note is 1, each "C ♯" note is 2, each "D" note is 3, etc. Create a table of transition probabilities from a piece of music, and then generate "new" melodies in the same way that `textsim.m` generates "new" sentences. (Observe that this procedure can be automated using standard MIDI files as input.)

Because this method derives the multiletter probabilities directly from a text, there is no need to compile transition probabilities for other languages. Using Vergil's *Aeneid* (with m=3) gives

```
Aenere omnibus praeviscrimus habes ergemio nam inquae enies
Media tibi troius antis igna volae subilius ipsis dardatuli
Cae sanguina fugis ampora auso magnum patrix quis ait longuin
```

which is not real Latin. Similarly,

```
Que todose remosotro enga tendo en guinada y ase aunque lo
Se dicielos escubra la no fuerta pare la paragales posa derse
Y quija con figual se don que espedios tras tu pales del arrecermos
```

is not Spanish (the input file was Cervante's *Don Quijote*, also with m=3), and

```
Seule sontagne trait homarcher de la t au onze le quance matices
Maississait passepart penaient la ples les au cherche de je
Chamain peut accide bien avaien rie se vent puis il nez pande
```

is not French.[6]

The input file to the program `textsim.m` is a MATLAB `.mat` file that is preprocessed to remove excessive line breaks, spaces, and capitalization using `textman.m`, which is why there is no punctuation in these examples. A large assortment of text files is available for downloading at the website of Project Gutenberg (at http://promo.net/pg/).

Text, in a variety of languages, retains some of the character of its language with correlations of 3 to 5 letters (21–35 bits, when coded in ASCII). Thus, messages written in those languages are *not* independent, except possibly at lengths greater than this. A result from probability theory suggests that if the letters are clustered into blocks that are longer than the correlation, then the blocks may be (nearly) independent. This is one strategy to pursue when designing codes that seek to optimize performance. Section 14.5 will explore some practical ways to attack this problem, but the next two sections establish a measure of performance such that it is possible to know how close to the optimal any given code lies.

14.3 Entropy

This section extends the concept of information from a single symbol to a sequence of symbols. As defined by Shannon,[7] the information in a symbol is inversely proportional to its probability of occurring. Since messages are composed of sequences of symbols, it is important to be able to talk concretely about the average flow of information. This is called the *entropy* and is formally defined as

$$H(x) = \sum_{i=1}^{N} p(x_i) I(x_i)$$

$$= \sum_{i=1}^{N} p(x_i) \log \left(\frac{1}{p(x_i)} \right)$$

$$= -\sum_{i=1}^{N} p(x_i) \log(p(x_i)), \tag{14.5}$$

where the symbols are drawn from an alphabet x_i, each with probability $p(x_i)$. $H(x)$ sums the information in each symbol, weighted by the probability of that symbol. Those

---

[6] The source was *Le Tour du Monde en Quatre Vingts Jours*, a translation of Jules Verne's *Around the World in Eighty Days*.

[7] Actually, Hartley was the first to use this as a measure of information in his 1928 paper in the *Bell Systems Technical Journal* called "Transmission of Information".

familiar with probability and random variables will recognize this as an expectation. Entropy[8] is measured in bits per symbol, and so gives a measure of the average amount of information transmitted by the symbols of the source. Sources with different symbol sets and different probabilities have different entropies. When the probabilities are known, the definition is easy to apply.

Example 14.3

Consider the $N = 3$ symbol set defined in Example 14.1. The entropy is

$$H(x) = \frac{1}{2} \cdot 1 + \frac{1}{4} \cdot 2 + \frac{1}{4} \cdot 2 = 1.5 \text{ bits/symbol}.$$

Problem

14.10. Reconsider the fair die of Problem 14.1. What is its entropy?

Example 14.4

Suppose that the message $\{x_1, x_3, x_2, x_1\}$ is received from a source characterized by

1. $N = 4$, $p(x_1) = 0.5$, $p(x_2) = 0.25$, $p(x_3) = p(x_4) = 0.125$. The total information is

$$\sum_{i=1}^{4} I(x_i) = \sum_{i=1}^{4} \log\left(\frac{1}{p(x_i)}\right) = 1 + 2 + 3 + 3 = 9 \text{ bits}.$$

The entropy of the source is

$$H(x) = \frac{1}{2} \cdot 1 + \frac{1}{4} \cdot 2 + \frac{1}{8} \cdot 3 + \frac{1}{8} \cdot 3 = 1.75 \text{ bits/symbol}.$$

2. $N = 4$, $p(x_i) = 0.25$ for all i. The total information is $I(x_i) = 2 + 2 + 2 + 2 = 8$. The entropy of the source is

$$H(x) = \frac{1}{4} \cdot 2 + \frac{1}{4} \cdot 2 + \frac{1}{4} \cdot 2 + \frac{1}{4} \cdot 2 = 2 \text{ bits/symbol}.$$

---

[8] Warning: though the word is the same, this is not the same as the notion of entropy that is familiar from physics.

FIGURE 14.2 Entropy
of a binary signal with
probabilities p and
$1 - p$.

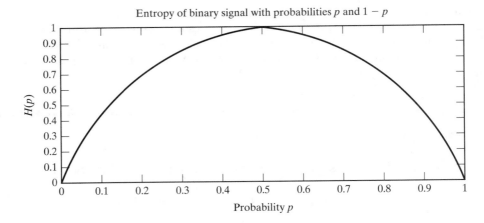

Entropy of binary signal with probabilities p and $1 - p$

Messages of the same length from the first source give less information than those from the second source. Hence, sources with the same number of symbols but different probabilities can have different entropies. The key is to design a system to maximize entropy since this will have the largest throughput, or largest average flow of information. But how can this be achieved?

First, consider the simple case in which there are two symbols in the alphabet, x_1 with probability p, and x_2 with probability $1 - p$. (Think of a coin that is weighted so as to give heads with higher probability than tails.) Applying the definition (14.5) shows that the entropy is

$$H(p) = -p\log(p) - (1 - p)\log(1 - p).$$

This is plotted as a function of p in Figure 14.2. For all allowable values of p, $H(p)$ is positive. As p approaches either zero or one, $H(p)$ approaches zero, which represent the symmetric cases in which either x_1 occurs all the time or x_2 occurs all the time, and no information is conveyed. $H(p)$ reaches its maximum in the middle, at $p = 0.5$. For this example, entropy is maximized when both symbols are equally likely.

Problem

14.11. Show that $H(p)$ is maximized at $p = 0.5$ by taking the derivative and setting it equal to zero.

The next result shows that an N-symbol source cannot have entropy larger than $\log(N)$, and that this bound is achieved when all the symbols are equally likely. Mathematically, $H(x) \leq \log(N)$, which is demonstrated by showing that $H(x) - \log(N) \leq 0$. First,

$$H(x) - \log(N) = \sum_{i=1}^{N} p(x_i)\log\left(\frac{1}{p(x_i)}\right) - \log(N)$$

$$= \sum_{i=1}^{N} p(x_i)\log\left(\frac{1}{p(x_i)}\right) - \sum_{i=1}^{N} p(x_i)\log(N),$$

since $\sum_{i=1}^{N} p(x_i) = 1$. Gathering terms, this can be rewritten

$$H(x) - \log(N) = \sum_{i=1}^{N} p(x_i) \left[\log \left(\frac{1}{p(x_i)} \right) - \log(N) \right]$$

$$= \sum_{i=1}^{N} p(x_i) \log \left(\frac{1}{Np(x_i)} \right),$$

and changing the base of the logarithm (using $\log(z) \equiv \log_2(z) = \log_2(e) \ln(z)$, where $\ln(z) \equiv \log_e(z)$), gives

$$H(x) - \log(N) = \log(e) \sum_{i=1}^{N} p(x_i) \ln \left(\frac{1}{Np(x_i)} \right).$$

If all symbols are equally likely, $p(x_i) = 1/N$, then $\frac{1}{Np(x_i)} = 1$ and $\ln \left(\frac{1}{Np(x_i)} \right) = \ln(1) = 0$. Hence $H(x) = \log(N)$. On the other hand, if the symbols are not equally likely, then the inequality $\ln(z) \le z - 1$ (which holds for $z \ge 0$) implies that

$$H(x) - \log(N) \le \log(e) \sum_{i=1}^{N} p(x_i) \left[\frac{1}{Np(x_i)} - 1 \right]$$

$$= \log(e) \left[\sum_{i=1}^{N} \frac{1}{N} - \sum_{i=1}^{N} p(x_i) \right]$$

$$= \log(e) [1 - 1] = 0. \tag{14.6}$$

Rearranging (14.6) gives the desired bound on the entropy, that $H(x) \le \log(N)$. This says that, all else being equal, it is preferable to choose a code in which each symbol occurs with the same probability. Indeed, Example 14.5 provides a concrete source for which the equal probability case has higher entropy than the unequal case.

Section 14.2 showed how letters in the text of natural languages do not occur with equal probability. Thus, naively using the letters will not lead to an efficient transmission. Rather, the letters must be carefully translated into equally probable symbols in order to increase the entropy. A method for accomplishing this translation is given in Section 14.5, but Section 14.4 examines the limits of attainable performance when transmitting symbols across a noisy (but otherwise perfect) channel.

14.4 Channel Capacity

Section 14.1 showed how much information (measured in bits) is contained in a given symbol, and Section 14.3 generalized this to the average amount of information contained in a sequence or set of symbols (measured in bits per symbol). In order to be useful in a communication system, however, the data must move from one place to another. What is the maximum amount of information that can pass through a channel in a given amount of time? The main result of this section is that the capacity of the channel defines the maximum possible flow of information through the channel. The capacity is a function of the bandwidth of the channel and of the amount of noise in the system, and it is measured in bits per second.

If the data is coded into $N = 2$ equally probable bits, then the entropy is $H_2 = 0.5 \log(2) + 0.5 \log(2) = 1$ bit per symbol. Why not increase the number of bits per symbol? This would allow representing more information. Doubling to $N = 4$, the entropy increases to $H_4 = 2$. In general, when using N bits, the entropy is $H_N = \log(N)$. By increasing N without bound, the entropy can be increased without bound! But is it really possible to send an infinite amount of information?

When doubling the size of N, one of two things must happen. Either the distance between the levels must decrease, or the power must increase. For instance, it is common to represent the binary signal as ± 1 and the four-level signal as $\pm 1, \pm 3$. In this representation, the distance between neighboring values is constant, but the power in the signal has increased. Recall that the power in a discrete signal $x[k]$ is

$$\lim_{T \to \infty} \frac{1}{T} \sum_{k=1}^{T} x^2[k].$$

For a binary signal with equal probabilities, this is $P_2 = \frac{1}{2}(1^2 + (-1)^2) = 1$. The four-level signal has power $P_4 = \frac{1}{4}(1^2 + (-1)^2 + 3^2 + (-3)^2) = 5$. To normalize the power to unity for the four-level signal, calculate the value x such that $\frac{1}{4}(x^2 + (-x)^2 + (3x)^2 + (-3x)^2) = 1$, which is $x = \sqrt{1/5}$. Figure 14.3 shows how the values of the N-level signal become closer together as N increases, when the power is held constant.

FIGURE 14.3 When the power is equal, the values of the N-level signal grow closer as N increases.

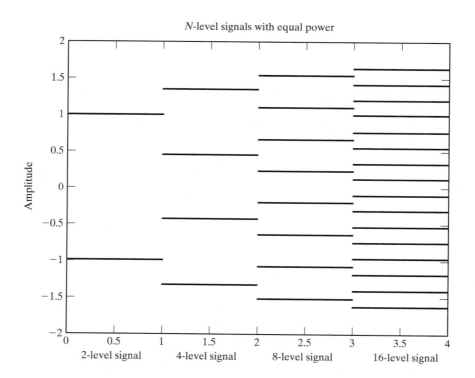

Now it will be clearer why it is not really possible to send an infinite amount of information in a single symbol. For a given transmitter power, the amplitudes become closer together for large N, and the sensitivity to noise increases. Thus, when there is noise (and some is inevitable), the four-level signal is more prone to errors than the two-level signal. Said another way, a higher signal-to-noise ratio[9] (SNR) is needed to maintain the same probability of error in the four-level signal as in the two-level signal.

Consider the situation in terms of the bandwidth required to transmit a given set of data containing M bits of information. From the Nyquist sampling theorem of Section 6.1, data can be sent through a channel of bandwidth B at a maximum rate of $2B$ symbols per second. If these symbols are coded into two binary levels, then M symbols must be sent. If the data are transmitted with four levels (by assigning pairs of binary digits to each four-level symbol), then only $M/2$ symbols are required. Thus the multilevel signal can operate at half the data rate of the binary signal. Said another way, the four-level signal requires only half the bandwidth of the two-level signal.

The previous two paragraphs show the tradeoff between signal-to-noise ratio and bandwidth. To maintain the same probability of error, larger bandwidth allows smaller SNR; larger SNR allows the use of a narrower frequency band. Quantifying this tradeoff was one of Shannon's greatest contributions.

While the details of a formal proof of the channel capacity are complex, the result is believable when thought of in terms of the relationship between the distance between the levels in a source alphabet and the average amount of noise that the system can tolerate. A digital signal with N levels has a maximum information rate $C = \frac{\log(N)}{T}$, where T is the time interval between transmitted symbols. C is the *capacity* of the channel, and has units of bits per second. This can be expressed in terms of the bandwidth B of the channel by recalling Nyquist's sampling theorem, which says that a maximum of $2B$ pulses per second can pass through the channel. Thus the capacity can be rewritten

$$C = 2B \log(N) \text{ bits per second.}$$

To include the effect of noise, observe that the power of the received signal is $S + P$ (where S is the power of the signal and P is the power of the noise). Accordingly, the average amplitude of the received signal is $\sqrt{S+P}$ and the average amplitude of the noise is $\sqrt{P}$. The average distance d between levels is twice the average amplitude divided by the number of levels (minus one), and so $d = \frac{2\sqrt{S+P}}{N-1}$. Many errors will occur in the transmission unless the distance between the signal levels is separated by at least twice the average amplitude of the noise, that is, unless

$$d = \frac{2\sqrt{S+P}}{N-1} > 2\sqrt{P}.$$

Rearranging this implies that $N-1$ must be no larger than $\frac{\sqrt{S+P}}{\sqrt{P}}$. The actual bound (as Shannon shows) is that $N \approx \frac{\sqrt{S+P}}{\sqrt{P}}$, and using this value gives

$$C = 2B \log\left(\frac{\sqrt{S+P}}{\sqrt{P}}\right) = B \log\left(1 + \frac{S}{P}\right) \text{ bits per second.} \qquad (14.7)$$

---

[9] As the term suggests, SNR is the ratio of the energy (or power) in the signal to the energy (or power) in the noise.

Observe that, if either the bandwidth or the SNR is increased, so is the channel capacity. For white noise, as the bandwidth increases, the power in the noise increases, the SNR decreases, and so the channel capacity does not become infinite. For a fixed channel capacity, it is easy to trade off bandwidth against SNR. For example, suppose a capacity of 1000 bits per second is required. Using a bandwidth of 1 KHz, we find that the signal and the noise can be of equal power. As the allowed bandwidth is decreased, the ratio $\frac{S}{P}$ increases rapidly:

Bandwidth	$\frac{S}{P}$
1000 Hz	1
500 Hz	3
250 Hz	15
125 Hz	255
100 Hz	1023

Shannon's result can now be stated succinctly. Suppose that there is a source producing information at a rate of R bits per second and a channel of capacity C. If $R < C$ (where C is defined as in (14.7)) then there exists a way to represent (or code) the data so that it can be transmitted with arbitrarily small error. Otherwise, the probability of error is strictly positive.

This is tantalizing and frustrating at the same time. The channel capacity defines the ultimate goal beyond which transmission systems cannot go, yet it provides no recipe for how to achieve the goal. The next sections describe various methods of representing or coding the data that assist in approaching this limit in practice.

The following MATLAB program explores a noisy system. A sequence of four-level data is generated by calling the pam.m routine. Noise is then added with power specified by p, the number of errors caused by this amount of noise is calculated in err.

---

noisychan.m: generate 4-level data and add noise

```
m=1000;                              % length of data sequence
p=1/15; s=1.0;                       % power of noise and signal
x=pam(m,4,s);                        % generate 4-PAM input with power 1...
l=sqrt(1/5);                         %      ...with amp levels 1
n=sqrt(p)*randn(1,m);                % generate noise with power p
y=x+n;                               % output of system adds noise to data
qy=quantalph(y,[-3*1,-1,1,3*1]);     % quantize output to [-3*1,-1,1,3*1]
err=sum(abs(sign(qy'-x)))/m;         % percent transmission errors
```

Typical outputs of noisychan.m are shown in Figure 14.4. Each plot shows the input sequence (the four solid horizontal lines), the input plus the noise (the cloud of small dots), and the error between the input and quantized output (the dark stars). Thus the dark stars that are not at zero represent errors in transmission. The noise $\mathcal{P}^\dagger$ in the right-hand case is the maximum noise allowable in the plausibility argument used to derive (14.7), which relates the average amplitudes of the signal plus the noise to the number of levels in the signal. For $\mathcal{S} = 1$ (the same conditions as in Problem 14.12(a)), the noise was chosen to be independent and normally distributed with power $\mathcal{P}^\dagger$ to ensure that $4 = \frac{\sqrt{1+\mathcal{P}^\dagger}}{\sqrt{\mathcal{P}^\dagger}}$.

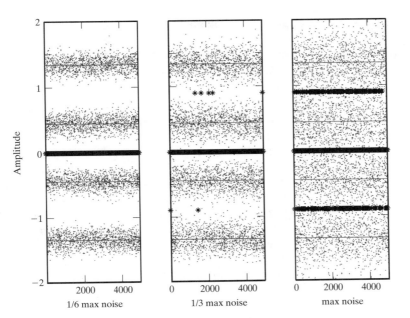

FIGURE 14.4 Each plot shows a four-level PAM signal (the four solid lines), the signal plus noise (the scattered dots), and the error between the data and the quantized output (the dark stars). The noise in the right-hand plot was at the Shannon limit $N \approx \frac{\sqrt{S + \mathcal{P}^\dagger}}{\sqrt{\mathcal{P}^\dagger}}$, in the middle plot the noise was at one-third the power, and in the left-hand plot the noise was at one-sixth the power.

The middle plot used a noise with power $\mathcal{P}^\dagger/3$ and the left-hand plot had noise power $\mathcal{P}^\dagger/6$. As can be seen from the plots, there were essentially no errors when using the smallest noise, a handful of errors in the middle, and about 6% errors when the power of the noise matched the Shannon capacity. Thus, this naive transmission of four-level data (i.e., with no coding) has many more errors than the Shannon limit suggests.

Problems

14.12. Find the amplitudes of the N-level (equally spaced) signal with unity power when
  (a) $N = 4$.
  (b) $N = 6$.
  (c) $N = 8$.

14.13. Use `noisychan.m` to compare the noise performance of two-level, four-level, and six-level transmissions.
  (a) Modify the program to generate two- and six-level signals.
  (b) Make a plot of the noise power versus the percentage of errors for two, four, and six levels.

14.14. Use `noisychan.m` to compare the power requirements for two-level, four-level, and six-level transmissions. Fix the noise power at p=0.01, and find the error probability for four-level transmission. Experimentally find the power S that is required to make the two-level and six-level transmissions have the same probability of error. Can you think of a way to calculate this?

14.15. Consider the (asymmetric, nonuniformly spaced) alphabet consisting of the symbols $-1, 1, 3, 4$.
  (a) Find the amplitudes of this 4-level signal with unity power.

(b) Use `noisychan.m` to examine the noise performance of this transmission by making a plot of the noise power versus percentage of errors.

(c) Compare this alphabet to 4-PAM with the standard alphabet $\pm 1, \pm 3$. Which would you prefer?

There are two different problems that can keep a transmission system from reaching the Shannon limit. The first is that the source may not be coded with maximum entropy, and this will be discussed next in Section 14.5. The second is when different symbols experience different amounts of noise. Recall that the plausibility argument for the channel capacity rested on the idea of the average noise. When symbols encounter anything less than the average noise, then all is well, since the average distance between levels is greater than the average noise. But errors occur when symbols encounter more than the average amount of noise. (This is why there are so many errors in the right-hand plot of Figure 14.4.) Good coding schemes try to ensure that all symbols experience (roughly) the average noise. This can be accomplished by grouping the symbols into clusters or blocks that distribute the noise evenly among all the symbols in the block. Such error coding is discussed in Section 14.6.

14.5 Source Coding

The results from Section 14.3 suggest that, all else being equal, it is preferable to choose a code in which each symbol occurs with the same probability. But what if the symbols occur with widely varying frequencies? Recall that this was shown in Section 14.2 for English and other natural languages. There are two basic approaches. The first aggregates the letters into clusters, and provides a new (longer) code word for each cluster. If properly chosen, the new code words can occur with roughly the same probability. The second approach uses variable-length code words, assigning short codes to common letters like "e" and long codes to infrequent letters like "x." Perhaps the most common variable-length code was that devised by Morse for telegraph operators, which used a sequence of "dots" and "dashes" (along with silences of various lengths) to represent the letters of the alphabet.

Before discussing how source codes can be constructed, consider an example using the $N = 4$ code from Example 14.4(a) in which $p(x_1) = 0.5$, $p(x_2) = 0.25$, and $p(x_3) = p(x_4) = 0.125$. As shown earlier, the entropy of this source is 1.75 bits/symbol, which means that there must be some way of coding the source so that, on average, 1.75 bits are used for each symbol. The naive approach to this source would use 2 bits for each symbol, perhaps assigning

$$x_1 \leftrightarrow 11, \quad x_2 \leftrightarrow 10, \quad x_3 \leftrightarrow 01, \quad \text{and } x_4 \leftrightarrow 00. \tag{14.8}$$

An alternative representation is

$$x_1 \leftrightarrow 1, \quad x_2 \leftrightarrow 01, \quad x_3 \leftrightarrow 001, \quad \text{and } x_4 \leftrightarrow 000, \tag{14.9}$$

where more probable symbols use fewer bits, and less probable symbols require more. For instance, the string

$$x_1, x_2, x_1, x_4, x_3, x_1, x_1, x_2$$

(in which each element appears with the expected frequency) is coded as

$$10110010001101.$$

This requires 14 bits to represent the 8 symbols. The average is $14/8 = 1.75$ bits per symbol, and so this coding is as good as possible, since it equals the entropy. In contrast, the naive code of (14.8) requires 16 bits to represent the 8 symbols for an average of 2 bits per symbol. One feature of the variable length code in (14.9) is that there is never any ambiguity about where it starts, since any occurrence of a 1 corresponds to the end of a symbol. The naive code requires knowledge of where the first symbol begins. For example, the string $01 - 10 - 11 - 00 - 1\_$ is very different from $\_0 - 11 - 01 - 10 - 01$ even though they contain the same bits in the same order. Codes for which the start and end are immediately recognizable are called *instantaneous* or *prefix* codes.

Since the entropy defines the smallest number of bits that can be used to encode a source, it can be used to define the *efficiency* of a code

$$\text{efficiency} = \frac{\text{entropy}}{\text{average number of bits per symbol used in code}}. \tag{14.10}$$

Thus the efficiency of the naive code (14.8) is $1.75/2 = 0.875$ while the efficiency of the variable rate code (14.9) is 1. Shannon's source coding theorem says that if an independent source has entropy H, then there exists a prefix code in which the average number of bits per symbol is between H and $H + 1$. Moreover, there is no uniquely decodable code that has smaller average length. Thus, if N symbols (each with entropy H) are compressed into less than NH bits, information is lost, while information need not be lost if $N(H + 1)$ bits are used. Shannon has defined the goal towards which all codes aspire, but provides no way of finding good codes for any particular case.

Fortunately, D. A. Huffman proposed an organized procedure to build variable-length codes that are as efficient as possible. Given a set of symbols and their probabilities, the procedure is as follows:

1. List the symbols in order of decreasing probability. These are the original "nodes."
2. Find the two nodes with the smallest probabilities, and combine them into one new node, with probability equal to the sum of the two. Connect the new nodes to the old ones with "branches" (lines).
3. Continue combining the pairs of nodes with the smallest probabilities. (If there are ties, pick any of the tied symbols).
4. Place a 0 or a 1 along each branch. The path from the rightmost node to the original symbol defines a binary list, which is the code word for that symbol.

This procedure is probably easiest to understand by working through an example. Consider again the $N = 4$ code from Example 14.4(a) in which the symbols have probabilities $p(x_1) = 0.5$, $p(x_2) = 0.25$, and $p(x_3) = p(x_4) = 0.125$. Following the foregoing procedure leads to the chart shown in Figure 14.5. In the first step, x_3 and x_4 are combined to form a new node with probability equal to 0.25 (the sum $p(x_3) + p(x_4)$). Then this new node is combined with x_2 to form a new node with probability 0.5. Finally, this is combined with x_1 to form the rightmost node. Each branch is now labelled. The

FIGURE 14.5 The Huff-
man code for the
source defined in
Example 14.4(a) can
be read directly from
this chart, which is
constructed using the
procedure (1) to (4) in
the text.

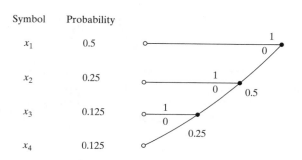

convention used in Figure 14.5 is to place a 1 on the top and a 0 on the bottom (assigning the binary digits in another order just relabels the code). The Huffman code for
this source can be read from the chart. Reading from the right-hand side, x_1 corresponds to 1, x_2 corresponds 01, x_3 to 001 and x_4 to 000. This is the same code as in
(14.9).

The Huffman procedure with a consistent branch labelling convention always leads
to a prefix code because all the symbols end the same (except for the maximal length
symbol x_4). More importantly, it always leads to a code that has average length very
near the optimal.

Problems

14.16. Consider the source with $N = 5$ symbols with probabilities $p(x_1) = 1/16$,
$p(x_2) = 1/8$, $p(x_3) = 1/4$, $p(x_4) = 1/16$, and $p(x_5) = 1/2$.
   (a) What is the entropy of this source?
   (b) Build the Huffman chart.
   (c) Show that the Huffman code is

$$x_1 \leftrightarrow 0001, \quad x_2 \leftrightarrow 001, \quad x_3 \leftrightarrow 01, \quad x_4 \leftrightarrow 0000, \quad \text{and } x_5 \leftrightarrow 1.$$

   (d) What is the efficiency of this code?
   (e) If this source were encoded naively, how many bits per symbol are needed?
   What is the efficiency?

14.17. Consider the source with $N = 4$ symbols with probabilities $p(x_1) = 0.3$,
$p(x_2) = 0.3$, $p(x_3) = 0.2$, and $p(x_4) = 0.2$.
   (a) What is the entropy of this source?
   (b) Build the Huffman code.
   (c) What is the efficiency of this code?
   (d) If this source were encoded naively, how many bits per symbol are needed?
   What is the efficiency?

14.18. Build the Huffman chart for the source defined by the 26 English letters (plus
"space") and their frequency in the *Wizard of Oz* as given in (14.4).

The MATLAB program `codex.m` demonstrates how a variable length code can be
encoded and decoded. The first step generates a 4-PAM sequence with the probabilities used in Example 14.4(a). In the code, the symbols are assigned numerical values
$\{\pm 1, \pm 3\}$. The symbols, their probabilities, the numerical values, and the variable length

Huffman code are as follows:

symbol	probability	value	Huffman code
x_1	0.5	$+1$	1
x_2	0.25	-1	01
x_3	0.125	$+3$	001
x_4	0.125	-3	000

This Huffman code was derived in Figure 14.5. For a length m input sequence, the second step replaces each symbol value with the appropriate binary sequence, and places the output in the vector cx.

---

codex.m: step 2: encode the sequence using Huffman code

```
j=1;
for i=1:m
    if x(i)==+1, cx(j:j)=[1];        j=j+1; end
    if x(i)==-1, cx(j:j+1)=[0,1];    j=j+2; end
    if x(i)==+3, cx(j:j+2)=[0,0,1];  j=j+3; end
    if x(i)==-3, cx(j:j+2)=[0,0,0];  j=j+3; end
end
```

The third step carries out the decoding. Assuming the encoding and decoding have been done properly, then cx is transformed into the output y, which should be the same as the original sequence x.

---

codex.m: step 3: decode the variable length sequence

```
j=1; i=1;
while i<=length(cx)
    if     cx(i:i)==[1],       y(j)=+1; i=i+1; j=j+1;
    elseif cx(i:i+1)==[0,1],   y(j)=-1; i=i+2; j=j+1;
    elseif cx(i:i+2)==[0,0,1], y(j)=+3; i=i+3; j=j+1;
    elseif cx(i:i+2)==[0,0,0], y(j)=-3; i=i+3; j=j+1; end
end
```

Indeed, running the program codex.m (which contains all three steps) gives a perfect decoding.

Problems

14.19. Mimicking the code in codex.m, create a Huffman encoder and decoder for the source defined in Example 14.16.

14.20. Use codex.m to investigate what happens when the probabilities of the source alphabet change.

(a) Modify step 1 of the program so that the elements of the input sequence have probabilities

$$x_1 \leftrightarrow 0.1, \ x_2 \leftrightarrow 0.1, \ x_3 \leftrightarrow 0.1, \ \text{and} \ x_4 \leftrightarrow 0.7. \qquad (14.11)$$

(b) Without changing the Huffman encoding to account for these changed probabilities, compare the average length of the coded data vector cx with the average length of the naive encoder (14.8). Which does a better job compressing the data?

(c) Modify the program so that the elements of the input sequence all have the same probability, and answer the same question.

(d) Build the Huffman chart for the probabilities defined in (14.11).

(e) Implement this new Huffman code and compare the average length of the coded data cx to the previous results. Which does a better job compressing the data?

14.21. Using codex.m, implement the Huffman code from Problem 14.18. What is the length of the resulting data when applied to the text of the *Wizard of Oz*? What rate of data compression has been achieved?

Source coding is used to reduce the redundancy in the original data. If the letters in the *Wizard of Oz* were independent, then the Huffman coding in Problem 14.21 would be optimal: no other coding method could achieve a better compression ratio. But the letters are not independent. More sophisticated schemes would consider not just the raw probabilities of the letters, but the probabilities of pairs of letters, or of triplets, or more. As suggested by the redundancy studies in Section 14.2, there is a lot that can be gained by exploiting higher order relationships between the symbols.

Problems

14.22. "Zipped" files (usually with a .zip extension) are a popular form of data compression for text (and other data) on the web. Download a handful of .zip files. Note the file size when the data is in its compressed form and the file size after decompressing ("unzipping") the file. How does this compare with the compression ratio achieved in Problem 14.21?

14.23. Using the routine writetext.m (this file, which can be found on the CD, uses the MATLAB command fwrite), write the *Wizard of Oz* text to a file OZ.doc. Use a compression routine (uuencode on a Unix or Linux machine, zip on a Windows machine, or *stuffit* on a Mac) to compress OZ.doc. Note the file size when the file is in its compressed form, and the file size after decompressing. How does this compare with the compression ratio achieved in Problem 14.21?

14.6 Channel Coding

The job of channel or error-correcting codes is to add some redundancy to a signal before it is transmitted so that it becomes possible to detect when errors have occurred and to correct them, when possible.

Perhaps the simplest technique is to send each bit three times. Thus, in order to transmit a 0, the sequence 000 is sent. In order to transmit 1, 111 is sent. This is the *encoder*. At the receiver, there must be a *decoder*. There are eight possible sequences that can be received, and a "majority rules" decoder assigns

$$000 \leftrightarrow 0 \quad 001 \leftrightarrow 0 \quad 010 \leftrightarrow 0 \quad 100 \leftrightarrow 0$$
$$101 \leftrightarrow 1 \quad 110 \leftrightarrow 1 \quad 011 \leftrightarrow 1 \quad 111 \leftrightarrow 1. \tag{14.12}$$

This encoder/decoder can identify and correct any isolated single error and so the transmission has smaller probability of error. For instance, assuming no more than one error per block, if 101 was received, then the error must have occurred in the middle bit, while if 110 was received, then the error must have been in the third bit. But the majority rules coding scheme is costly: three times the number of symbols must be transmitted, which reduces the bit rate by a factor of three. Over the years, many alternative schemes have been designed to reduce the probability of error in the transmission, without incurring such a heavy penalty.

Linear block codes are popular because they are easy to design, easy to implement, and because they have a number of useful properties. With $n > k$, an (n, k) linear code operates on sets of k symbols, and transmits a length n code word for each set. Each code is defined by two matrices: the k by n *generator* matrix G, and the $n - k$ by n *parity check* matrix H. In outline, the operation of the code is as follows:

1. Collect k symbols into a vector $\mathbf{x} = \{x_1, x_2, \ldots, x_k\}$.
2. Transmit the length n code word $\mathbf{c} = \mathbf{x}G$.
3. At the receiver, the vector $\mathbf{y}$ is received. Calculate $\mathbf{y}H^T$.
4. If $\mathbf{y}H^T = 0$, then no errors have occurred.
5. When $\mathbf{y}H^T \neq 0$, errors have occurred. Look up $\mathbf{y}H^T$ in a table of "syndromes," which contains a list of all possible received values and the most likely code word to have been transmitted, given the error that occurred.
6. Translate the corrected code word back in to the vector $\mathbf{x}$.

The simplest way to understand this is to work through an example in detail.

14.6.1 A (5,2) Binary Linear Block Code

To be explicit, consider the case of a $(5, 2)$ binary code with generator matrix

$$G = \begin{bmatrix} 1 & 0 & 1 & 0 & 1 \\ 0 & 1 & 0 & 1 & 1 \end{bmatrix} \tag{14.13}$$

and parity check matrix

$$H^T = \begin{bmatrix} 1 & 0 & 1 \\ 0 & 1 & 1 \\ 1 & 0 & 0 \\ 0 & 1 & 0 \\ 0 & 0 & 1 \end{bmatrix}. \tag{14.14}$$

This code bundles the bits into pairs, and the four corresponding code words are

$$x_1 = 00 \leftrightarrow c_1 = x_1 G = 00000,$$

$$x_2 = 01 \leftrightarrow c_2 = x_2 G = 01011,$$

$$x_3 = 10 \leftrightarrow c_3 = x_3 G = 10101,$$

and

$$x_4 = 11 \leftrightarrow c_4 = x_4 G = 11110.$$

There is one subtlety. The arithmetic used in the calculation of the code words (and indeed throughout the linear block code method) is not standard. Because the input source is binary, the arithmetic is also binary. Binary addition and multiplication are shown in Table 14.1. The operations of binary arithmetic may be more familiar as *exclusive OR* (binary addition), and *logical AND* (binary multiplication).

In effect, at the end of every calculation, the answer is taken modulo 2. For instance, in standard arithmetic, $x_4 G = 11112$. The correct code word c_4 is found by reducing each calculation modulo 2. In MATLAB, this is done with `mod(x4*g,2)` where `x4=[1,1]` and `g` is defined as in (14.13). In modulo 2 arithmetic, 1 represents any odd number and 0 represents any even number. This is also true for negative numbers so that, for instance, $-1 = +1$ and $-4 = 0$.

After transmission, the received signal $\mathbf{y}$ is multiplied by H^T. If there were no errors in transmission, then $\mathbf{y}$ is equal to one of the four code words c_i. With H defined as in (14.14), $c_1 H^T = c_2 H^T = c_3 H^T = c_4 H^T = 0$, where the arithmetic is binary, and where 0 means the zero vector of size 1 by 3 (in general, 1 by $(n-k)$). Thus $\mathbf{y} H^T = 0$ and the received signal is one of the code words.

However, when there are errors, $\mathbf{y} H^T \neq 0$, and the value can be used to determine the most likely error to have occurred. To see how this works, rewrite

$$\mathbf{y} = \mathbf{c} + (\mathbf{y} - \mathbf{c}) \equiv \mathbf{c} + \mathbf{e},$$

where $\mathbf{e}$ represents the error(s) that have occurred in the transmission. Note that

$$\mathbf{y} H^T = (\mathbf{c} + \mathbf{e}) H^T = \mathbf{c} H^T + \mathbf{e} H^T = \mathbf{e} H^T,$$

since $\mathbf{c} H^T = 0$. The value of $\mathbf{e} H^T$ is used by looking in the syndrome Table 14.2. For example, suppose that the symbol $x_2 = 01$ is transmitted using the code $c_2 = 01011$. But an error occurs in transmission so that $\mathbf{y} = 11011$ is received. Multiplication by the parity check matrix gives $\mathbf{y} H^T = \mathbf{e} H^T = 101$. Looking this up in the syndrome table shows that the most likely error was 10000. Accordingly, the most likely code word to have been transmitted was $\mathbf{y} - \mathbf{e} = 11011 - 10000 = 01011$, which is indeed the correct code word c_2.

On the other hand, if more than one error occurred in a single symbol, then the (5,2) code cannot necessarily find the correct code word. For example, suppose that the symbol $x_2 = 01$ is transmitted using the code $c_2 = 01011$ but that two errors occur in transmission so that $\mathbf{y} = 00111$ is received. Multiplication by the parity check matrix gives $\mathbf{y} H^T = \mathbf{e} H^T = 111$. Looking this up in the syndrome table shows that the most likely error was 10010. Accordingly, the most likely symbol to have been transmitted

Table 14.1.   Modulo 2 Arithmetic

+	0	1		·	0	1
0	0	1		0	0	0
1	1	0		1	0	1

Table 14.2. Syndrome Table for the binary (5, 2) code with generator matrix (14.13) and parity check matrix (14.14)

Syndrome $\mathbf{e}H^T$	Most likely error $\mathbf{e}$
000	00000
001	00001
010	00010
011	01000
100	00100
101	10000
110	11000
111	10010

was $\mathbf{y} - \mathbf{e} = 00111 + 10010 = 10101$, which is the code word c_3 corresponding to the symbol x_3, and not c_2.

The syndrome table can be built as follows. First, take each possible single error pattern, that is, each of the $n = 5$ $\mathbf{e}$'s with exactly one 1, and calculate $\mathbf{e}H^T$ for each. As long as the columns of H are nonzero and distinct, each error pattern corresponds to a different syndrome. To fill out the remainder of the table, take each of the possible double errors (each of the $\mathbf{e}$'s with exactly two 1's) and calculate $\mathbf{e}H^T$. Pick two that correspond to the remaining unused syndromes. Since there are many more possible double errors $n(n-1) = 20$ than there are syndromes ($2^{n-k} = 8$), these are beyond the ability of the code to correct.

The MATLAB program `blockcode52.m` shows details of how this encoding and decoding proceeds. The first part defines the relevant parameters of the (5, 2) binary linear block code: the generator g, the parity check matrix h, and the syndrome table syn. The rows of syn are ordered so that the binary digits of eH^T can be used to directly index into the table.

---

blockcode52.m: Part 1: Definition of (5,2) binary linear block code the generator and parity check matrices

```
g=[1 0 1 0 1;
   0 1 0 1 1];
h=[1 0 1 0 0;
   0 1 0 1 0;
   1 1 0 0 1];
% the four code words cw=x*g (mod 2)
x(1,:)=[0 0]; cw(1,:)=mod(x(1,:)*g,2);
x(2,:)=[0 1]; cw(2,:)=mod(x(2,:)*g,2);
x(3,:)=[1 0]; cw(3,:)=mod(x(3,:)*g,2);
x(4,:)=[1 1]; cw(4,:)=mod(x(4,:)*g,2);
% the syndrome table
syn=[0 0 0 0 0;
     0 0 0 0 1;
     0 0 0 1 0;
     0 1 0 0 0;
     0 0 1 0 0;
     1 0 0 0 0;
     1 1 0 0 0;
     1 0 0 1 0];
```

The second part carries out the encoding and decoding process. The variable p specifies the chance that bit errors will occur in the transmission. The code words c are constructed using the generator matrix. The received signal is multiplied by the parity check matrix h to give the syndrome, which is then used as an index into the syndrome table (matrix) syn. The resulting "most likely error" is subtracted from the received signal, and this is the "corrected" code word that is translated back into the message. Because the code is linear, code words can be translated back into the message using an "inverse" matrix[10] and there is no need to store all the code words. This becomes important when there are millions of possible code words, but when there are only four it is not crucial. The translation is done in blockcode52.m in the for j loop with by searching.

---

blockcode52.m: Part 2: encoding and decoding data

```
p=.1;                                % probability of bit flip
m=10000;                             % length of message
dat=0.5*(sign(rand(1,m)-0.5)+1);     % m random 0s and 1s
for i=1:2:m
  c=mod([dat(i) dat(i+1)]*g,2);      % build code word
  for j=1:length(c)
    if rand<p, c(j)=-c(j)+1; end     % flip bits with prob p
  end
  y=c;                               % received signal
  eh=mod(y*h',2);                    % multiply by parity check h'
  ehind=eh(1)*4+eh(2)*2+eh(3)+1;     % turn syndrome into index
  e=syn(ehind,:);                    % error from syndrome table
  y=mod(y-e,2);                      % add e to correct errors
  for j=1:max(size(x))               % recover message from code words
    if y==cw(j,:),  z(i:i+1)=x(j,:); end
  end
end
err=sum(abs(z-dat))                  % how many errors occurred
```

Running blockcode52.m with the default parameters of 10% bit errors and length m=10000 will give about 400 errors, a rate of about 4%. Actually, as will be shown in the next section, the performance of this code is slightly better than these numbers suggest, because it is also capable of detecting certain errors that it cannot correct, and this feature is not implemented in blockcode52.m.

Problems

14.24. Use blockcode52.m to investigate the performance of the binary (5, 2) code. Let p take on a variety of values $p = 0.001, 0.01, 0.02, 0.05, 0.1, 0.2, 0.5$ and plot the percentage of errors as a function of the percentage of bits flipped.

14.25. This exercise compares the performance of the (5,2) block code in a more "real-istic" setting and provides a good warm-up exercise for the receiver to be built in Chapter 15. The program nocode52.m (all MATLAB files are available on the CD) provides a template with which you can add the block coding into a "real" transmitter and receiver pair. Observe, in particular, that the block coding is placed after the translation of the text into binary but before the translation into 4-PAM (for transmission). For efficiency, the text is encoded using text2bin.m

---

[10] This is explored in the context of blockcode52.m in Problem 14.26.

(recall Example 8.2). At the receiver, the process is reversed: the raw 4-PAM data is translated into binary, then decoded using the (5,2) block decoder, and finally translated back into text (using `bin2text.m`) where you can read it. Your task in this problem is to experimentally verify the gains possible when using the (5,2) code. First, merge the programs `blockcode52.m` and `nocode52.m`. Measure the number of errors that occur as noise is increased (the variable `varnoise` scales the noise). Make a plot of the number of errors as the variance increases. Compare this with the number of errors that occur as the variance increases when no coding is used (i.e., running `nocode52.m` without modification).

14.26. Use the matrix `ginv=[1 1;1 0 ;0 0;1 0;0 1];` to replace the `for j` loop in `blockcode52.m`. Observe that this reverses the effect of constructing the code words from the x since `cw*ginv=x (mod 2)`.

14.27. Implement the simple majority rules code described in (14.12).
  (a) Plot the percentage of errors after coding as a function of the number of symbol errors.
  (b) Compare the performance of the majority rules code to the (5, 2) block code.
  (c) Compare the data rate required by the majority rules code to that required by the (5, 2) code, and to the naive (no coding) case.

14.6.2 Minimum Distance of a Linear Code

In general, linear codes work much like the example in the previous section, although the generator matrix, parity check matrix, and the syndrome table are unique to each code. The details of the arithmetic may also be different when the code is not binary. Two examples will be given later. This section discusses the general performance of linear block codes in terms of the *minimum distance* of a code, which specifies how many errors the code can detect and how many errors it can correct.

A code C is a collection of code words c_i which are n-vectors with elements drawn from the same alphabet as the source. An encoder is a rule that assigns a k-length message to each codeword.

Example 14.5

The code words of the (5, 2) binary code are 00000, 01011, 10101, and 11110, which are assigned to the four input pairs 00, 01, 10, and 11 respectively.

The *Hamming distance*[11] between any two elements in C is equal to the number of places in which they disagree. For instance, the distance between 00000 and 01011 is three, which is written $d(00000, 01011) = 3$. The distance between 1001 and 1011 is

---

[11] Named after R. Hamming, who also created the Hamming blip as a windowing function. *Telecommunication Breakdown* adopted the blip in previous chapters as a convenient pulse shape.

$d(1001, 1011) = 1$. The *minimum distance* of a code C is the smallest distance between any two code words. In symbols,

$$d_{min} = \min_{i \neq j} d(c_i, c_j),$$

where $c_i \in C$.

Problems

14.28. Show that the minimum distance of the (5, 2) binary linear block code is $d_{min} = 3$.

14.29. Write down all code words for the majority rules code (14.12). What is the minimum distance of this code?

14.30. A code C has four elements {0000, 0101, 1010, 1111}. What is the minimum distance of this code?

Let $D_i(t)$ be the "decoding sets" of all possible received signals that are less than t away from c_i. For instance, the majority rules code has two code words, and hence two decoding sets. With $t = 1$, these are

$$D_1(1) = \{000, \ 001, \ 100, \ 010\} \text{ and } D_2(1) = \{111, \ 110, \ 011, \ 101\}. \tag{14.15}$$

When any of the elements in $D_1(1)$ are received, the code word $c_1 = 0$ is used; when any of the elements in $D_2(1)$ are received, the code word $c_2 = 1$ is used. For $t = 0$, the decoding sets are

$$D_1(0) = \{000\} \text{ and } D_2(0) = \{111\}. \tag{14.16}$$

In this case, when 000 is received, c_1 is used; when 111 is received, c_2 is used. When the received bits are in neither of the D_i, an error is detected, though it cannot be corrected. When $t > 1$, the $D_i(t)$ are not disjoint and so cannot be used for decoding.

Problems

14.31. What are the $t = 0$ decoding sets for the four-element code in Problem 14.30? Are the $t = 1$ decoding sets disjoint?

14.32. Write down all possible disjoint decoding sets for the (5, 2) linear binary block code.

One use of decoding sets lies in their relationship with d_{min}. If $2t < d_{min}$, then the decoding sets are disjoint. Suppose that the code word c_i is transmitted over a channel, but that c (which is obtained by changing at most t components of c_i) is received. Then c still belongs to the correct decoding set D_i, and is correctly decoded. This is an error-correction code that handles up to t errors.

Now suppose that the decoding sets are disjoint with $2t + s < d_{min}$, but that $t < d(c, c_i) \leq t + s$. Then c is not a member of any decoding set. Such an error cannot be corrected by the code, though it is detected. The following example shows how the ability to detect errors and the ability to correct them can be traded off.

Example 14.6

Consider again the majority rules code C with two elements {000, 111}. This code has $d_{min} = 3$ and can be used as follows:

1. $t = 1$, $s = 0$. In this mode, using decoding sets (14.15), code words could suffer any single error and still be correctly decoded. But if two errors occurred, the message would be incorrect.
2. $t = 0$, $s = 2$. In this mode, using decoding sets (14.16), the code word could suffer up to two errors and the error would be detected, but there would be no way to correct it with certainty.

Example 14.7

Consider the code C with two elements {0000000, 1111111}. Then, $d_{min} = 7$. This code can be used in the following ways:

1. $t = 3$, $s = 0$. In this mode, the code word could suffer up to three errors and still be correctly decoded. But if four errors occurred, the message would be incorrect.
2. $t = 2$, $s = 2$. In this mode, if the codeword suffered up to two errors, then it would be correctly decoded. If there were three or four errors, then the errors are detected, but because they cannot be corrected with certainty, no (incorrect) message is generated.

Thus, the minimum distance of a code is a resource that can be allocated between error detection and error correction. How to trade these off is a system design issue. In some cases, the receiver can ask for a symbol to be retransmitted when an error occurs (for instance, in a computer modem or when reading a file from disk), and it may be sensible to allocate d_{min} to detecting errors. In other cases (such as broadcast), it is more common to focus on error correction.

The discussion in this section so far is completely general; that is, the definition and results on minimum distance apply to any code of any size, whether linear or nonlinear. There are two problems with large nonlinear codes:

- It is hard to specify codes with large d_{min}.
- Implementing coding and decoding can be expensive in terms of memory and computational power.

To emphasize this, consider a code that combines binary digits into clusters of 56 and codes these clusters using 64 bits. Such a code requires about 10^{101} code words. Considering that the estimated number of elementary particles in the universe is about

Table 14.3.   Syndrome Table for the binary (7, 3) code.

Syndrome $\mathbf{e}H^T$	Most likely error $\mathbf{e}$
0000	0000000
0001	0000001
0010	0000010
0100	0000100
1000	0001000
1101	0010000
1011	0100000
0111	1000000
0011	0000011
0110	0000110
1100	0001100
0101	0011000
1010	0001010
1001	0010100
1110	0111000
1111	0010010

10^{80}, this is a problem. When the code is linear, however, it is not necessary to store all the code words; they can be generated as needed. This was noted in the discussion of the (5, 2) code of the previous section. Moreover, finding the minimum distance of a linear code is also easy, since d_{min} is equal to the smallest number of nonzero coordinates in any code word (not counting the zero code word). Thus d_{min} can be calculated directly from the definition by finding the distances between all the code words, or by finding the code word that has the smallest number of 1's. For instance, in the (5, 2) code, the two elements 01011 and 10101 each have exactly three nonzero terms.

14.6.3 Some More Codes

This section gives two examples of (n, k) linear codes. If the generator matrix G has the form

$$G = [I_k | P],\tag{14.17}$$

where I_k is the k by k identity matrix and P is some k by $n - k$ matrix, then

$$[I_k | P] \left[\begin{array}{c} -P \\ --- \\ I_{n-k} \end{array} \right] = 0,\tag{14.18}$$

where the 0 is the k by $n-k$ matrix of all zeroes. Hence, define $H = [-P | I_{n-k}]$. Observe that the (5,2) code is of this form, since in binary arithmetic $-1 = +1$ and so $-P = P$.

Example 14.8

A (7,3) binary code has generator matrix

$$G = \begin{bmatrix} 1 & 0 & 0 & 0 & 1 & 1 & 1 \\ 0 & 1 & 0 & 1 & 0 & 1 & 1 \\ 0 & 0 & 1 & 1 & 1 & 0 & 1 \end{bmatrix}$$

and parity check matrix

$$H^T = \begin{bmatrix} 0 & 1 & 1 & 1 \\ 1 & 0 & 1 & 1 \\ 1 & 1 & 0 & 1 \\ - & - & - & - \\ 1 & 0 & 0 & 0 \\ 0 & 1 & 0 & 0 \\ 0 & 0 & 1 & 0 \\ 0 & 0 & 0 & 1 \end{bmatrix}.$$

The syndrome Table 14.3 is built by calculating which error pattern is most likely (i.e., has the fewest bits flipped) for each given syndrome eH^T. This code has $d_{min} = 4$, and hence the code can correct any 1-bit errors, 7 (out of 21) possible 2-bit errors, and 1 of the many 3-bit errors.

Problem

14.33. Using the code from `blockcode.m`, implement the binary (7,3) linear block code. Compare its performance and efficiency with the (5,2) code and to the majority rules code.
   (a) For each code, plot the percentage p of bit flips in the channel versus the percentage of bit flips in the decoded output.
   (b) For each code, what is the average number of bits transmitted for each bit in the message?

Sometimes, when the source alphabet is not binary, the elements of the code words are also not binary. In this case, using the binary arithmetic of Table 14.1 is inappropriate. For example, consider a source alphabet with 5 symbols labelled 0, 1, 2, 3, 4. Arithmetic operations for these elements are addition and multiplication modulo 5, which are defined in Table 14.4. These can be implemented in MATLAB using the `mod` function. For some source alphabets, the appropriate arithmetic operations are not modulo operations, and in these cases, it is normal to simply define the desired operations via tables like 14.1 and 14.4.

Table 14.4.   Modulo 5 Arithmetic

+	0	1	2	3	4
0	0	1	2	3	4
1	1	2	3	4	0
2	2	3	4	0	1
3	3	4	0	1	2
4	4	0	1	2	3

·	0	1	2	3	4
0	0	0	0	0	0
1	0	1	2	3	4
2	0	2	4	1	3
3	0	3	1	4	2
4	0	4	3	2	1

Example 14.9

A (6,4) code using a $q = 5$ element source alphabet has generator matrix

$$G = \begin{bmatrix} 1 & 0 & 0 & 0 & 4 & 4 \\ 0 & 1 & 0 & 0 & 4 & 3 \\ 0 & 0 & 1 & 0 & 4 & 2 \\ 0 & 0 & 0 & 1 & 4 & 1 \end{bmatrix}$$

and parity check matrix

$$H^T = \begin{bmatrix} -4 & -4 \\ -4 & -3 \\ -4 & -2 \\ -4 & -1 \\ 1 & 0 \\ 0 & 1 \end{bmatrix} = \begin{bmatrix} 1 & 1 \\ 1 & 2 \\ 1 & 3 \\ 1 & 4 \\ 1 & 0 \\ 0 & 1 \end{bmatrix},$$

since, in mod 5 arithmetic, $-4 = 1$, $-3 + 2$, $-2 = 3$, and $-1 = 4$. Observe that these fit in the general form of (14.17) and (14.18). The syndrome Table 14.5 lists the $q^{n-k} = 5^{6-4} = 25$ syndromes and corresponding errors.

The code in this example corrects all one-symbol errors (and no others).

Problems

14.34. Find all the code words in the $q = 5$ (6,4) linear block code from Example 14.9.

14.35. What is the minimum distance of the $q = 5$ (6,4) linear block code from Example 14.9?

14.36. Mimicking the code in `blockcode52.m`, implement the $q = 5$ (6,4) linear block code from Example 14.9. Compare its performance with the (5,2) and (7,3) binary codes in terms of
(a) performance in correcting errors
(b) data rate
Be careful: how can a $q = 5$ source alphabet be compared fairly with a binary alphabet? Should the comparison be in terms of percentage of bit errors or percentage of symbol errors?

**Table 14.5.   Syndrome Table
for the $q = 5$ source alphabet
(6, 4) code.**

Syndrome $\mathbf{e}H^T$	Most likely error $\mathbf{e}$
00	000000
01	000001
10	000010
14	000100
13	001000
12	010000
11	100000
02	000002
20	000020
23	000200
21	002000
24	020000
22	200000
03	000003
30	000030
32	000300
34	003000
31	030000
33	300000
04	000004
40	000040
41	000400
42	004000
43	040000
44	400000

14.7 Encoding a Compact Disc

The process of writing to and reading from a compact disc is involved. The essential idea in optical media is that a laser beam bounces off the surface of the disc. If there is a pit, then the light travels a bit further than if there is no pit. The distances are controlled so that the extra time required by the round trip corresponds to a phase shift of 180 degrees. Thus, the light travelling back interferes destructively if there is a pit, while it reinforces constructively if there is no pit. The strength of the beam is monitored to detect a 0 (a pit) or a 1 (no pit).

While the complete system can be made remarkably accurate, the reading and writing procedures are prone to errors. This is a perfect application for error correcting codes! The encoding procedure is outlined in Figure 14.6. The original signal is digitized at 44, 100 samples per second in each of two stereo channels. Each sample is 16 bits, and the effective data rate is 1.41 Mbps (mega bits per second). The Cross Interleave Reed–Solomon Code (CIRC) encoder (described shortly) has an effective rate of about

3/4, and its output is at 1.88 Mbps. Then control and timing information is added, which contains the track and subtrack numbers that allow CD tracks to be accessed rapidly. The "EFM" (Eight-to-Fourteen Module) is an encoder that spreads the audio information in time by changing each possible 8-bit sequence into a predefined 14-bit sequence so that each one is separated by at least two (and at most ten) zeros. This is used to help the tracking mechanism. Reading errors on a CD often occur in clusters (a small scratch may be many hundreds of bits wide) and interleaving distributes the errors so that they can be corrected more effectively. Finally, a large number of synchronization bits are added. These are used by the control mechanism of the laser to tell it where to shine the beam in order to find the next bits. The final encoded data rate is 4.32 Mbps. Thus, about 1/3 of the bits on the CD are actual data, and about 2/3 of the bits are present to help the system function and to detect (and/or correct) errors when they occur.

The CIRC encoder consists of two special linear block codes called Reed–Solomon codes (which are named after their inventors). Both use $q = 256$ (8-bit) symbols, and each 16-bit audio sample is split into two code words. The first code is a (32,28) linear code with $d_{min} = 5$, and the second code is a linear (28,24) code, also with $d_{min} = 5$. These are nonbinary and use special arithmetic operations defined by the "Galois Field" with 256 symbols. The encoding is split into two separate codes so that an interleaver can be used between them. This spreads out the information over a larger range and helps to spread out the errors (making them easier to detect and/or correct).

FIGURE 14.6 CDs can be used for audio or for data. The encoding procedure is the same, though decoding may be done differently for different applications.

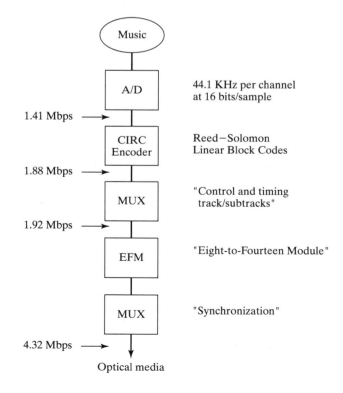

The encoding process on the CD is completely specified, but each manufacturer can implement the decoding as they wish. Accordingly, there are many choices. For instance, the Reed–Solomon codes can be used to correct two errors each, or to detect up to five errors. When errors are detected, a common strategy is to interpolate the audio, which may be transparent to the listener as long as the error rate is not too high. Manufacturers may also choose to mute the audio when the error rate is too high. For data purposes, the controller can also ask that the data be reread. This may allow correction of the error when it was caused by mistracking or some other transitory phenomenon, but will not be effective if the cause is a defect in the medium.

14.8 For Further Reading

The paper that started information theory is still a good read half a century after its initial publication.

- C. E. Shannon, "A mathematical theory of communication," *Bell System Technical Journal*, vol. 27, pp. 379–423 and 623–656, July and October, 1948.

PART

V

The Integration Layer

The last layer is the final project of Chapter 15 which integrates all the fixes of the adaptive component layer (recall page 175) into the receiver structure of the idealized system layer (from page 57) to create a fully functional digital receiver. The well-fabricated receiver is robust to distortions such as those caused by noise, multipath interference, timing inaccuracies, and oscillator mismatches.

15 Mix'n'match® Receiver Design

Make it so.

Captain Picard

This chapter describes a software-defined radio design project called $\mathcal{M}^6$, the *Mix 'n' Match Mostly Marvelous Message Machine*. The $\mathcal{M}^6$ transmission standard is specified so that the receiver can be designed using the building blocks of the preceding chapters. The DSP portion of the $\mathcal{M}^6$ can be simulated in MATLAB by combining the functions and subroutines from the examples and exercises of the previous chapters.

The input to the digital portion of the $\mathcal{M}^6$ receiver is a sampled signal at intermediate frequency (IF) that contains several simultaneous messages, each transmitted in its own frequency band. The original message is text that has been converted into symbols drawn from a 4-PAM constellation, and the pulse shape is a square-root raised cosine. The sample frequency can be less than twice the highest frequency in the analog IF signal, but it must be sufficiently greater than the inverse of the transmitted symbol period to be twice the bandwidth of the baseband signal. The successful $\mathcal{M}^6$ MATLAB program will demodulate, synchronize, equalize, and decode the signal, so it is a "fully operational" software-defined receiver (although it is not intended to work in "real time"). The receiver must overcome multiple impairments. There may be phase noise in the transmitter oscillator. There may be an offset between the frequency of the oscillator in the transmitter and the frequency of the oscillator in the receiver. The pulse clocks in the transmitter and receiver may differ. The transmission channel may be noisy. Other users in spectrally adjacent bands may be actively transmitting at the same time. There may be intersymbol interference caused by multipath channels.

The next section describes the transmitter, the channel, and the analog front-end of the receiver. Then Section 15.2 makes several generic observations about receiver design, and proposes a methodology for the digital receiver design. The final section describes the receiver design challenge that serves as the culminating design experience of this book. Actually building the $\mathcal{M}^6$ receiver, however, is left to you. You will know that your receiver works when you can recover the mystery message hidden inside the received signal.

15.1 How the Received Signal Is Constructed

Receivers cannot be designed in a vacuum; they must work in tandem with a particular transmitter. Sometimes, a communication system designer gets to design both ends of the

313

system. More often, however, the designer works on one end or the other, with the goal of making the signal in the middle meet some standard specifications. The standard for the $\mathcal{M}^6$ is established on the transmitted signal, and consists, in part, of specifications on the allowable bandwidth and on the precision of its carrier frequency. The standard also specifies the source constellation, the modulation, and the coding schemes to be used. The front-end of the receiver provides some bandpass filtering, downconversion to IF, and automatic gain control prior to the sampler.

This section describes the construction of the sampled IF signal that must be processed by the $\mathcal{M}^6$ receiver. The system that generates the analog received signal is shown in block diagram form in Figure 15.1. The front end of the receiver that turns this into a sampled IF signal is shown in Figure 15.2.

The original message in Figure 15.1 is a character string of English text. Each character is converted into a 7-bit binary string according to the ASCII conversion format (e.g., the letter "a" is 1100001 and the letter "M" is 1001101), as in Example 8.2. The bit string is coded using the (5,2) linear block code specified in `blockcode52.m` which associates a 5-bit code with each pair of bits. The output of the block code is then partitioned into pairs that are associated with the four integers of a 4-PAM alphabet ±1 and ±3 via the mapping

$$
\begin{array}{rcl}
11 & \rightarrow & +3 \\
10 & \rightarrow & +1 \\
01 & \rightarrow & -1 \\
00 & \rightarrow & -3
\end{array}
\tag{15.1}
$$

as in Example 8.1. Thus, if there are n letters, there are $7n$ (uncoded) bits, $7n(\frac{5}{2})$ coded bits, and $7n(\frac{5}{2})(\frac{1}{2})$ 4-PAM symbols. These mappings are familiar from Section 8.1, and are easy to use with the help of the MATLAB functions `bin2text.m` and

FIGURE 15.1 Received signal generator.

FIGURE 15.2 Receiver front end.

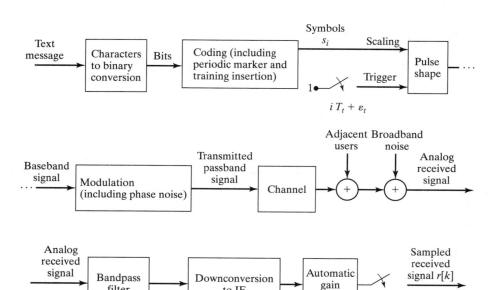

`text2bin.m`. Problem 14.25 provides several hints to help implement the $\mathcal{M}^6$ encoding, and the MATLAB function `nocode.m` outlines the necessary transformations from the original text into a sequence of 4-PAM symbols $s[i]$.

In order to decode the message at the receiver, the recovered symbols must be properly grouped and the start of each group must be located. To aid this frame synchronization, a marker sequence is inserted in the symbol stream at the start of every block of 100 letters (at the start of every 875 symbols). The header/training sequence that starts each frame is given by the phrase

A0Oh well whatever Nevermind

which codes into 245 4-PAM symbols and is assumed to be known at the receiver. This marker text string can be used as a training sequence by the adaptive equalizer. The unknown message begins immediately after each training segment. Thus, the $\mathcal{M}^6$ symbol stream is a coded message periodically interrupted by the same marker/training clump.

As indicated in Figure 15.1, pulses are initiated at intervals of T_t seconds, and each is scaled by the 4-PAM symbol value. This translates the discrete-time symbol sequence $s[i]$ (composed of the coded message interleaved with the marker/training segments) into a continuous-time signal

$$s(t) = \sum_i s[i]\delta(t - iT_t - \epsilon_t).$$

The actual transmitter symbol period T_t is required to be within 0.01 percent of a nominal $\mathcal{M}^6$ symbol period $T = 6.4$ microseconds. The transmitter symbol period clock is assumed to be steady enough that the timing offset ϵ_t and its period T_t are effectively time-invariant over the duration of a single frame.

Details of the $\mathcal{M}^6$ transmission specifications are given in Table 15.1. The pulse-shaping filter $P(f)$ is a square-root raised cosine filter symmetrically truncated to eight symbol periods. The rolloff factor β of the pulse-shaping filter is fixed within some range and is known at the receiver, though it could take on different values with different transmissions. The (half-power) bandwidth of the square-root raised cosine pulse could be as large as ≈ 102 kHz for the nominal T. With double sideband modulation, the pulse shape bandwidth doubles so that each passband FDM signal will need a bandwidth at least 204 kHz wide.

The channel may be near ideal (i.e., a unit gain multisymbol delay) or it may have significant intersymbol interference. In either case, the impulse response of the channel is unknown at the receiver, though an upper bound on its delay spread may be available. There are also disturbances that may occur during the transmission. These may be wideband noise with flat power spectral density or they may be narrowband interferers, or both. They are unknown at the receiver.

The achieved intermediate frequency is required to be within 0.01 percent of its assigned value. The carrier phase $\theta(t)$ is unknown to the receiver and may vary over time, albeit slowly. This means that the phase of the intermediate frequency signal presented to the receiver sampler may also vary.

The bandpass filter before the downconverter in the front-end of the receiver in Figure 15.2 partially attenuates adjacent 204 kHz wide FDM user bands. The automatic

Table 15.1. $\mathcal{M}^6$ **Signal Specifications**

Symbol source alphabet	$\pm1, \pm3$
Assigned intermediate frequency	2 MHz
Nominal symbol period	6.4 microseconds
SRRC pulse shape rolloff factor	$\beta \in [0.1, 0.3]$
FDM user slot allotment	204 kHz
Truncated width of SRRC pulse shape	8 transmitter clock periods
Frame marker/training sequence	A0Oh well whatever Nevermind
Frame marker sequence period	1120 symbols
Time-varying IF carrier phase	Lowpass filtered white noise
Transmitter IF offset	Fixed, less than 0.01% of assigned value
Transmitter timing offset	Fixed
Transmitter symbol period offset	Fixed, less than 0.01% of assigned value
Intersymbol interference	Maximum delay spread = 7 symbols
Sampler frequency	850 kHz

FIGURE 15.3 DSP portion of software-defined receiver.

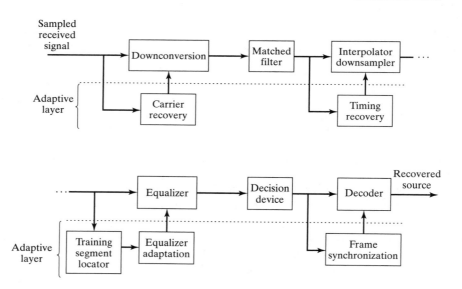

gain control is presumed locked and fixed over each transmission. The free-running sampler frequency of 850 kHz is well above twice the 102 kHz baseband bandwidth of the user of interest. This is necessary for the baseband analog signal interpolator used in the timer in the DSP portion of the receiver in Figure 15.3. However, the sampler frequency is not above twice the highest frequency of the IF signal. This means that the sampled received signal has replicated the spectrum at the output of the front-end analog downconverter lowpass filter to frequencies between zero and IF.

15.2 A Design Methodology for the $\mathcal{M}^6$ Receiver

Before describing the specific design requirements that must be met by a successful $\mathcal{M}^6$ receiver, this section makes some generic remarks about a systematic approach to receiver design. There are four generic stages:

1. *Choose the order in which the basic operations of the receiver occur.*
2. *Select components and methods that can perform the basic operations in an ideal setting.*
3. *Select adaptive elements that allow the receiver to continue functioning when there are impairments.*
4. *Verify that the performance requirements are met.*

While it may seem as though each stage requires that choices made in the preceding stages be fixed, in reality, difficulties encountered at one stage in the design process may require a return to (and different choices to be made in) earlier stages. As will soon become clear, the $\mathcal{M}^6$ problem specification has basically (pre)resolved the design issues of the first two stages.

15.2.1 Stage One: Ordering the Pieces

The first stage is to select the basic components and the order in which they occur. The design layout first established in Figure 2.11 (and reappearing in the schematic of the DSP portion of the receiver in Figure 15.3) suggests one feasible structure. As the signal enters the receiver it is downconverted (with carrier recovery), matched filtered, interpolated (with timing recovery), equalized (adaptively), quantized, and decoded (with frame synchronization). This classical ordering, while popular, is not the only (nor necessarily the best) way to recover the message from the noisy, ISI-distorted, FDM–PAM–IF received signal. However, it offers a useful foundation for assessing the relative benefits and costs of alternative receiver configurations. Also, we know for sure that the $\mathcal{M}^6$ receiver can be built this way. Other configurations may work, but we have not tested them.[1]

How was this ordering of components chosen? The authors have consulted with, worked for, talked about (and argued with) engineers working on a number of receiver systems including HDTV (high definition television), DSL, and AlohaNet. The ordering of components in Figures 2.11 and 15.3 represents an amalgamation of ideas from these (and other) systems. Sometimes it is easy to argue why a particular order is good, sometimes it is a matter of preference or personal experience, and sometimes the choice is based on factors outside the engineer's control.[2]

For example, the carrier recovery algorithms of Chapter 10 are not greatly affected by noise or intersymbol interference (as was shown in Problems 10.31 and 10.35). Thus carrier recovery can be done before equalization, and this is the path we have followed. But it need not be done in this order.[3] Another example is the placement of the timing recovery element. The algorithms of Chapter 12 operate at baseband, and hence the timing recovery in Figure 15.3 is placed after the demodulation. But there are passband timing recovery algorithms that could have been used to reverse the order of these two operations.

---

[1] If this sounds like a challenge, rest assured it is. Research continues worldwide, making compilation of a complete handbook of receiver designs and algorithms a Sisyphean task. The creation of "new" algorithms with minor variations that exploit a particular application-specific circumstance is a popular pastime of communication engineers. Perhaps you too will come up with a unique approach!

[2] For instance, the company might have a patent on a particular method of timing recovery and using any other method might require royalty payments.

[3] For instance, in the QAM radio of *A Digital Quadrature Amplitude Modulation Radio*, available in the CD, the blocks appear in a different order.

15.2.2 Stage Two: Selecting Components

Choices for the second design stage are relatively set as well. Since the sampling is done at a sub-Nyquist rate f_s (relative to the IF frequency f_I), the spectrum of the analog received signal is replicated every f_s. The integer n for which $f^\dagger = |f_I - nf_s|$ is smallest defines the nominal frequency $f^\dagger$ from which further downconversion is needed. Recall that such downconversion by sampling was discussed in Section 6.2. Using different specifications, the $\mathcal{M}^6$ sampling frequency f_s may be above the Nyquist frequency associated with the IF frequency f_I.[4]

The most common method of downconversion is to use mixing followed by an FIR lowpass filter. This will be followed by an FIR matched filter, an interpolator–decimator for downsampling, and a symbol-spaced FIR equalizer that adapts its coefficients based on the training data contained in the transmission. The output of the equalizer is quantized to the nearest 4-PAM symbol value, translated back into binary, decoded (using the (5,2) block decoder) and finally turned back into readable text.

Given adequate knowledge of the operating environment (the SNR in the received signal, the carrier frequency and phase, the clock period and symbol timing, and the marker location), the designer-selected parameters within these components can be set to recover the message. This was, in fact, the strategy followed in the idealized receiver of Chapter 9. Said another way, the choices in stages one and two are presumed to admit an acceptable answer if properly tuned. Component selections at this point (including specification of the fixed lowpass filter in the downconverter and the fixed matched filter preceding the interpolator/downsampler) can be confirmed by simulations of the ISI-free ideal/full-knowledge setting. Thus, the upper half of Figure 15.3 is specified by stage two activities.

15.2.3 Stage Three: Anticipating Impairments

In the third design stage, the choices are less constrained. Elements of the third stage are shown in the lower half of the receiver schematic (the "adaptive layer" of Figure 15.3) and include the selection of algorithms for carrier, timing, frame synchronization, and equalizer adaptation. There are several issues to consider.

One of the primary stage three activities is algorithm selection—which performance function to use in each block. For example, should the $\mathcal{M}^6$ receiver use a phase-locked loop, a Costas loop, or a decision-directed method for carrier recovery? Is a dual loop needed to provide adequate carrier tracking, or will a single loop suffice? What performance function should be used for the equalizer? Which algorithm is best for the timing recovery? Is simple correlation suitable to locate the training and marker segment?

Once the specific methods have been chosen, it is necessary to select specific variables and parameters within the algorithms. This is a traditional aspect of engineering design that is increasingly dominated by computer-aided design, simulation, and visualization tools. For example, error surfaces and eye diagrams can be used to compare the performance of the various algorithms in particular settings. They can be used to help determine which technique is more effective for the application at hand.

As software-aided design packages proliferate, the need to understand the computational mechanics underlying a particular design becomes less of a barrier. For instance,

---

[4] Indeed, changing parameters such as this allows an instructor to create new transmission "standards" for each class!

Telecommunication Breakdown has relied exclusively on the filter design algorithms built into MATLAB. But the specification of the filter (its shape, cutoff frequencies, computational complexity, and filter length) cannot be left to MATLAB. The more esoteric the algorithm, the less transparent is the process of selecting design parameters. Thus, *Telecommunication Breakdown* has devoted considerable space to the design and operation of adaptive elements.

But, even assuming that the tradeoffs associated with each of the individual components are clear, how can everything be integrated together to succeed at a multifaceted design objective such as the $\mathcal{M}^6$ receiver?

15.2.4 Sources of Error and Tradeoffs

Even when a receiver is fully operational, it may not decode every symbol precisely. There is always a chance of error. Perhaps part of the error is due to a frequency mismatch, part of the error is due to noise in the channel, part of the error is due to a nonoptimal timing offset, etc. This section (and the next) suggest a general strategy for allocating "part of" the error to each component. Then, as long as the sum of all the partial errors does not exceed the maximum allowable error, there is a good chance that the complete receiver will work according to its specifications.

The approach is to choose a method of measuring the amount of error, for instance, the average of the squared recovery error. Each individual component can be assigned a threshold, and its parameters can be adjusted so that it does not contribute more than its share to the total error. Assuming that the accumulation of the errors from various sources is additive, the complete receiver will have no larger error than the concatenation of all its parts. This additivity assumption is effectively an assumption that the individual pieces of the system do not interact with each other. If they do (or when they do), then the threshold allotments may need to be adjusted.

There are many factors that contribute to the recovery error, including the following:

- Residual interference from adjacent FDM bands (caused by imperfect bandpass filtering before downconversion and imperfect lowpass filtering after downconversion).

- AGC jitter (caused by the deviation in the instantaneous signal from its desired average and scaled by the stepsize in the AGC element).

- Quantization noise in the sampler (caused by coarseness in the magnitudes of the quantizer).

- Round-off noise in filters (caused by wordlength limitations in filter parameters and filter algebra).

- Residual interference from the doubly upconverted spectrum (caused by imperfect lowpass filtering after downconversion).

- Carrier phase jitter (occurs physically as a system impairment and is caused by the stepsize in the carrier recovery element).

- Timing jitter (occurs physically as a system impairment and is caused by the stepsize in the timing recovery element).

- Residual mean squared error left by the equalizer (even an infinitely long linear equalizer cannot remove all recovery error in the presence of simultaneous channel noise and ISI).

- Equalizer parameter jitter (caused by the nonvanishing stepsize in the adaptive equalizer).

- Noise enhancement by the equalizer (caused by ISI that requires large equalizer gains, such as a deep channel null at frequencies that also include noise).

Because MATLAB implements all calculations in floating point arithmetic, the quantization and round-off noise in the simulations is imperceptible. The project setup presumes that the AGC has no jitter. A well-designed and sufficiently long lowpass filter in the downconverter can effectively remove the interference from outside the user band of interest. The in-band interference from sloppy adjacent FDM signals should be considered part of the in-band channel noise. This leaves carrier phase, timing jitter, imperfections in the equalizer, tap jitter, and noise gain. All of these are potentially present in the $\mathcal{M}^6$ software-defined digital radio.

In all of the cases in which error is due to the jiggling of the parameters in adaptive elements (in the estimation of the sampling instants, the phase errors, the equalizer taps), the errors are proportional to the stepsize used in the algorithm. Thus, the (asymptotic) recovery error can be made arbitrarily small by reducing the appropriate stepsize. The problem is that, if the stepsize is too small, the element takes longer to converge. If the time to convergence of the element is too long (for instance, longer than the complete message), then the error is increased. Accordingly, there is some optimal stepsize that is large enough to allow rapid convergence yet small enough to allow acceptable error. An analogous trade-off arises with the choice of the length of the equalizer. Increasing its length reduces the size of the residual error. But as the length grows, so does the amount of tap jitter.

Such tradeoffs are common in any engineering design task. The next section suggests a method of quantifying the tradeoffs to help make concrete decisions.

15.2.5 Tuning and Testing

The testing and verification stage of receiver design is not a simple matter because there are so many things that can go wrong. (There is so much stuff that can happen!) Of course, it is always possible to simply build a prototype and then test to see if the specifications are met. Such a haphazard approach may result in a working receiver, but then again, it may not. Surely there is a better way! This section suggests a commonsense approach that is not uncommon among practicing engineers. It represents a "practical" compromise between excessive analysis (such as one might find in some advanced communication texts) and excessive trial and error (such as "try something and cross your fingers").

The idea is to construct a simulator that can create a variety of test signals that fall within the $\mathcal{M}^6$ specification. The parameters within the simulator can then be changed one at a time, and their effect noted on various candidate receivers. By systematically varying the test signals, the worst components of the receiver can be identified and then replaced. As the tests proceed, the receiver gradually improves. As long as the complete set of test signals accurately represents the range of situations that will be encountered in operation, the testing will lead to a successful design.

Given the particular stage one and two design choices for the $\mathcal{M}^6$ receiver, the previous section outlined the factors that may degrade the performance of the receiver. The following steps suggest some detailed tests that may facilitate the design process:

- Step 1: *Tuning the Carrier Recovery* As shown in Chapter 10, any of the carrier recovery algorithms are capable of locating a fixed phase offset in a receiver in which everything else is operating optimally. Even when there is noise or ISI, the best settings for the frequency and phase of the demodulation sinusoid are those that match the frequency and phase of the carrier of the IF signal. For the $\mathcal{M}^6$ receiver, there are two issues that must be considered. First, the $\mathcal{M}^6$ specification allows the frequency to be (somewhat) different from its nominal value. Is a dual-loop structure needed? Or can a single loop adequately track the expected variations? Second, the transmitter phase may be jittering.

  The user choosable features of the carrier recovery algorithms are the LPF and the algorithm stepsize, both of which influence the speed at which the estimates can change. Since the carrier recovery scheme needs to track a time-varying phase, the stepsize cannot be chosen too small. Since a large stepsize increases the error due to phase jitter, it cannot be chosen too large. Thus, an acceptable stepsize will represent a compromise.

  To conduct a test to determine the stepsize (and LPF) requires creating test signals that have a variety of off-nominal frequency offsets and phase jitters. A simple way to model phase jitter is to add a lowpass filtered version of zero-mean white noise to a nominal value. The quality of a particular set of parameters can then be measured by averaging (over all the test signals) the mean squared recovery error. Choosing the LPF and stepsize parameters to make this error as small as possible gives the "best" values. This average error provides a measure of the portion of the total error that is due to the carrier recovery component in the receiver.

- Step 2: *Tuning the Timing Recovery* As shown in Chapter 12, there are several algorithms that can be used to find the best timing instants in the ideal setting. When the channel impairment consists purely of additive noise, the optimal sampling times remain unchanged, though the estimates will likely be more noisy. As shown by Example 12.3, and in Figure 12.12, however, when the channel contains ISI, the answer returned by the algorithms differs from what might be naively expected.

  There are two parts to the experiments at this step. The first is to locate the best timing recovery parameter for each test signal. (This value will be needed in the next step to assess the performance of the equalizer.) The second part is to find the mean squared recovery error due to jitter of the timing recovery algorithm.

  The first part is easy. For each test signal, run the chosen timing recovery algorithm until it converges. The convergent value gives the timing offset (and indirectly specifies the ISI) to which the equalizer will need to respond. (If it jiggles excessively, then decrease the stepsize.)

  Assessing the mean squared recovery error due to timing jitter can be done much like the measurement of jitter for the carrier recovery: measure the average error that occurs over each test signal when the algorithm is initialized at its optimum answer; then average over all the test signals. The answer may be affected by the various parameters of the algorithm: the δ that determines the approximation to the derivative, the $\mathtt{l}$ parameter that specifies the time support of the interpolation, and the stepsize (these variable names are from the first part of the timing recovery algorithm `clockrecdp.m` on page 233.)

  In operation, there may also be slight inaccuracies in the specification of the clock period. When the clock period at the transmitter and receiver differ, the stepsize must be large enough so that the timing estimates can follow the changing period. (Recall the discussion surrounding Example 12.4.) Thus, again, there is a tension between a large stepsize needed to track rapid changes and a small stepsize to minimize the effect of the jitter on the mean squared recovery error. In a more complex environment, in which clock phases might be varying, it might be necessary to follow a procedure more like that considered in step 1.

• Step 3: *Tuning the Equalizer* After choosing the equalizer method (as specified by the performance function), there are a number of parameters that must be chosen and decisions that must be made in order to implement the linear equalizer. These are

 – the order of the equalizer (number of taps),

 – initializing the equalizer,

 – selecting the training signal delay (if using the training signal), and

 – choosing the stepsize.

As in the previous steps, it is a good idea to create a collection of test signals using a simulation of the transmitter. To test the performance of the equalizer, the test signals should contain a variety of ISI channels and/or additive interferences.

As suggested in Chapter 13, in a high SNR scenario the T-spaced equalizer tries to implement an approximation of the inverse of the ISI channel. If the channel is mild, with all its roots well away from the unit circle, then its inverse may be fairly short. But if the channel has zeros that are near the unit circle, then its FIR inverse may need to be quite long. While much can be said about this, a conservative guideline is that the equalizer should be from two to five times longer than the maximum anticipated channel delay spread.

One subtlety that arises in making this decision and in consequent testing is that any channel ISI that is added into a simulation may appear differently at the receiver because of the sampling. This effect was discussed at length in Section 13.1, where it was shown how the effective digital model of the channel includes the timing offset. Thus (as mentioned in the previous step) assessing the "actual" channel to which the equalizer will adapt requires knowing the timing offset that will be found by the timing recovery. Fortunately, in the $\mathcal{M}^6$ receiver structure of Figure 15.3, the timing recovery algorithm operates independently of the equalizer, and so the optimal value can be assessed beforehand.

For most of the adaptive equalizers in Chapter 13, the center spike initialization is used. This was justified in Section 13.4 (see page 261) as a useful method of initialization. Only if there is some concrete a prior knowledge of the channel characteristics would other initializations be used.

The problem of finding an appropriate delay was discussed in Section 13.2.3, where the least squares solution was recomputed for each possible delay. The delay with the smallest error was the best. In a real receiver, it will not be possible to do an extensive search, and so it is necessary to pick some delay. The $\mathcal{M}^6$ receiver uses correlation to locate the marker sequence and this can be used to locate the time index corresponding to the first training symbol. This location plus half the length of the equalizer should correspond closely to the desired delay. Of course, this value may change depending on the particular ISI (and channel lengths) used in a given test signal. Choose a value that, over the complete set of test signals, provides a reasonable answer.

The remaining designer-selected variable is stepsize. As with all adaptive methods, there is a tradeoff inherent in stepsize selection: making it too large can result in excessive jitter or algorithm instability, while making it too small can lead to an unacceptably long convergence time. A common technique is to select the largest stepsize consistent with achievement of the component's assigned asymptotic performance threshold.

- Step 4: *Frame Synchronization* Any error in locating the first symbol of each four-symbol block can completely garble the reconstructed text. The frame synchronizer operates on the output of the quantizer, which should contain few errors once the equalizer, timing recovery, and phase recovery have converged. The success of frame synchronization relies on the peakiness of the correlation of the marker/training sequence. The chosen marker/training sequence "A0Oh well whatever Nevermind" should be long enough that there are few false spikes when correlating to find the start of the message within each block. To test software written to locate the marker, feed it a sample symbol string assembled according to the specifications described in the previous section as if the downconverter, clock timing, equalizer, and quantizer had recovered the transmitted symbol sequence perfectly.

Finally, after tuning each component separately, it is necessary to confirm that when all the pieces of the system are operating simultaneously, there are no excessive negative interactions. Hopefully, little further tuning will prove necessary to complete a successful design. The next section has more specifics about the $\mathcal{M}^6$ receiver design.

15.3 The $\mathcal{M}^6$ Receiver Design Challenge

The analog front end of the receiver in Figure 15.2 takes the signal from an antenna, amplifies it, and crudely bandpass filters it to (partially) suppress frequencies outside the desired user's frequency band. An analog converter modulates the received signal (approximately) down to the nominal intermediate frequency f_I at 2 MHz. The output of the analog downconverter is set by an automatic gain controller to fit the range of the sampler. The output of the AGC is sampled at intervals of $T_s = 850$ kHz to give $r[k]$, which provides a "Nyquist" bandwidth of 425 kHz that is ample for a 102 kHz baseband user bandwidth. The sampled received signal $r[k]$ from Figure 15.2 is the input to the DSP portion of the receiver in Figure 15.3.

The following comments on the components of the digital receiver in Figure 15.3 help characterize the design task:

- The downconversion to baseband uses the sampler frequency f_s, the known intermediate frequency f_I, and the current phase estimates to determine the mixer frequency needed to demodulate the signal. The $\mathcal{M}^6$ receiver may use any of the phase tracking algorithms of Chapter 10. A second loop may also help with frequency offset.

- The lowpass filtering in the demodulator should have a bandwidth of roughly 102 kHz, which will cover the selected source spectrum but reject components outside the frequency band of the desired user.

- The interpolator/downsampler implements the reduction in data rate to T-spaced values. This block must also implement the timing synchronization, so that the time between samples after timing recovery is representative of the true spacing of the samples at the transmitter. You are free to implement this in any of the ways discussed in Chapter 12.

- Since there could be a significant amount of intersymbol interference due to channel dynamics, an equalizer is essential. Any one will do. A trained equalizer requires finding the start of the marker/training segment while a blind equalizer may converge more slowly.

- The decision device is a quantizer defined to reproduce the known alphabet of the $s[i]$ by a memoryless nearest-element decision.

- At the final step, the decoding from `blockcode52.m` in conjunction with `bin2text.m` can be used to reconstruct the original text. This also requires a frame synchronization that finds and removes the start block consisting of marker plus training, which is most likely implemented using a correlation technique.

The software-defined radio should have the following user-selectable variables that can readily be set at the start of processing of the received block of data:

- rolloff factor β for the square-root raised cosine pulse shape,
- initial phase offset,
- initial timing offset, and
- initial equalizer parameterization.

The following are some suggestions:

- Build your own transmitter in addition to a digital receiver simulation. This will enable you to test your receiver as described in the methodology proposed in the preceding section over a wider range of conditions than just the cases available on the CD. Also, building a transmitter will increase your understanding of the composition of the received signal.

- Try to break your receiver. See how much noise can be present in the received signal before accurate (e.g., less than 1% symbol errors) demodulation seems impossible. Find the fastest change in the carrier phase that your receiver can track, even with a bad initial guess.

- In order to facilitate more effective debugging while building the project, implementation of a debug mode in the receiver is recommended. The information of interest will be plots of the time histories of pertinent signals as well as timing information (e.g., a graph of matched filter average output power versus receiver symbol timing offset). One convenient way to add this feature to your MATLAB receiver would be to include a debug flag as an argument that produces these plots when the flag is activated.

- When debugging adaptive components, use a test with initialization at the right answer and zero stepsize to determine if the problem is in the adaptation or in the fixed component structure. An initialization very near the desired answer with a small stepsize will reveal that the adaptive portion is working properly if the adaptive parameter trajectory remains in the close vicinity of the desired answer. A rapid divergence may indicate that the update has the wrong sign or that the stepsize is way too large. An aimless wandering that drifts away from the vicinity of the desired answer represents a more subtle problem that requires reconsideration of the algorithm code and/or its suitability for the circumstance at hand.

Several test files that contain a "mystery signal" with a quote from a well known book are available on the CD. They are labelled `easyN.mat`, `mediumN.mat`, and `hardN.mat`.[5] These have been created with a variety of different rolloff factors, carrier frequencies, phase noises, ISI, interferers, and symbol timing offsets. We encourage the adventurous reader to try to "receive" these secret signals. Solve the mystery. Break it down.

15.4 For Further Reading

An overview of a practical application of *software-defined radio* emphasizing the redefinability of the DSP portion of the receiver can be found in

- B. Bing and N. Jayant, "A cellphone for all standards," *IEEE Spectrum*, pg 34–39, May 2002.

The field of "software radio" erupted with a special issue of the *IEEE Communications Magazine* in May 1995. This was called a "landmark special issue" in an editorial in the more recent

- J. Mitola, III, V. Bose, B. M. Leiner, T. Turletti and D. Tennenhouse, Ed., *IEEE Journal on Selected Areas in Communications* (Special Issue on Software Radios), vol. 17, April 1999.

For more information on the technological context and the relevance of software implementations of communications systems, see

- E. Buracchini, "The Software Radio Concept," *IEEE Communications Magazine*, vol. 38, pp. 138–143, September 2000

and papers from the (occasional) special section in the *IEEE Communications Magazine* on topics in software and DSP in radio. For much more, see

- J. H. Reed, *Software Radio: A Modern Approach to Radio Engineering*, Prentice Hall, 2002

which overlaps in content (if not style) with the first half of *Telecommunication Breakdown*.

Two recommended monographs that include more attention than most to the methodology of the same slice of digital receiver design as we consider here are

- J. A. C. Bingham, *The Theory and Practice of Modem Design*, Wiley Interscience, 1988. (especially Chapter 5)

---

[5] One student remarked that these should have been called `hardN.mat`, `harderN.mat`, and `complete-lyridiculousN.mat`. Nonetheless, a well crafted $\mathcal{M}^6$ receiver can recover the hidden messages.

- H. Meyr, M. Moeneclaey, and S. A. Fechtel, *Digital Communication Receivers: Synchronization, Channel Estimation, and Signal Processing*, Wiley Interscience, 1998 (especially Section 4.1).

A similar design challenge can be found in the extension to QAM found in the CD in *A Digital Quadrature Amplitude Modulation (QAM) Radio*.

APPENDIX

A

Transforms, Identities, and Formulas

Just because some of us can read and write and do a little math, that doesn't mean we deserve to conquer the Universe.
—Kurt Vonnegut, *Hocus Pocus*, 1990

This appendix gathers together all of the math facts used in the text. They are divided into six categories:

- Trigonometric identities
- Fourier transforms and properties
- Energy and power
- Z-transforms and properties
- Integral and derivative formulas
- Matrix algebra

So, with no motivation or interpretation, just labels, here they are:

A.1 Trigonometric Identities

- **Euler's relation**

$$e^{\pm jx} = \cos(x) \pm j\sin(x) \qquad (A.1)$$

- **Exponential definition of a cosine**

$$\cos(x) = \frac{1}{2}\left(e^{jx} + e^{-jx}\right) \qquad (A.2)$$

- **Exponential definition of a sine**

$$\sin(x) = \frac{1}{2j}\left(e^{jx} - e^{-jx}\right) \qquad (A.3)$$

- **Cosine squared**

$$\cos^2(x) = \frac{1}{2}\left(1 + \cos(2x)\right) \qquad (A.4)$$

327

- **Sine squared**

$$\sin^2(x) = \frac{1}{2}(1 - \cos(2x)) \tag{A.5}$$

- **Sine and Cosine as phase shifts of each other**

$$\sin(x) = \cos\left(\frac{\pi}{2} - x\right) = \cos\left(x - \frac{\pi}{2}\right) \tag{A.6}$$

$$\cos(x) = \sin\left(\frac{\pi}{2} - x\right) = -\sin\left(x - \frac{\pi}{2}\right) \tag{A.7}$$

- **Sine–cosine product**

$$\sin(x)\cos(y) = \frac{1}{2}\left[\sin(x - y) + \sin(x + y)\right] \tag{A.8}$$

- **Cosine–cosine product**

$$\cos(x)\cos(y) = \frac{1}{2}\left[\cos(x - y) + \cos(x + y)\right] \tag{A.9}$$

- **Sine–sine product**

$$\sin(x)\sin(y) = \frac{1}{2}\left[\cos(x - y) - \cos(x + y)\right] \tag{A.10}$$

- **Odd symmetry of the sine**

$$\sin(-x) = -\sin(x) \tag{A.11}$$

- **Even symmetry of the cosine**

$$\cos(-x) = \cos(x) \tag{A.12}$$

- **Cosine angle sum**

$$\cos(x \pm y) = \cos(x)\cos(y) \mp \sin(x)\sin(y) \tag{A.13}$$

- **Sine angle sum**

$$\sin(x \pm y) = \sin(x)\cos(y) \pm \cos(x)\sin(y) \tag{A.14}$$

A.2 Fourier Transforms and Properties

- **Definition of Fourier transform**

$$W(f) = \int_{-\infty}^{\infty} w(t)e^{-j2\pi ft}dt \tag{A.15}$$

- **Definition of Inverse Fourier transform**

$$w(t) = \int_{-\infty}^{\infty} W(f)e^{j2\pi ft}df \tag{A.16}$$

- **Fourier transform of a sine**

$$\mathcal{F}\{A\sin(2\pi f_0 t + \phi)\} = j\frac{A}{2}\left[-e^{j\phi}\delta(f - f_0) + e^{-j\phi}\delta(f + f_0)\right] \qquad (A.17)$$

- **Fourier transform of a cosine**

$$\mathcal{F}\{A\cos(2\pi f_0 t + \phi)\} = \frac{A}{2}\left[e^{j\phi}\delta(f - f_0) + e^{-j\phi}\delta(f + f_0)\right] \qquad (A.18)$$

- **Fourier transform of impulse**

$$\mathcal{F}\{\delta(t)\} = 1 \qquad (A.19)$$

- **Fourier transform of rectangular pulse**

  With

$$\Pi\left(\frac{t}{T}\right) = \left[\begin{array}{cc} 1 & -T/2 \leq t \leq T/2 \\ 0 & \text{otherwise} \end{array}\right. , \qquad (A.20)$$

$$\mathcal{F}\{\Pi\left(\frac{t}{T}\right)\} = T\frac{\sin(\pi f T)}{\pi f T} \equiv T\,\text{sinc}(f T). \qquad (A.21)$$

- **Fourier transform of sinc function**

$$\mathcal{F}\{\text{sinc}(2Wt)\} = \frac{1}{2W}\Pi\left(\frac{f}{2W}\right) \qquad (A.22)$$

- **Fourier transform of raised cosine**

  With

$$w(t) = 2f_0\left(\frac{\sin(2\pi f_0 t)}{2\pi f_0 t}\right)\left[\frac{\cos(2\pi f_\Delta t)}{1 - (4f_\Delta t)^2}\right], \qquad (A.23)$$

$$\mathcal{F}\{w(t)\} = \begin{cases} 1 & |f| < f_1 \\ \frac{1}{2}\left(1 + \cos\left[\frac{\pi(|f| - f_1)}{2f_\Delta}\right]\right) & f_1 < |f| < B \\ 0 & |f| > B \end{cases} , \qquad (A.24)$$

  with the *rolloff factor* defined as $\beta = f_\Delta/f_0$.

- **Fourier transform of square-root raised cosine (SRRC)**

  With

$$w(t) = \begin{cases} \dfrac{1}{\sqrt{T}}\dfrac{\sin(\pi(1-\beta)t/T) + (4\beta t/T)\cos(\pi(1+\beta)t/T)}{(\pi t/T)(1 - (4\beta t/T)^2)} & t \neq 0,\ t \neq \pm\frac{T}{4\beta} \\ \dfrac{1}{\sqrt{T}}(1 - \beta + (4\beta/\pi)) & t = 0 \\ \dfrac{\beta}{\sqrt{2T}}\left[\left(1 + \dfrac{2}{\pi}\right)\sin\left(\dfrac{\pi}{4\beta}\right) + \left(1 - \dfrac{2}{\pi}\right)\cos\left(\dfrac{\pi}{4\beta}\right)\right] & t = \pm\frac{T}{4\beta} \end{cases} ,$$

$$\qquad (A.25)$$

$$\mathcal{F}\{w(t)\} = \begin{cases} 1 & |f| < f_1 \\ \left[\frac{1}{2} \left(1 + \cos\left[\frac{\pi(|f| - f_1)}{2f_\Delta} \right] \right) \right]^{1/2} & f_1 < |f| < B \\ 0 & |f| > B \end{cases} . \quad \text{(A.26)}$$

- **Fourier transform of periodic impulse sampled signal**

    With

    $$\mathcal{F}\{w(t)\} = W(f),$$

    and

    $$w_s(t) = w(t) \sum_{k=-\infty}^{\infty} \delta(t - kT_s), \quad \text{(A.27)}$$

    $$\mathcal{F}\{w_s(t)\} = \frac{1}{T_s} \sum_{n=-\infty}^{\infty} W(f - (n/T_s)). \quad \text{(A.28)}$$

- **Fourier transform of a step**

    With

    $$w(t) = \begin{cases} A & t > 0 \\ 0 & t < 0 \end{cases},$$

    $$\mathcal{F}\{w(t)\} = A \left[\frac{\delta(f)}{2} + \frac{1}{j2\pi f} \right]. \quad \text{(A.29)}$$

- **Fourier transform of ideal $\pi/2$ phase shifter (Hilbert transformer) filter impulse response**

    With

    $$w(t) = \begin{cases} \frac{1}{\pi t} & t > 0 \\ 0 & t < 0 \end{cases},$$

    $$\mathcal{F}\{w(t)\} = \begin{cases} -j & f > 0 \\ j & f < 0 \end{cases}. \quad \text{(A.30)}$$

- **Linearity property**

    With

    $$\mathcal{F}\{w_i(t)\} = W_i(f),$$

    $$\mathcal{F}\{aw_1(t) + bw_2(t)\} = aW_1(f) + bW_2(f). \quad \text{(A.31)}$$

- **Duality property**

    With

    $$\mathcal{F}\{w(t)\} = W(f),$$

    $$W(t) = w(-f). \quad \text{(A.32)}$$

- **Cosine modulation frequency shift property**
  With

$$\mathcal{F}\{w(t)\} = W(f),$$

$$\mathcal{F}\{w(t)\cos(2\pi f_c t + \theta)\} = \frac{1}{2}\left[e^{j\theta}W(f - f_c) + e^{-j\theta}W(f + f_c)\right]. \quad (A.33)$$

- **Exponential modulation frequency shift property**
  With

$$\mathcal{F}\{w(t)\} = W(f),$$

$$\mathcal{F}\{w(t)e^{j2\pi f_0 t}\} = W(f - f_0). \quad (A.34)$$

- **Complex conjugation (symmetry) property**
  If $w(t)$ is real valued,

$$W^*(f) = W(-f), \quad (A.35)$$

where the superscript $*$ denotes complex conjugation (i.e., $(a + jb)^* = a - jb$). In particular, $|W(f)|$ is even and $\angle W(f)$ is odd.

- **Symmetry property for real signals**
  Suppose $w(t)$ is real.

$$\text{If } w(t) = w(-t), \text{ then } W(f) \text{ is real.} \quad (A.36)$$

$$\text{If } w(t) = -w(-t), \text{ then } W(f) \text{ is purely imaginary.} \quad (A.37)$$

- **Time shift property**
  With

$$\mathcal{F}\{w(t)\} = W(f),$$

$$\mathcal{F}\{w(t - t_0)\} = W(f)e^{-j2\pi f t_0}. \quad (A.38)$$

- **Differentiation property**
  With

$$\mathcal{F}\{w(t)\} = W(f),$$

$$\frac{dw(t)}{dt} = j2\pi f W(f). \quad (A.39)$$

- **Convolution ↔ multiplication property**
  With

$$\mathcal{F}\{w_i(t)\} = W_i(f),$$

$$\mathcal{F}\{w_1(t) * w_2(t)\} = W_1(f)W_2(f) \quad (A.40)$$

and

$$\mathcal{F}\{w_1(t)w_2(t)\} = W_1(f) * W_2(f), \quad (A.41)$$

where the convolution operator "$*$" is defined via

$$x(\alpha) * y(\alpha) \equiv \int_{-\infty}^{\infty} x(\lambda)y(\alpha - \lambda)d\lambda. \tag{A.42}$$

- **Parseval's theorem**

With

$$\mathcal{F}\{w_i(t)\} = W_i(f),$$

$$\int_{-\infty}^{\infty} w_1(t)w_2^*(t)dt = \int_{-\infty}^{\infty} W_1(f)W_2^*(f)df. \tag{A.43}$$

- **Final value theorem**

With $\lim_{t\to-\infty} w(t) = 0$ and $w(t)$ bounded,

$$\lim_{t\to\infty} w(t) = \lim_{f\to 0} j2\pi f W(f), \tag{A.44}$$

where $\mathcal{F}\{w(t)\} = W(f)$.

A.3 Energy and Power

- **Energy of a continuous time signal $s(t)$ is**

$$E(s) = \int_{-\infty}^{\infty} s^2(t)dt \tag{A.45}$$

if the integral is finite.

- **Power of a continuous time signal $s(t)$ is**

$$P(s) = \lim_{T\to\infty} \frac{1}{T} \int_{-T/2}^{T/2} s^2(t)dt \tag{A.46}$$

if the limit exists.

- **Energy of a discrete time signal $s[k]$ is**

$$E(s) = \sum_{-\infty}^{\infty} s^2[k] \tag{A.47}$$

if the sum is finite.

- **Power of a discrete time signal $s[k]$ is**

$$P(s) = \lim_{N\to\infty} \frac{1}{2N} \sum_{k=-N}^{N} s^2[k] \tag{A.48}$$

if the limit exists.

- **Power Spectral Density**

  With input and output transforms $X(f)$ and $Y(f)$ of a linear filter with impulse response transform $H(f)$ (such that $Y(f) = H(f)X(f)$),

  $$\mathcal{P}_y(f) = \mathcal{P}_x(f)\ |H(f)|^2, \qquad (A.49)$$

  where the *power spectral density* (PSD) is defined as

  $$\mathcal{P}_x(f) = \lim_{T \to \infty} \frac{|X_T(f)|^2}{T} \quad \text{(Watts/Hz)}, \qquad (A.50)$$

  where $\mathcal{F}\{x_T(t)\} = X_T(f)$ and

  $$x_T(t) = x(t)\ \Pi\left(\frac{t}{T}\right), \qquad (A.51)$$

  where $\Pi(\cdot)$ is the rectangular pulse (A.20).

A.4 Z-Transforms and Properties

- **Definition of the Z-transform**

  $$X(z) = \mathcal{Z}\{x[k]\} = \sum_{k=-\infty}^{\infty} x[k]z^{-k} \qquad (A.52)$$

- **Time-shift property**

  With

  $$\mathcal{Z}\{x[k]\} = X(z),$$
  $$\mathcal{Z}\{x[k - \Delta]\} = z^{-\Delta}X(z). \qquad (A.53)$$

- **Linearity property**

  With

  $$\mathcal{Z}\{x_i[k]\} = X_i(z),$$
  $$\mathcal{Z}\{ax_1[k] + bx_2[k]\} = aX_1(z) + bX_2(z). \qquad (A.54)$$

- **Final Value Theorem for z-transforms**

  If $X(z)$ converges for $|z| > 1$ and all poles of $(z - 1)X(z)$ are inside the unit circle, then

  $$\lim_{k \to \infty} x[k] = \lim_{z \to 1}(z - 1)X(z). \qquad (A.55)$$

A.5 Integral and Derivative Formulas

- **Sifting property of impulse**

  $$\int_{-\infty}^{\infty} w(t)\delta(t - t_0)dt = w(t_0) \qquad (A.56)$$

- **Schwarz's inequality**

$$\left| \int_{-\infty}^{\infty} a(x)b(x)dx \right|^2 \leq \left\{ \int_{-\infty}^{\infty} |a(x)|^2 dx \right\} \left\{ \int_{-\infty}^{\infty} |b(x)|^2 dx \right\} \qquad \text{(A.57)}$$

and equality occurs only when $a(x) = kb^*(x)$, where superscript $*$ indicates complex conjugation (i.e., $(a + jb)^* = a - jb$).

- **Leibniz's rule**

$$\frac{d\left[\int_{a(x)}^{b(x)} f(\lambda, x)d\lambda \right]}{dx} = f(b(x), x)\frac{db(x)}{dx} - f(a(x), x)\frac{da(x)}{dx} + \int_{a(x)}^{b(x)} \frac{\partial f(\lambda, x)}{\partial x}d\lambda$$
$$\text{(A.58)}$$

- **Chain rule of differentiation**

$$\frac{dw}{dx} = \frac{dw}{dy}\frac{dy}{dx} \qquad \text{(A.59)}$$

- **Derivative of a product**

$$\frac{d}{dx}(wy) = w\frac{dy}{dx} + y\frac{dw}{dx} \qquad \text{(A.60)}$$

- **Derivative of signal raised to a power**

$$\frac{d}{dx}(y^n) = ny^{n-1}\frac{dy}{dx} \qquad \text{(A.61)}$$

- **Derivative of cosine**

$$\frac{d}{dx}(\cos(y)) = -(\sin(y))\frac{dy}{dx} \qquad \text{(A.62)}$$

- **Derivative of sine**

$$\frac{d}{dx}(\sin(y)) = (\cos(y))\frac{dy}{dx} \qquad \text{(A.63)}$$

A.6 Matrix Algebra

- **Transpose transposed**

$$(A^T)^T = A \qquad \text{(A.64)}$$

- **Transpose of a product**

$$(AB)^T = B^T A^T \qquad \text{(A.65)}$$

- **Transpose and inverse commutativity**
  If A^{-1} exists,

$$\left(A^T\right)^{-1} = \left(A^{-1}\right)^T.$$ (A.66)

- **Inverse identity**
  If A^{-1} exists,

$$A^{-1}A = AA^{-1} = I.$$ (A.67)

APPENDIX

B *Simulating Noise*

> Anyone who considers arithmetic methods of producing random digits is, of course, in a state of sin.
>
> —John von Neumann

Noise generally refers to unwanted or undesirable signals that disturb or interfere with the operation of a system. There are many sources of noise. In electrical systems, there may be coupling with the power lines, lightning, bursts of solar radiation, or thermal noise. Noise in a transmission system may arise from atmospheric disturbances, from other broadcasts that are not well shielded, from unreliable clock pulses or inexact frequencies used to modulate signals.

Whatever the physical source, there are two very different kinds of noise: narrowband and broadband. Narrowband noise consists of a thin slice of frequencies. With luck, these frequencies will not overlap the frequencies that are crucial to the communication system. When they do not overlap, it is possible to build filters that reject the noise and pass only the signal, analogous to the filter designed in Section 7.2.2 to remove certain frequencies from the gong waveform. When running simulations or examining the behavior of a system in the presence of narrowband noise, it is common to model the narrowband noise as a sum of sinusoids.

Broadband noise contains significant amounts of energy over a large range of frequencies. This is problematic because there is no obvious way to separate the parts of the noise that lie in the same frequency regions as the signals from the signals themselves. Often, stochastic or probabilistic models are used to characterize the behavior of systems under uncertainty. The simpler approach employed here is to model the noise in terms of its spectral content. Typically, the noise v will also be assumed to be uncorrelated with the signal w, in the sense that R_{wv} of (8.3) is zero. The remainder of this section explores mathematical models of (and computer implementations for simulations of) several kinds of noise that are common in communications systems.

The simplest broadband noise is one that contains "all" frequencies in equal amounts. By analogy with white light, which contains all frequencies of visible light, this is called *white* noise. Most random number generators, by default, give (approximately) white noise. For example, the following MATLAB code uses the function `randn` to create a vector with N normally distributed (or Gaussian) random numbers.

---

randspec.m spectrum of random numbers

```
N=2^16;                             % how many random #'s
Ts=0.001; t=Ts*(1:N);              % define a time vector
ssf=(-N/2:N/2-1)/(Ts*N);          % frequency vector
x=randn(1,N);                      % N random numbers
fftx=fft(x);                       % spectrum of random numbers
```

Running `randspec.m` gives a plot much like that shown in Figure B.1, though details may change because the random numbers are different each time the program is run.

The random numbers themselves fall mainly between ±4, though most are less than ±2. The average (or mean) value is

$$m = \frac{1}{N} \sum_{k=1}^{N} x[k] \tag{B.1}$$

and is very close to zero, as can be verified by calculating

```
m=sum(x)/length(x).
```

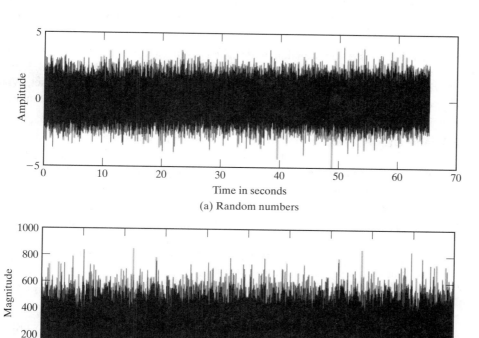

FIGURE B.1 A random signal and its (white) spectrum.

(a) Random numbers

(b) Spectrum of random numbers

The variance (the width, or spread of the random numbers) is defined by

$$v = \frac{1}{N} \sum_{k=1}^{N} (x[k] - m)^2 \tag{B.2}$$

and can easily be calculated with the MATLAB code

```
v=sum((x-m).*(x-m))/length(x).
```

For `randn`, this is very close to 1.0. When the mean is zero, this variance is the same as the power. Hence, if `m=0`, `v=pow(x)` also gives the variance.

The spectrum of a numerically generated white noise sequence typically appears as in the bottom plot of Figure B.1. Observe the symmetry in the spectrum (which occurs because the random numbers are real valued). In principle, the spectrum is flat (all frequencies are represented equally), but in reality, any given time the program is run, some frequencies appear slightly larger than others. In the figure, there is such a spurious peak near 275 Hz, and a couple more near 440 Hz. Verify that each time the program is run, these spurious peaks are at different frequencies.

Problems

B.1. Use `ranspec.m` to investigate the spectrum when different numbers of random values are chosen. Try $N = 10, 100, 2^{10}, 2^{18}$. For each of the values N, locate any spurious peaks. When the same program is run again, do they occur at the same frequencies?[1]

B.2. MATLAB's `randn` function is designed so that the mean is always (approximately) zero and the variance is (approximately) unity. Consider a signal defined by `w = a randn + b`; that is, the output of `randn` is scaled and offset. What are the mean and variance of `w`? Hint: Use (B.1) and (B.2). What values must `a` and `b` have to create a signal that has mean 1.0 and variance 5.0?

B.3. Another MATLAB function to generate random numbers is `rand`, which creates numbers between 0 and 1. Try the code `x=rand(1,N)-0.5` in `randspec.m`, where the 0.5 causes `x` to have zero mean. What are the mean and the variance of `x`? What does the spectrum of `rand` look like? Is it also "white"? What happens if the 0.5 is removed? Explain what you see.

B.4. Create two different white signals $w[k]$ and $v[k]$ that are at least $N = 2^{16}$ elements long.
   (a) For each j between -100 and $+100$, find the cross-correlation $R_{wv}[j]$ between $w[k]$ and $v[k]$.
   (b) Find the autocorrelations $R_w[j]$ and $R_v[j]$. What value(s) of j give the largest autocorrelation?

Though many types of noise may have a wide bandwidth, few are truly white. A common way to generate random sequences with (more or less) any desired spectrum is to pass white noise through a linear filter with a specified passband. The output then

---

[1] MATLAB allows control over whether the "random" numbers are the same each time using the "seed" option in the calls to the random number generator. Details can be found in the help files for `rand` and `randn`.

has a spectrum that coincides with the passband of the filter. For example, the following program creates such "colored" noise by passing white noise through a bandpass filter that attenuates all frequencies but those between 100 and 200 Hz:

---

randcolor.m generating a colored noise spectrum

```
N=2^16;                              % how many random #'s
Ts=0.001; nyq=0.5/Ts;                % sampling interval and nyquist rate
ssf=(-N/2:N/2-1)/(Ts*N);             % frequency vector
x=randn(1,N);                        % N random numbers
fbe=[0 100 110 190 200 nyq]/nyq;     % definition of desired filter
damps=[0 0 1 1 0 0];                 % desired amplitudes
fl=70;                               % filter size
b=remez(fl,fbe,damps);               % design the impulse response
y=filter(b,1,x);                     % filter x with impulse response b
```

Plots from a typical run of `randcolor.m` are shown in Figure B.2, which illustrates the spectrum of the white input and the spectrum of the colored output. Clearly, the bandwidth of the output noise is (roughly) between 100 and 200 Hz.

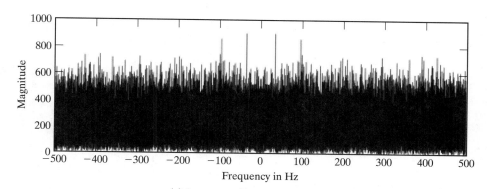

(a) Spectrum of input random numbers

FIGURE B.2 A white input signal (top) is passed through a bandpass filter, creating a noisy signal with bandwidth between 100 and 200 Hz.

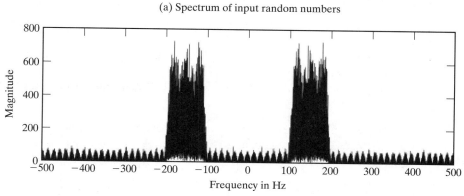

(b) Spectrum of output random numbers

Problems

B.5. Create a noisy signal that has no energy below 100 Hz. It should then have (linearly) increasing energy from 100 Hz to the Nyquist rate at 500 Hz.

    (a) Design an appropriate filter using `remez`. Verify its frequency response by using `freqz`.

    (b) Generate a white noise and pass it through the filter. Plot the spectrum of the input and the spectrum of the output.

B.6. Create two noisy signals $w[k]$ and $v[k]$ that are $N = 2^{16}$ elements long. The bandwidths of both $w[k]$ and $v[k]$ should lie between 100 and 200 Hz as in `randcolor.m`.

    (a) For each j between -100 and $+100$, find the cross-correlation $R_{wv}[j]$ between $w[k]$ and $v[k]$.

    (b) Find the autocorrelations $R_w[j]$ and $R_v[j]$. What value(s) of j give the largest autocorrelation?

    (c) Are there any similarities between the two autocorrelations?

    (d) Are there any similarities between these autocorrelations and the impulse response b of the bandpass filter?

C Envelope of a Bandpass Signal

You know that the Radio wave is sent across, "transmitted" from the transmitter, to the receiver through the ether. Remember that ether forms a part of everything in nature—that is why Radio waves travel everywhere, through houses, through the earth, through the air.
—*Fundamental Principles of Radio: Certified Radio-tricians Course,*
National Radio Institute, Washington DC, 1914

The envelope of a signal is a curve that smoothly encloses the signal, as shown in Figure C.1. An envelope detector is a circuit (or computer program) that outputs the envelope when the signal is applied at its input.

In early analog radios, envelope detectors were used to help recover the message from the modulated carrier, as discussed in Section 5.1. One simple design includes a diode, capacitor, and resistor arranged as in Figure C.2. The oscillating signal arrives from an antenna. When the voltage is positive, current passes through the diode, and charges the capacitor. When the voltage is negative, the diode blocks the current, and the capacitor discharges through the resistor. The time constants are chosen so that the charging of the capacitor is quick (so that the output follows the upward motion of the signal), but the discharging is relatively slow (so that the output decays slowly from its peak value). Typical output of such a circuit is shown by the jagged line in Figure C.1, a reasonable approximation to the actual envelope.

It is easy to approximate the action of an envelope detector. The essence of the method is to apply a static nonlinearity (analogous to the diode in the circuit) followed by a lowpass filter (the capacitor and resistor). For example, the MATLAB code in AMlarge.m on page 81 extracted the envelope using an absolute value nonlinearity and a LPF, and this method is also used in envsig.m.

---

envsig.m: "envelope" of a bandpass signal

```
time=.33; Ts=1/10000;                    % sampling interval and time
t=0:Ts:time; lent=length(t);             % define a "time" vector
fc=1000; c=cos(2*pi*fc*t);               % define signal as fast wave
fm=10; w=cos(2*pi*fm*t).*exp(-5*t)+0.5;  % times slow decaying wave
x=c.*w;                                  % with offset
fbe=[0 0.05 0.1 1]; damps=[1 1 0 0]; fl=100;  % low pass filter design
b=remez(fl,fbe,damps);                   % impulse response of LPF
envx=(pi/2)*filter(b,1,abs(x));          % find envelope full rectify
```

FIGURE C.1 The enve-
lope of a signal out-
lines the extremes in a
smooth manner.

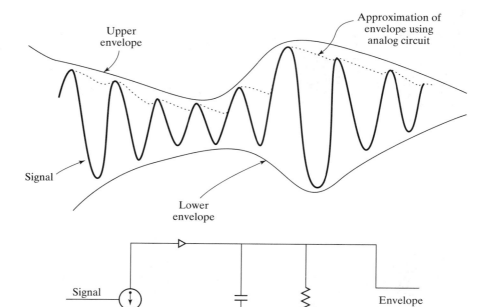

FIGURE C.2 A circuit
that extracts the enve-
lope from a signal.

Suppose that a pure sine wave is input into this envelope detector. Then the output of the LPF would be the average of the absolute value of the sine wave (the integral of the absolute value of a sine wave over a period is $\frac{2}{\pi}$). The factor $\frac{\pi}{2}$ in the definition of envx accounts for this factor so that the output rides on crests of the wave. The output of envsig.m is shown in Figure C.3(a), where the envelope signal envx follows the outline of the narrow bandwidth passband signal x, though with a slight delay. This delay is caused by the linear filter, and can be removed by shifting the envelope curve by the group delay of the filter. This is fl/2, half the length of the lowpass filter when designed using the remez command.

A more formal definition of envelope uses the notion of in-phase and quadrature components of signals to reexpress the original bandpass signal $x(t)$ as the product of a complex sinusoid and a slowly varying envelope function

$$x(t) = \text{Re}\{g(t)e^{j2\pi f_c t}\}. \tag{C.1}$$

The function $g(t)$ is called the *complex envelope* of $x(t)$, and f_c is the carrier frequency in Hz.

To see this is always possible, consider Figure C.4. The input $x(t)$ is assumed to be a narrowband signal centered near f_c (with support between $f_c - B$ and $f_c + B$ for some small B). Multiplication by the two sine waves modulates this to a pair of signals centered at baseband and at $2f_c$. The LPF removes all but the baseband, and so the spectra of both $x_c(t)$ and $x_s(t)$ are contained between $-B$ to B. Modulation by the final two sinusoids returns the baseband signals to a region around f_c, and adding them together gives exactly the signal $x(t)$. Thus, Figure C.4 represents an identity. It is useful

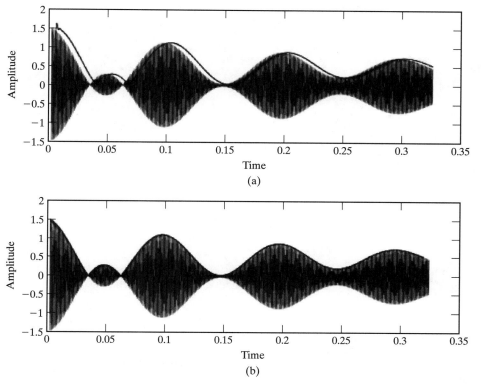

FIGURE C.3 The envelope smoothly outlines the contour of the signal. (a) shows the output of `envsig.m`, while (b) shifts the output to account for the delay caused by the linear filter.

(a)

(b)

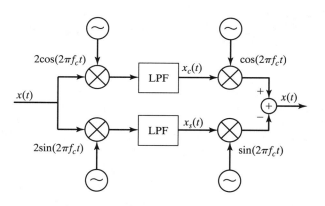

FIGURE C.4 The envelope can be written in terms of the two baseband signals $x_c(t)$ (the in-phase component) and $x_s(t)$ (the quadrature component). Assuming the lowpass filters are perfect, this represents an identity; $x(t)$ at the input equals $x(t)$ at the output.

because it allows any passband signal to be expressed in terms of two baseband signals, which are called the in-phase and quadrature components of the signal.

Symbolically, the signal $x(t)$ can be written

$$x(t) = x_c(t) \cos(2\pi f_c t) - x_s(t) \sin(2\pi f_c t),$$

where

$$x_c(t) = \text{LPF}\{2x(t) \cos(2\pi f_c t)\}$$
$$x_s(t) = -\text{LPF}\{2x(t) \sin(2\pi f_c t)\}.$$

Applying Euler's identity (A.1) then shows that the envelope $g(t)$ can be expressed in terms of the in-phase and quadrature components as

$$g(t) = \sqrt{x_c^2(t) + x_s^2(t)}.$$

Any physical (real valued) bandlimited waveform can be represented as in (C.1) and so it is possible to represent many of the standard modulation schemes in a unified notation.

For example, consider the case in which the complex envelope is a scaled version of the message waveform. (i.e., $g(t) = A_c w(t)$). Then

$$x(t) = \text{Re}\{A_c w(t) e^{j2\pi f_c t}\}.$$

Using $e^{\pm jx} = \cos(x) \pm j\sin(x)$,

$$x(t) = \text{Re}\{w(t)[A_c\cos(2\pi f_c t) + jA_c\sin(2\pi f_c t)]\}$$
$$= w(t)A_c\cos(2\pi f_c t),$$

which is the same as AM with suppressed carrier from Section 5.2.

AM with large carrier can also be written in the form of (C.1) with $g(t) = A_c[1+w(t)]$. Then

$$x(t) = \text{Re}\{A_c[1 + w(t)]e^{j2\pi f_c t}\}$$
$$= \text{Re}\{A_c e^{j2\pi f_c t} + A_c w(t)e^{j2\pi f_c t}\}$$
$$= A_c\cos(2\pi f_c t) + w(t)A_c\cos(2\pi f_c t).$$

The envelope $g(t)$ is real in both of these cases when $w(t)$ is real.

When the envelope $g(t) = x(t) + jy(t)$ is complex valued, then $x(t)$ in (C.1) becomes

$$x(t) = \text{Re}\{(x(t) + jy(t))e^{j2\pi f_c t}\}.$$

With $e^{jx} = \cos(x) + j\sin(x)$,

$$x(t) = \text{Re}\{x(t)\cos(2\pi f_c t) + jx(t)\sin(2\pi f_c t) + jy(t)\cos(2\pi f_c t) + j^2 y(t)\sin(2\pi f_c t)\}$$
$$= x(t)\cos(2\pi f_c t) - y(t)\sin(2\pi f_c t).$$

This is the same as quadrature modulation of Section 5.3.

Problems

C.1. Replace the `filter` command with the `filtfilt` command and rerun `envsig.m`. Observe the effect of the delay. Read the MATLAB help file for `filtfilt`, and try to adjust the programs so that the outputs coincide. Hint: You will need to change the filter parameters as well as the decay of the output.

C.2. Replace the absolute value nonlinearity with a rectifying nonlinearity

$$\bar{x}(t) = \begin{cases} x(t) & t \geq 0 \\ 0 & t < 0 \end{cases},$$

which more closely simulates the action of a diode. Mimic the code in `envsig.m` to create an envelope detector. What is the appropriate constant that must be used to make the output smoothly touch the peaks of the signal?

C.3. Use `envsig.m` and the following code to find the envelope of a signal

```
xc=filter(b,1,2*x.*cos(2*pi*fc*t));    % in-phase component
xs=filter(b,1,2*x.*sin(2*pi*fc*t));    % quadrature component
envx=abs(xc+xs);                        % envelope of signal
```

Can you see how to write these three lines of code in one (complex valued) line?

C.4. For those who have access to the MATLAB signal processing toolbox, an even simpler syntax for the complex envelope is

```
envx=abs(hilbert(x));
```

Can you figure out why the "Hilbert transform" is useful for calculating the envelope?

C.5. Find a signal $x(t)$ for which all the methods of envelope detection fail to provide a convincing "envelope." Hint: Try signals that are *not* narrowband.

APPENDIX

D

Relating the Fourier Transform to the DFT

Mathematics compares the most diverse phenomena and discovers the secret analogies that unite them.

—Joseph Fourier

Most people are quite familiar with "time domain" thinking: a plot of voltage versus time, how stock prices vary as the days pass, a function that grows (or shrinks) over time. One of the most useful tools in the arsenal of an electrical engineer is the idea of transforming a problem into the frequency domain. Sometimes this transformation works wonders; what at first seemed intractable is now obvious at a glance. Much of this appendix is about the process of making the transformation from time into frequency, and back again from frequency into time. The primary mathematical tool is the Fourier transform (and its discrete-time counterparts).

D.1 The Fourier Transform and Its Inverse

By definition, the Fourier transform of a time function $w(t)$ is

$$W(f) = \int_{-\infty}^{\infty} w(t)e^{-j2\pi ft}dt, \tag{D.1}$$

which appeared earlier in Equation (2.1). The Inverse Fourier transform is

$$w(t) = \int_{-\infty}^{\infty} W(f)e^{j2\pi ft}df. \tag{D.2}$$

Observe that the transform is an integral over all time, while the inverse transform is an integral over all frequency; the transform converts a signal from time into frequency, while the inverse converts from frequency into time. Because the transform is invertible, it does not create or destroy information. Everything about the time signal $w(t)$ is contained in the frequency signal $W(f)$ and vice versa.

The integrals (D.1) and (D.2) do not always exist; they may fail to converge or they may become infinite if the signal is bizarre enough. Mathematicians have catalogued exact conditions under which the transforms exist, and it is a reasonable engineering assumption that any signal encountered in practice fulfills these conditions.

346

Perhaps the most useful property of the Fourier transform (and its inverse) is its linearity. Suppose that $w(t)$ and $v(t)$ have Fourier transforms $W(f)$ and $V(f)$, respectively. Then superposition suggests that the function $s(t) = aw(t) + bv(t)$ should have transform $S(f) = aW(f) + bV(f)$, where a and b are any complex numbers. To see that this is indeed the case, observe that

$$S(f) = \int_{-\infty}^{\infty} s(t)e^{-j2\pi ft}dt$$

$$= \int_{-\infty}^{\infty} (aw(t) + bv(t))e^{-j2\pi ft}dt$$

$$= a\int_{-\infty}^{\infty} w(t)e^{-j2\pi ft}dt + b\int_{-\infty}^{\infty} v(t)e^{-j2\pi ft}dt$$

$$= aW(f) + bV(f).$$

What does the transform mean? Unfortunately, this is not immediately apparent from the definition. One common interpretation is to think of $W(f)$ as describing how to build the time signal $w(t)$ out of sine waves (more accurately, out of complex exponentials). Conversely, $w(t)$ can be thought of as the unique time waveform that has the frequency content specified by $W(f)$.

Even though the time signal is usually a real-valued function, the transform $W(f)$ is, in general, complex valued due to the complex exponential $e^{-j2\pi ft}$ appearing in the definition. Thus $W(f)$ is a complex number for each frequency f. The *magnitude spectrum* is a plot of the magnitude of the complex numbers $W(f)$ as a function of f, and the *phase spectrum* is a plot of the angle of the complex numbers $W(f)$ as a function f.

D.2 The DFT and the Fourier Transform

This section derives the DFT as a limiting approximation of the Fourier transform, showing the relationship between the continuous-time and discrete-time transforms.

The Fourier transform cannot be applied directly to a waveform that is defined only on a finite interval $[0, T]$. But any finite length signal can be extended to infinite length by assuming it is zero outside of $[0, T]$. Accordingly, consider the windowed waveform

$$w_w(t) = \left\{ \begin{array}{ll} w(t) & 0 \leq t \leq T \\ 0 & t \text{ otherwise} \end{array} \right\} = w(t)\Pi\left(\frac{t - (T/2)}{T}\right),$$

where Π is the rectangular pulse (2.7). The Fourier transform of this windowed (finite support) waveform is

$$W_w(f) = \int_{t=-\infty}^{\infty} w_w(t)e^{-j2\pi ft}dt = \int_{t=0}^{T} w(t)e^{-j2\pi ft}dt.$$

Approximating the integral at $f = n/T$ and replacing the differential dt with Δt ($= T/N$) allows it to be approximated by the sum

$$\left.\int_0^T w(t)e^{-2j\pi ft}\,dt\right|_{f=n/T} \approx \sum_{k=0}^{N-1} w(k\Delta t)e^{-j2\pi(n/T)k(T/N)}\Delta t$$

$$= \Delta t \sum_{k=0}^{N-1} w(k\Delta t)e^{-j(2\pi/N)nk},$$

where the substitution $t \approx k\Delta t$ is used. Identifying $w(k\Delta t)$ with $w[k]$ gives

$$W_w(f)|_{f=n/T} \approx \Delta t\, W[n].$$

As before, T is the time window of the data record, and N is the number of data points. Δt ($= T/N$) is the time between samples (or the time resolution), which is chosen to satisfy the Nyquist rate so that no aliasing will occur. T is selected for a desired frequency resolution $\Delta f = 1/T$; that is, T must be chosen large enough so that Δf is small enough. For a frequency resolution of 1 Hz, a second of data is needed. For a frequency resolution of 1 KHz, 1 msec of data is needed.

Suppose N is to be selected so as to achieve a time resolution $\Delta t = 1/\alpha f^\dagger$, where $\alpha > 2$ causes no aliasing (i.e., the signal is bandlimited to $f^\dagger$). Suppose T is specified to achieve a frequency resolution $1/T$ that is β times the signal's highest frequency, so $T = 1/(\beta f^\dagger)$. Then the (required) number of data points N, which equals the ratio of the time window T to the time resolution Δt, is α/β.

For example, consider a waveform that is zero for all time before $-\frac{T_d}{2}$, when it becomes a sine wave lasting until time $\frac{T_d}{2}$. This "switched sinusoid" can be modelled as

$$w(t) = \Pi\left(\frac{t}{T_d}\right)A\sin(2\pi f_0 t) = \Pi\left(\frac{t}{T_d}\right)A\cos(2\pi f_0 t - \pi/2).$$

From (2.8), the transform of the pulse is

$$\mathcal{F}\left\{\Pi\left(\frac{t}{T_d}\right)\right\} = T_d\frac{\sin(\pi f T_d)}{\pi f T_d}.$$

Using the frequency translation property, the transform of the switched sinusoid is

$$W(f) = A\left(\frac{1}{2}\right)\left[e^{-j\pi/2}T_d\frac{\sin(\pi(f-f_0)T_d)}{\pi(f-f_0)T_d} + e^{j\pi/2}T_d\frac{\sin(\pi(f+f_0)T_d)}{\pi(f+f_0)T_d}\right],$$

which can be simplified (using $e^{-j\pi/2} = -j$ and $e^{j\pi/2} = j$) to

$$W(f) = \frac{jAT_d}{2}\left[\frac{\sin(\pi(f+f_0)T_d)}{\pi(f+f_0)T_d} - \frac{\sin(\pi(f-f_0)T_d)}{\pi(f-f_0)T_d}\right]. \tag{D.3}$$

This transform can be approximated numerically, as in the following program `switchsin.m`. Assume the total time window of the data record of $N = 1024$ samples is $T = 8$ seconds and that the underlying sinusoid of frequency $f_0 = 10$ Hz is switched on for only the first $T_d = 1$ seconds.

---

switchsin.m spectrum of a switched sine

```
Td=1;                            % pulse width [-Td/2,Td/2]
N=1024;                          % number of data points
f=10;                            % frequency of sine
T=8;                             % total time window
trez=T/N; frez=1/T;             % time and freq. resolution
w=zeros(size(1:N));             % vector for full data record
w(N/2-Td/(trez*2)+1:N/2+Td/(2*trez))=sin(trez*(1:Td/trez)*2*pi*f);
dftmag=abs(fft(w));             % magnitude of spectrum of w
spec=trez*[dftmag((N/2)+1:N),dftmag(1:N/2)];
ssf=frez*[-(N/2)+1:1:(N/2)];
plot(trez*[-length(w)/2+1:length(w)/2],w,'-');  % plot (a)
plot(dftmag,'-');                % plot (b)
plot(ssf,spec,'-');              % plot (c)
```

Plots of the key variables are shown in Figure D.1. The switched sinusoid w is shown plotted against time, and the "raw" spectrum dftmag is plotted as a function of its index. The proper magnitude spectrum spec is plotted as a function of frequency, and the final plot shows a zoom into the low frequency region. In this case, the time resolution is $\Delta t = T/N = 0.0078$ seconds and the frequency resolution is $\Delta f = 1/T = 0.125$ Hz. The largest allowable f_0 without aliasing is $N/2T = 64$ Hz.

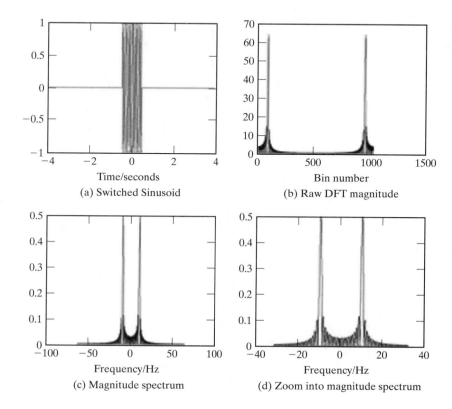

(a) Switched Sinusoid

(b) Raw DFT magnitude

(c) Magnitude spectrum

(d) Zoom into magnitude spectrum

FIGURE D.1 Spectrum of the switched sinusoid calculated using the DFT. (a) the time waveform, (b) the raw magnitude data, (c) the magnitude spectrum, (d) zoomed into the magnitude spectrum.

Problem

D.1. Rerun the preceding program with `T=16`, `Td=2`, and `f=5`. Comment on the location and the width of the two spectral lines. Can you find particular values so that the peaks are extremely narrow? Can you relate the locations of these narrow peaks to (D.3)?

APPENDIX

E *Power Spectral Density*

One way of classifying and measuring signals and systems is by their power (or energy), and the amount of power (or energy) in various frequency regions. This section defines the power spectral density, and shows how it can be used to measure the power in signals, to measure the correlation within a signal, and to talk about the gain of a linear system. In *Telecommunication Breakdown*, power spectral density is used mainly in Chapter 11 in the discussion of the design of matched filtering.

The (time) energy of a signal was defined in (A.45) as the integral of the signal squared, and Parseval's theorem (A.43) guarantees that this is the same as the total energy measured in frequency

$$E = \int_{-\infty}^{\infty} |W(f)|^2 df,$$

where $W(f) = \mathcal{F}\{w(t)\}$ is the Fourier transform of $w(t)$.

When E is finite, $w(t)$ is called an energy waveform. But E is infinite for many common signals in communications such as the sine wave and the sinc functions. In this case, the power, as defined in (A.46), is

$$P = \lim_{T \to \infty} \frac{1}{T} \int_{t=-T/2}^{T/2} |w(t)|^2 dt, \qquad \text{(E.1)}$$

which is the average of the energy, and which can be used to measure the signal. Signals for which P is nonzero and finite are called power waveforms.

Define the truncated waveform

$$w_T(t) = w(t) \, \Pi\left(\frac{t}{T}\right),$$

where $\Pi(\cdot)$ is the rectangular pulse (2.7) that is 1 between $-T/2$ and $T/2$, and is zero elsewhere. When $w(t)$ is real valued, (E.1) can be rewritten

$$P = \lim_{T \to \infty} \frac{1}{T} \int_{t=-\infty}^{\infty} w_T^2(t) \, dt.$$

Parseval's theorem (A.43) shows that this is the same as

$$P = \lim_{T \to \infty} \frac{1}{T} \int_{f=-\infty}^{\infty} |W_T(f)|^2 \, df = \int_{f=-\infty}^{\infty} \left(\lim_{T \to \infty} \frac{|W_T(f)|^2}{T} \right) df,$$

where $W_T(f) = \mathcal{F}\{w_T(t)\}$. The *power spectral density* (PSD) is then defined as

$$\mathcal{P}_w(f) = \lim_{T \to \infty} \frac{|W_T(f)|^2}{T} \quad \text{(Watts/Hz)},$$

which allows the total power to be written

$$P = \int_{f=-\infty}^{\infty} \mathcal{P}_w(f) \, df. \tag{E.2}$$

Observe that the PSD is always real and nonnegative. When $w(t)$ is real valued, then the PSD is symmetric, $\mathcal{P}_w(f) = \mathcal{P}_w(-f)$.

The PSD can be used to reexpress the autocorrelation function (the correlation of $w(t)$ with itself),

$$R_w(\tau) = \lim_{T \to \infty} \frac{1}{T} \int_{-T/2}^{T/2} w(t)w(t + \tau) \, dt,$$

in the frequency domain. This is the continuous-time counterpart of the cross-correlation (8.3) with $w = v$. First, replace τ with $-\tau$. Now the integrand is a convolution, and so the Fourier transform is the product of the spectra. Hence,

$$\mathcal{F}\{R_w(\tau)\} = \mathcal{F}\{R_w(-\tau)\} = \mathcal{F}\{ \lim_{T \to \infty} \frac{1}{T} w(t) * w(t)\}$$

$$= \lim_{T \to \infty} \frac{1}{T} \mathcal{F}\{w(t) * w(t)\}$$

$$= \lim_{T \to \infty} \frac{1}{T} |W(f)|^2 = \mathcal{P}_w(f).$$

Thus, the Fourier transform of the autocorrelation function of $w(t)$ is equal to the power spectral density of $w(t)$,[1] and it follows that

$$P = \int_{-\infty}^{\infty} \mathcal{P}_w(f) \, df = R_w(0),$$

which says that the total power is equal to the autocorrelation at $\tau = 0$.

The PSD can also be used to quantify the power gain of a linear system. Recall that the output $y(t)$ of a linear system is given by the convolution of the impulse response $h(t)$ with the input $x(t)$. Since convolution in time is the same as multiplication in frequency, $Y(f) = X(f)H(f)$. Assuming that $H(f)$ has finite energy, the PSD of y is

$$\mathcal{P}_y(f) = \lim_{T \to \infty} \frac{1}{T} |Y_T(f)|^2 \tag{E.3}$$

---

[1] This is known as the Wiener–Khintchine theorem, and it formally requires that $\int_{-\infty}^{\infty} \tau R_w(\tau)d\tau$ be finite; that is, the correlation between $w(t)$ and $w(t + \tau)$ must die away as τ gets large.

$$= \lim_{T \to \infty} \frac{1}{T} |X_T(f)|^2 |H(f)|^2 = \mathcal{P}_x(f) |H(f)|^2, \tag{E.4}$$

where $y_T(t) = y(t)\Pi\left(\frac{t}{T}\right)$ and $x_T(t) = x(t)\Pi\left(\frac{t}{T}\right)$ are truncated versions of $y(t)$ and $x(t)$. Thus, the PSD of the output is precisely the PSD of the input times the magnitude of the frequency response (squared), and the power gain of the linear system is exactly $|H(f)|^2$ for each frequency f.

Relating Difference Equations to Frequency Response and Intersymbol Interference

This appendix presents background material that is useful when designing equalizers. The first tool can be thought of as a variation on the Fourier transform called the $\mathcal{Z}$-transform, which is used to represent the channel model in a concise way. The frequency response of these models can easily be derived using a simple graphical technique that also provides insight into the inverse model. This can be useful in visualizing equalizer design as in Chapter 13. Finally, the "open eye" criterion provides a way of determining how good the design is.

F.1 $\mathcal{Z}$-Transforms

Fundamental to any digital signal is the idea of the unit delay, a time delay T of exactly one sample interval. There are several ways to represent this mathematically, and this section uses the $\mathcal{Z}$-Transform, which is closely related to a discrete version of the Fourier transform. Define the variable z to represent a (forward) time shift of one sample interval. Thus, $z\,u(kT) \equiv u((k+1)T)$. The inverse is the backward time shift $z^{-1}u(kT) \equiv u((k-1)T)$. These are most commonly written without explicit reference to the sampling rate as

$$z\,u[k] = u[k+1] \text{ and } z^{-1}u[k] = u[k-1].$$

For example, the FIR filter

$$u[k] + 0.6u[k-1] - 0.91u[k-2]$$

can be rewritten in terms of the time delay operator z as

$$(1 + 0.6z^{-1} - 0.91z^{-2})u[k].$$

Formally, the $\mathcal{Z}$-transform is defined much like the transforms of Chapter 5. The $\mathcal{Z}$-transform of a sequence $y[k]$ is

$$Y(z) = \mathcal{Z}\{y[k]\} = \sum_{k=-\infty}^{\infty} y[k]z^{-k}. \tag{F.1}$$

Though it may not at first be apparent, this definition corresponds to the intuitive idea of a unit delay. The $\mathcal{Z}$-transform of a delayed sequence is

$$\mathcal{Z}\{y[k-\Delta]\} = \sum_{k=-\infty}^{\infty} y[k-\Delta]z^{-k}.$$

Applying the change of variable $k - \Delta = j$ (so that $k = j + \Delta$), this can be rewritten

$$\mathcal{Z}\{y[k-\Delta]\} = \sum_{j+\Delta=-\infty}^{\infty} y[j]z^{-(j+\Delta)}$$

$$= z^{-\Delta} \sum_{j+\Delta=-\infty}^{\infty} y[j]z^{-j}$$

$$= z^{-\Delta}Y(z). \tag{F.2}$$

In words, the $\mathcal{Z}$-transform of the time-shifted sequence $y[k-\Delta]$ is $z^{-\Delta}$ times the $\mathcal{Z}$-transform of the original sequence $y[k]$. Observe the similarity between this property and the time delay property of Fourier transforms, equation (A.38). This similarity is no coincidence; formally substituting $z \leftrightarrow e^{j2\pi f}$ and $\Delta \leftrightarrow t_0$ turns (F.2) into (A.38).

In fact, most of the properties of the Fourier transform and the DFT have a counterpart in $\mathcal{Z}$-transforms. For instance, it is easy to show from the definition (F.1) that the $\mathcal{Z}$-transform is linear; that is,

$$\mathcal{Z}\{ay[k] + bu[k]\} = aY(z) + bU(z).$$

Similarly, the product of two $\mathcal{Z}$-transforms is given by the convolution of the time sequences (analogous to (7.2)), and the ratio of the $\mathcal{Z}$-transform of the output to the $\mathcal{Z}$-transform of the input is a (discrete-time) transfer function.

For instance, the simple two-tap finite impulse response difference equation,

$$y[k] = u[k] - bu[k-1], \tag{F.3}$$

can be represented in transfer function form by taking the $\mathcal{Z}$-transform of both sides of the equation, applying (F.2) and using linearity. Thus,

$$Y(z) \equiv \mathcal{Z}\{y[k]\} = \mathcal{Z}\{u[k] - bu[k-1]\}$$

$$= \mathcal{Z}\{u[k]\} - \mathcal{Z}\{bu[k-1]\}$$

$$= U(z) - bz^{-1}U(z)$$

$$= (1 - bz^{-1})U(z),$$

which can be solved algebraically for

$$H(z) = \frac{Y(z)}{U(z)} = 1 - bz^{-1} = \frac{z-b}{z}. \tag{F.4}$$

$H(z)$ is called the transfer function of the filter (F.3).

There are two types of singularities that a z-domain transfer function may have. *Poles* are those values of z that make the magnitude of the transfer function infinite. The transfer function in (F.4) has a pole at $z = 0$. *Zeros* are those values of z that make the magnitude of the frequency response equal to zero. The transfer function in (F.4) has one zero at $z = b$. There are always the same number of poles as there are zeros in a transfer function, though some may occur at infinite values of z. For example, the transfer function $\frac{z-a}{1}$ has one finite-valued zero at $z = a$ and a pole at $z = \infty$.

A z-domain discrete-time system transfer function is called *minimum phase* (*maximum phase*) if it is causal and all of its singularities are located inside (outside) the unit circle. If some singularities are inside and others outside the unit circle, the transfer function is *mixed phase*. If it is causal and all of the poles of the transfer function are strictly inside the unit circle (i.e., if all the poles have magnitudes less than unity), then the system is stable, and a bounded input always leads to a bounded output. For example, the FIR difference equation

$$y[k] = u[k] + 0.6u[k-1] - 0.91u[k-2]$$

has the transfer function

$$H(z) = \frac{Y(z)}{U(z)} = 1 + 0.6z^{-1} - 0.91z^{-2} = (1 - 0.7z^{-1})(1 + 1.3z^{-1}) = \frac{(z-0.7)(z+1.3)}{z^2}.$$

This is mixed phase and stable, with zeros at $z = 0.7$ and -1.3 and two poles at $z = 0$.

Problems

F.1. Use the definition of the $\mathcal{Z}$-transform to show that the transform is linear; that is, show that $\mathcal{Z}\{ay[k] + bu[k]\} = aY(z) + bU(z)$.

F.2. Find the z-domain transfer function of the system defined by $y[k] = b_1u[k] + b_2u[k-1]$.

    1. What are the poles of the transfer function?
    2. What are the zeros?
    3. For what values of b_1 and b_2 is the system stable?
    4. For what values of b_1 and b_2 is the system minimum phase?
    5. For what values of b_1 and b_2 is the system maximum phase?

F.3. Find the z-domain transfer function of the system defined by $y[k] = ay[k-1] + bu[k-1]$.
    (a) What are the poles of the transfer function?
    (b) What are the zeros?
    (c) For what values of a and b is the system stable?
    (d) For what values of a and b is the system minimum phase?
    (e) For what values of a and b is the system maximum phase?

F.2 Sketching the Frequency Response from the $\mathcal{Z}$-Transform

A complex number $\alpha = a + jb$ can be drawn in the complex plane as a vector from the origin to the point (a, b). Figure F.1 is a graphical illustration of the difference between two complex numbers $\beta - \alpha$, which is equal to the vector drawn from α to β. The

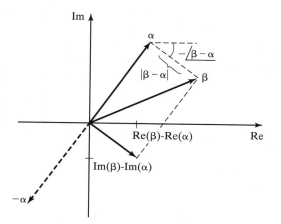

FIGURE F.1 Graphical calculation of the difference between two complex numbers.

magnitude is the length of this vector, and the angle is measured counterclockwise from the horizontal drawn to the right of α to the direction of $\beta - \alpha$, as shown.

As with Fourier transforms, the discrete-time transfer function in the z-domain can be used to describe the gain and phase that a sinusoidal input of frequency ω (in radians/second) will experience when passing through the system. With transfer function $H(z)$, the frequency response can be calculated by evaluating the magnitude of the complex number $H(z)$ at all points on the unit circle, that is, at all $z = e^{j\omega T}$. (T has units of seconds/sample.)

For example, consider the transfer function $H(z) = z - a$. At $z = e^{j0T} = 1$ (zero frequency), $H(z) = 1 - a$. As the frequency increases (as ω increases), the "test point" $z = e^{j\omega T}$ moves along the unit circle (think of this as the β in Figure F.1). The value of the frequency response at the test point $H(e^{j\omega T})$ is the difference between this β and the zero of $H(z)$ at $z = a$ (which corresponds to α in Figure F.1).

Suppose that $0 < a < 1$. Then the distance from the test point to the zero is smallest when $z = 1$, and increases continuously as the test point moves around the circle, reaching a maximum at $\omega T = \pi$ radians. Thus, the frequency response is highpass. The phase goes from $0°$ to $180°$ as ωT goes from 0 to π. On the other hand, if $-1 < a < 0$, then the system is lowpass.

More generally, consider any polynomial transfer function

$$a_N z^N + a_{N-1} z^{N-1} + \cdots + a_2 z^2 + a_1 z + a_0.$$

This can be factored into a product of N (possibly complex) roots:

$$H(z) = g \Pi_{i=1}^{N} (z - \gamma_i).$$

Accordingly, the magnitude of this FIR transfer function at any value z is the product of the magnitudes of the distances from z to the zeros. For any test point on the unit circle, the magnitude is equal to the product of all the distances from the test point to the zeros. An example is shown in Figure F.2, where a transfer function has three zeros. Two "test points" are shown at frequencies corresponding to (approximately) 15 degrees and 80 degrees. The magnitude at the first test point is equal to the product of the lengths a_1, a_2, and a_3, while the magnitude at the second is $b_1 b_2 b_3$. Qualitatively, the frequency response begins at some value and slowly decreases in magnitude until it

FIGURE F.2 Suppose a transfer function has three zeros. At any frequency "test point" (specified in radians around the unit circle), the magnitude of the transfer function is the product of the distances from the test point to the zeros.

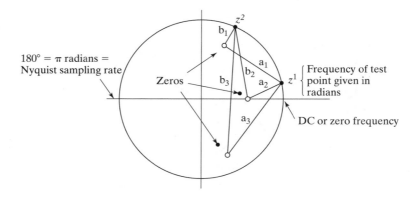

nears the second test point. After this, it rises. Accordingly, this transfer function is a "notch" filter.

Problem

F.4. Consider the transfer function $(z - a)(z - b)$ with $1 > a > 0$ and $0 > b > -1$. Sketch the magnitude of the frequency response, and show that it has a bandpass shape over the range of frequencies between 0 and π radians.

As another example, consider a ring of equally-spaced zeros in the complex z-plane. The resulting frequency response magnitude will be relatively flat because no matter where the test point is taken on the unit circle, the distances to the zeros in the ring of zeros is roughly the same. As the number of zeros in the ring decreases (increases), scallops in the frequency response magnitude will become more (less) pronounced. This is true if the ring of transfer function zeros is inside or outside the unit circle. Of course, the phase curves will be different in the two cases.

Problems

F.5. Sketch the frequency response of $H(z) = z - a$ when $a = 2$. Sketch the frequency response of $H(z) = z - a$ when $a = -2$.

F.6. Sketch the frequency responses of
  (a) $H(z) = (z - 1)(z - 0.5)$
  (b) $H(z) = z^2 - 2z + 1$
  (c) $H(z) = (z^2 - 2z + 1)(z + 1)$
  (d) $H(z) = g(z^n - 1)$ for $g = 0.1, 1.0, 10$ and $n = 2, 5, 25, 100$.

Of course, these frequency responses can also be evaluated numerically. For instance, the impulse response of the system described by $H(z) = 1 + 0.6z^{-1} - 0.91z^{-2}$ is the vector `h=[1 0.6 -0.91]`. Using the command `freqz(h)` draws the frequency response.

Problem

F.7. Draw the frequency response for each of the systems $H(z)$ in Problem F.6 using MATLAB.

If the transfer function included finite-valued poles, then the gain of the transfer function would be divided by the product of the distances from a test point on the unit circle to the poles. The counterclockwise angles from the positive horizontal at each pole location to the vector pointing from there to the test point on the unit circle would be subtracted in the overall phase formula. The point of this technique is not to carry out complex calculations better left to computers, but to learn to reason qualitatively using plots of the singularities of transfer functions.

F.3 Measuring Intersymbol Interference

The ideas of frequency response and difference equations can be used to interpret and analyze properties of the transmission system. When all aspects of the system operate well, quantizing the received signal to the nearest element of the symbol alphabet recovers the transmitted symbol. This requires (among other things) that there is no significant multipath interference. This section uses the graphical tool of the eye diagram to give a measure of the severity of the intersymbol interference. In Section 11.3, the eye diagram was introduced as a way to visualize the intersymbol interference caused by various pulse shapes. Here, the eye diagram is used to help visualize the effects of intersymbol interference caused by multipath channels such as (13.2).

For example, consider a binary ± 1 source $s[k]$ and a three-tap FIR channel model that produces the received signal $r[k]$

$$r[k] = b_0 s[k] + b_1 s[k - 1] + b_2 s[k - 2].$$

This is shown in Figure F.3, where the received signal is quantized using the sign operator in order to produce the binary sequence $y[k]$, which provides an estimate of the source. Depending on the values of the b_i, this estimate may or may not accurately reflect the source.

Suppose $b_1 = 1$ and $b_0 = b_2 = 0$. Then $r[k] = s[k - 1]$ and the output of the decision device is, as desired, a replica of a delayed version of the source (i.e., $y[k] = \text{sign}\{s[k-1]\} = s[k-1]$). Like the eye diagrams of Chapter 9, which are "open" whenever the intersymbol interference admits perfect reconstruction of the source message, the eye is said to be open.

If $b_0 = 0.5$, $b_1 = 1$, and $b_2 = 0$, then $r[k] = 0.5s[k] + s[k - 1]$. Consider the four possibilities: $(s[k], s[k - 1]) = (1, 1), (1, -1), (-1, 1)$, or $(-1, -1)$, for which $r[k]$ is 1.5, 0.5, -0.5, and -1.5, respectively. In each case, $\text{sign}\{r[k]\} = s[k - 1]$. The eye is still open.

Now consider $b_0 = 0.4$, $b_1 = 1$, and $b_2 = -0.2$. The eight possibilities for $(s[k], s[k - 1], s[k - 2])$ in

$$r[k] = 0.4s[k] + s[k - 1] - 0.2s[k - 2]$$

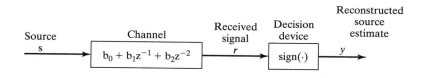

FIGURE F.3 Channel and Binary Decision Device.

are $(1, 1, 1)$, $(1, 1, -1)$, $(1, -1, 1)$, $(1, -1, -1)$, $(-1, 1, 1)$, $(-1, 1, -1)$, $(-1, -1, 1)$, and $(-1, -1, -1)$. The resulting choices for $r[k]$ are 1.2, 1.6, -0.8, -0.4, 0.4, 0.8, -1.6, -1.2, respectively, with the corresponding $s[k - 1]$ of $1,1,-1,-1,1,1,-1,-1$. For all of the possibilities, $y[k] = \text{sign}\{r[k]\} = s[k - 1]$. Furthermore, $y[k] \neq s[k]$ and $y[k] \neq s[k - 2]$ across the same set of choices. The eye is still open.

Now consider $b_0 = 0.5$, $b_1 = 1$, and $b_2 = -0.6$. The resulting $r[k]$ are 0.9, 2.1, -1.1, 0.1, -0.1, 1.1, -2.1, and -0.9, with $s[k - 1] = 1$, 1, -1, -1, 1, 1, -1, -1. Out of these eight possibilities, two cause $\text{sign}\{r[k]\} \neq s[k - 1]$. (Neither $s[k]$ or $s[k - 2]$ does better.) The eye is closed.

This can be explored in MATLAB using the program `openclosed.m`, which defines the channel in b and implements it using the `filter` command. After passing through the channel, the binary source becomes multivalued, taking on values $\pm b_1 \pm b_2 \pm b_3$. Typical outputs of `openclosed.m` are shown in Figure F.4 for channels b=[0.4 1 -0.2] and [0.5 1 -0.6]. In the first case, four of the possible values are above zero (when b_2 is positive) and four are below (when b_2 is negative). In the second case, there is no universal correspondence between the sign of the input data and the sign of the received data y. This is the purpose of the final `for` statement, which counts the number of errors that occur at each delay. In the first case, there is a delay that causes no errors at all. In the second case, there are always errors.

---

openclosed.m: draw eye diagrams

```
b=[0.4 1 -0.2];               % define channel
m=1000; s=sign(randn(1,m));   % binary input of length m
r=filter(b,1,s);              % output of channel
y=sign(r);                    % quantization
for sh=0:5    % error at different delays
 err(sh+1)=0.5*sum(abs(y(sh+1:end)-s(1:end-sh)));
end
```

In general for the binary case, if for some i

$$|b_i| > \sum_{j \neq i} |b_j|,$$

then such incorrect decisions cannot occur. The greatest distortion occurs at the boundary between the open and closed eyes. Let α be the index at which the impulse response has its largest coefficient (in magnitude), so $|b_\alpha| \geq |b_i|$ for all i. Define the *open eye measure* for a binary ± 1 input

$$\text{OEM} = 1 - \frac{\sum_{i \neq \alpha} |b_i|}{|b_\alpha|}.$$

For $b_0 = 0.4$, $b_1 = 1$, and $b_2 = -0.2$, OEM $= 1 - (0.6/1) = 0.4$. This value is how far from zero (i.e., crossing over to the other source alphabet value) the equalizer output is in worst case (as can be seen in Figure F.4). Thus, error-free behavior could be assured as long as all other sources of error are smaller than this OEM. For the channel $[0.5, 1, -0.6]$, the OEM is negative, and the eye is closed.

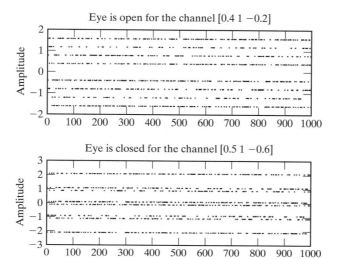

FIGURE F.4 Eye Diagrams for two channels

If the source is not binary, but instead takes on maximum ($s_{\max}$) and minimum ($s_{\min}$) magnitudes, then, as a worst-case measure,

$$\text{OEM} = 1 - \frac{(\sum_{i \neq \alpha} |b_i|)s_{\max}}{|b_\alpha|s_{\min}}.$$

As defined, OEM is always less than unity, with this value achieved only in the trivial case that all $|b_i|$ are zero for $i \neq \alpha$ and $|b_\alpha| > 0$. Thus,

- OEM > 0 is good (i.e., open eye)
- OEM < 0 is bad (i.e., closed eye)

The OEM provides a way of measuring the interference from a multipath channel. It does not measure the severity of other kinds of interference such as noise or in-band interferences caused by (say) other users.

Problems

F.8. Use `openclosed.m` to investigate the channels
   (a) b=[.3 -.3 .3 1 .1]
   (b) b=[.1 .1 .1 -.1 -.1 -.1 -.1]
   (c) b=[1 2 3 -10 3 2 1]
   For each channel, is the eye open? If so, what is the delay associated with the open eye? What is the OEM measure in each case?

F.9. Modify `openclosed.m` so that the received signal is corrupted by a (bounded uniform) additive noise with maximum amplitude s. How does the equivalent of Figure F.4 change? For what values of s do the channels in Problem F.8 have an open eye? For what values of s does the channel b=[.1 -.1 10 -.1] have an open eye? Hint: Use 2*s*(rand-0.5) to generate the noise.

F.10. Modify `openclosed.m` so that the input uses the source alphabet $\pm 1, \pm 3$. Are any of the channels in Problem F.8 open eye? Is the channel b=`[.1 -.1 10 -.1 ]` open eye? What is the OEM measure in each case?

When a channel has an open eye, all the intersymbol interference can be removed by the quantizer. But when the eye is closed, something more must be done. Opening a closed eye is the job of an equalizer, and is discussed at length in Chapter 13.

APPENDIX

G *Averages and Averaging*

Australian drinkers knocked back 336 beers each on average last year.
—Josh Whittington, *The Mercury*, 24 November, 1998, p. 12

There are two results in this appendix. The first section argues that averages (whether implemented as a simple sum, a moving average, or in recursive form) have an essentially "lowpass" character. This is used repeatedly in Chapters 6, 10, 12, and 13 to study the behavior of adaptive elements by simplifying the cost function to remove extraneous high frequency signals. The second result is that the derivative of an average (or a LPF) is *almost* the same as the average (or LPF) of the derivative. This approximation is formalized in (G.13) and is used throughout *Telecommunication Breakdown* to calculate the derivatives that occur in adaptive elements such as the phase locked loop, the automatic gain control, output energy maximization for timing recovery, and various equalization algorithms.

G.1 Averages and Filters

There are several kinds of averages. The *simple average* $\alpha[N]$ of a sequence of N numbers $\sigma[i]$ is

$$\alpha[N] = \text{avg}\{\sigma[i]\} = \frac{1}{N} \sum_{i=1}^{N} \sigma[i]. \tag{G.1}$$

For instance, the average temperature last year can be calculated by adding up the temperature on each day, and then dividing by the number of days.

When talking about averages over time, it is common to emphasize recent data and to de-emphasize data from the distant past. This can be done using a *moving average* of length P, which has a value at time k of

$$\alpha[k] = \text{avg}\{\sigma[i]\} = \frac{1}{P} \sum_{i=k-P+1}^{k} \sigma[i]. \tag{G.2}$$

This can also be implemented as a finite impulse response filter

$$\alpha[k] = \frac{1}{P}\sigma[k] + \frac{1}{P}\sigma[k-1] + \cdots + \frac{1}{P}\sigma[k-P+1]. \tag{G.3}$$

363

Instead of averaging the temperature over the whole year all at once, a moving average over a month ($P = 30$) finds the average over each consecutive 30 day period. This would show, for instance, that it is very cold in Wisconsin in the winter and hot in the summer. The simple annual average, on the other hand, would be more useful to the Wisconsin tourist bureau, since it would show a moderate yearly average of about 50 degrees.

Closely related to these averages is the recursive summer,

$$\alpha[i] = \alpha[i-1] + \mu\sigma[i] \text{ for } i = 1, 2, 3, \ldots, \tag{G.4}$$

which adds up each new element of the input sequence $\sigma[i]$, scaled by μ. Indeed, if the recursive filter (G.4) has $\mu = \frac{1}{N}$ and is initialized with $\alpha[0] = 0$, then $\alpha[N]$ is identical to the simple average in (G.1).

Writing these averages in the form of the filters (G.4) and (G.3) suggests the question: what kind of filters are these? The impulse response $h[k]$ of the moving average filter is

$$h[k] = \begin{bmatrix} 0 & k < 0 \\ \dfrac{1}{P} & 0 \le k < P \\ 0 & k \ge P \end{bmatrix},$$

which is essentially a "rectangle" shape in time. Accordingly, the frequency response is sinc shaped, from (A.20). Thus, the averaging "filter" passes very low frequencies and attenuates high frequencies. It thus has a lowpass character, though it is far from an ideal LPF.

The impulse response for the simple recursive filter (G.4) is

$$h[k] = \begin{bmatrix} 0 & k < 0 \\ \mu & k \ge 0 \end{bmatrix}.$$

This is also a "rectangle" that widens as k increases, which again represents a filter with a lowpass character. This can be seen using the techniques of Appendix F by observing that the transfer function of (G.4) has a single pole at 1 which causes the magnitude of the frequency response to decrease as the frequency increases. Thus, averages such as (G.1), moving average filters such as (G.3), and recursive filters such as (G.4) all have a "lowpass" character.

G.2 Derivatives and Filters

Averages and lowpass filters occur within the definitions of the performance functions associated with many adaptive elements. For instance, the AGC of Chapter 6, the phase tracking algorithms of Chapter 10, the timing recovery methods of Chapter 12, and the equalizers of Chapter 13 all involve LPFs, averages, or both. Finding the correct form for the adaptive updates requires taking the derivative of filtered and averaged signals. This section shows when it is possible to commute the two operations, that is, when the derivative of the filtered (averaged) signal is the same as a filtering (averaging) of the derivative. The derivative is taken with respect to some variable β, and the key to the commutativity is how β enters the filtering operation. Sometimes the derivative is taken with respect to time, sometimes it is taken with respect to a coefficient of the filter, and sometimes it appears as a parameter within the signal.

When the derivative is taken with respect to time, the LPF and/or average commute with the derivative; that is,

$$\text{LPF}\left\{\frac{d\alpha}{d\beta}\right\} = \frac{d}{d\beta}\text{LPF}\{\alpha\} \tag{G.5}$$

and

$$\text{avg}\left\{\frac{d\alpha}{d\beta}\right\} = \frac{d}{d\beta}\text{avg}\{\alpha\}, \tag{G.6}$$

where α is the signal and β represents time. This is a direct consequence of linearity; the LPF and the derivative are both linear operations. Since linear operations commute, so do the filters (averages) and the derivatives. This is demonstrated using the code in dlpf.m in which a random signal s is passed through an arbitrary linear system defined by the impulse response h. The derivative is approximated in dlpf.m using the diff function, and the calculation is done two ways: first by taking the derivative of the filtered signal, and then by filtering the derivative. Observe that the two methods give the same output after the filters have settled.

---

dlpf.m: differentiation and filtering commute

```
s=randn(1,100);           % generate random 'data'
h=randn(1,10);            % an arbitrary impulse response
dlpfs=diff(filter(h,1,s));   % take deriv of filtered input
lpfds=filter(h,1,diff(s));   % filter the deriv of input
dlpfs-lpfds               % compare the two
```

When the derivative is taken with respect to a coefficient (tap weight) of the filter, then (G.5) does not hold. For example, consider the time-invariant linear filter

$$\alpha[k] = \sum_{i=0}^{P-1} b_i \sigma[k-i],$$

which has impulse response $[b_{P-1}, \ldots, b_1, b_0]$. If the b_i are chosen so that $\alpha[k]$ represents a lowpass filtering of the $\sigma[k]$, then the notation

$$\alpha[k] = \text{LPF}\{\sigma[k]\}$$

is appropriate, while if $b_i = 1/P$, then $\alpha[k] = \text{avg}\{\sigma[k]\}$ might be more apropos. In either case, consider the derivative of $\alpha[k]$ with respect to a parameter b_j:

$$\frac{d\alpha[k]}{db_j} = \frac{d}{db_j}(b_0\sigma[k] + b_1\sigma[k-1] + \cdots + b_{P-1}\sigma[k-P+1])$$

$$= \frac{db_1\sigma[k]}{db_j} + \frac{db_2\sigma[k-1]}{db_j} + \cdots + \frac{db_{P-1}\sigma[k-P+1]}{db_j}. \tag{G.7}$$

Since all the terms $b_i\sigma[k-i]$ are independent of b_j for $i \neq j$, the terms $\frac{db_i\sigma[k-i]}{db_j}$ are zero. For $i = j$,

$$\frac{db_j\sigma[k-j]}{db_j} = \sigma[k-j]\frac{db_j}{db_j} = \sigma[k-j],$$

so

$$\frac{d\alpha[k]}{db_j} = \frac{d\text{LPF}\{\sigma[k]\}}{db_j} = \sigma[k-j]. \tag{G.8}$$

On the other hand, $\text{LPF}\left\{\frac{d\sigma[k]}{db_j}\right\} = 0$ because $\sigma[k]$ is not a function of b_j. The derivative and the filter do not commute and (G.5) (with $\beta = b_j$) does not hold.

An interesting and useful case is when the signal that is to be filtered is a function of β. Let $\sigma[\beta, k]$ be the input to the filter that is explicitly parameterized by both β and by time k. Then

$$\alpha[k] = \text{LPF}\{\sigma(\beta, k)\} = \sum_{i=0}^{P-1} b_i\sigma(\beta, k-i).$$

The derivative of $\alpha[k]$ with respect to β is

$$\frac{d\text{LPF}\{\sigma(\beta, k)\}}{d\beta} = \frac{d}{d\beta}\sum_{i=0}^{P-1} b_i\sigma(\beta, k-i) = \sum_{i=0}^{P-1} b_i\frac{d\sigma(\beta, k-i)}{d\beta} = \text{LPF}\left\{\frac{d\sigma(\beta, k)}{d\beta}\right\}.$$

Thus, (G.5) holds in this case.

Example G.1

This example is reminiscent of the phase tracking algorithms in Chapter 10. Let $\beta = \theta$ and $\sigma(\beta, k) = \sigma(\theta, k) = \sin(2\pi f kT + \theta)$. Then

$$\frac{d\text{LPF}\{\sigma(\theta, k)\}}{d\theta} = \frac{d}{d\theta}\sum_{i=0}^{P-1} b_i\sin(2\pi(k-i)T + \theta)$$

$$= \sum_{i=0}^{P-1} b_i\frac{d}{d\theta}\sin(2\pi(k-i)T + \theta)$$

$$= -\sum_{i=0}^{P-1} b_i\cos(2\pi(k-i)T + \theta)$$

$$= \text{LPF}\{\frac{d\sigma(\theta, k)}{d\theta}\}. \tag{G.9}$$

Example G.2

This example is reminiscent of the equalization algorithms that appear in Chapter 13 in which the signal $\sigma(\beta, k)$ is formed by filtering a signal $u[k]$ that is independent of β. To be precise, let $\beta = a_1$ and $\sigma(\beta, k) = \sigma(a_1, k) = a_0 u[k] + a_1 u[k-1] + a_2 u[k-2]$. Then

$$
\begin{aligned}
\frac{d\text{LPF}\{\sigma(a_1, k)\}}{da_1} &= \frac{d}{da_1} \sum_{i=0}^{P-1} b_i (a_0 u[k-i] + a_1 u[k-i-1] + a_2 u[k-i-2]) \\
&= \sum_{i=0}^{P-1} b_i \frac{d}{da_1} (a_0 u[k-i] + a_1 u[k-i-1] + a_2 u[k-i-2]) \\
&= \sum_{i=0}^{P-1} b_i u[k-i-1] \\
&= \text{LPF}\{\frac{d\sigma(a_1, k)}{da_1}\}.
\end{aligned}
\tag{G.10}
$$

The transition between the second and third equality mimics the transition from (G.7) to (G.8), with u playing the role of σ and a_1 playing the role of b_j.

G.3 Differentiation Is a Technique, Approximation Is an Art

When β (the variable with respect to which the derivative is taken) is not a function of time, then the derivatives can be calculated without ambiguity or approximation, as was done in the previous section. In most of the applications in *Telecommunication Breakdown*, however, the derivative is being calculated for the express purpose of adapting the parameter, that is, with the intent of changing β so as to maximize or minimize some performance function. In this case, the derivative is not straightforward to calculate, and it is often simpler to use an approximation.

To see the complication, suppose that the signal σ is a function of time k and the parameter β and that β is itself time dependent. Then it is more proper to use the notation $\sigma(\beta[k], k)$, and to take the derivative with respect to $\beta[k]$. If it were simply a matter of taking the derivative of $\sigma(\beta[k], k)$ with respect to $\beta[k]$, then there is no problem, since

$$
\frac{d\sigma(\beta[k], k)}{d\beta[k]} = \frac{d\sigma(\beta, k)}{d\beta}\bigg|_{\beta=\beta[k]}.
\tag{G.11}
$$

When taking the derivative of a filtered version of the signal $\sigma(\beta[k], k)$, however, all the terms are not exactly of this form. Suppose, for example, that

$$
\alpha[k] = \text{LPF}\{\sigma(\beta[k], k)\} = \sum_{i=0}^{P-1} b_i \sigma(\beta[k-i], k-i)
$$

is a filtering of σ and the derivative is to be taken with respect to $\beta[k]$:

$$\frac{d\alpha[k]}{d\beta[k]} = \sum_{i=0}^{P-1} b_i \frac{d\sigma(\beta[k-i], k-i)}{d\beta[k]}.$$

Only the first term in the sum has the form of (G.11). All others are of the form

$$\frac{d\sigma(\beta[k-i], k-i)}{d\beta[k]},$$

with $i \neq 0$. If there were no functional relationship between $\beta[k]$ and $\beta[k-i]$, then this derivative would be zero, and $\frac{d\alpha[k]}{d\beta[k]} = b_0 \frac{d\sigma(\beta,k)}{d\beta}|_{\beta=\beta[k]}$. But, of course, there generally is a functional relationship between β at different times, and proper evaluation of the derivative requires that this relationship be taken into account.

The situation that is encountered repeatedly throughout *Telecommunication Breakdown* occurs when $\beta[k]$ is defined by a small stepsize iteration; that is,

$$\beta[k] = \beta[k-1] + \mu\gamma(\beta[k-1], k-1), \tag{G.12}$$

where $\gamma(\beta[k-1], k-1)$ is some bounded signal (with bounded derivative) which may itself be a function of time $k-1$ and the state $\beta[k-1]$. A key feature of (G.12) is that μ is a user-choosable stepsize parameter. As will be shown, when a sufficiently small μ is chosen, the derivative $\frac{d\alpha[k]}{d\beta[k]}$ can be approximated efficiently as

$$\begin{aligned}
\frac{d\text{LPF}\{\sigma(\beta[k], k)\}}{d\beta[k]} &= \sum_{i=0}^{P-1} b_i \frac{d\sigma(\beta[k-i], k-i)}{d\beta[k]} \\
&\approx \sum_{i=0}^{P-1} b_i \frac{d\sigma(\beta[k-i], k-i)}{d\beta[k-i]} \\
&= \text{LPF}\left\{ \frac{d\sigma(\beta, k)}{d\beta}\bigg|_{\beta=\beta[k]} \right\},
\end{aligned} \tag{G.13}$$

which nicely recaptures the commutativity of the LPF and the derivative as in (G.5). A special case of (G.13) is to replace "LPF" with "avg."

The remainder of this section provides a detailed justification for this approximation and provides two detailed examples. Other examples appear throughout *Telecommunication Breakdown*.

As a first step, consider $\frac{d\sigma(\beta[k-1], k-1)}{d\beta[k]}$, which appears in the second term of the sum in (G.13). This term can be rewritten using the chain rule (A.59),

$$\frac{d\sigma(\beta[k-1], k-1)}{d\beta[k]} = \frac{d\sigma(\beta[k-1], k-1)}{d\beta[k-1]} \frac{d\beta[k-1]}{d\beta[k]}.$$

Rewriting (G.12) as $\beta[k-1] = \beta[k] - \mu\gamma(\beta[k-1], k-1)$ yields

$$\frac{d\sigma(\beta[k-1], k-1)}{d\beta[k]} = \frac{d\sigma(\beta[k-1], k-1)}{d\beta[k-1]} \frac{d(\beta[k] - \mu\gamma(\beta[k-1], k-1))}{d\beta[k]}$$

$$= \frac{d\sigma(\beta, k-1)}{d\beta}\bigg|_{\beta=\beta[k-1]} \left(1 - \mu\frac{d\gamma(\beta[k-1], k-1)}{d\beta[k]} \right). \tag{G.14}$$

Applying similar logic to the derivative of γ shows that

$$
\begin{aligned}
\frac{d\gamma(\beta[k-1], k-1)}{d\beta[k]} &= \frac{d\gamma(\beta[k-1], k-1)}{d\beta[k-1]} \frac{d\beta[k-1]}{d\beta[k]} \\
&= \frac{d\gamma(\beta[k-1], k-1)}{d\beta[k-1]} \frac{d(\beta[k] - \mu\gamma(\beta[k-1], k-1))}{d\beta[k]} \\
&= \left.\frac{d\gamma(\beta, k-1)}{d\beta}\right|_{\beta=\beta[k-1]} \left(1 - \mu\frac{d\gamma(\beta[k-1], k-1)}{d\beta[k]}\right).
\end{aligned}
$$

While this may appear at first glance to be a circular argument (since $\frac{d\gamma(\beta[k-1],k-1)}{d\beta[k]}$ appears on both sides), it can be solved algebraically as

$$
\frac{d\gamma(\beta[k-1], k-1)}{d\beta[k]} = \frac{\left.\frac{d\gamma(\beta,k-1)}{d\beta}\right|_{\beta=\beta[k-1]}}{1 + \mu\left.\frac{d\gamma(\beta,k-1)}{d\beta}\right|_{\beta=\beta[k-1]}} \equiv \frac{\gamma_0}{1 + \mu\gamma_0}, \tag{G.15}
$$

where

$$
\gamma_0 = \left.\frac{d\gamma(\beta, k-1)}{d\beta}\right|_{\beta=\beta[k-1]}. \tag{G.16}
$$

Substituting (G.15) back into (G.14) yields

$$
\frac{d\sigma(\beta[k-1], k-1)}{d\beta[k]} = \left.\frac{d\sigma(\beta, k-1)}{d\beta}\right|_{\beta=\beta[k-1]} \left(1 - \mu\frac{\gamma_0}{1 + \mu\gamma_0}\right).
$$

The point of this calculation is that, since the value of μ is chosen by the user, it can be made as small as needed to ensure that

$$
\frac{d\sigma(\beta[k-1], k-1)}{d\beta[k]} \approx \left.\frac{d\sigma(\beta, k-1)}{d\beta}\right|_{\beta=\beta[k-1]}.
$$

Following the same basic steps for the general delay term in (G.13) shows that

$$
\frac{d\sigma(\beta[k-n], k-n)}{d\beta[k]} = \left.\frac{d\sigma(\beta, k-n)}{d\beta}\right|_{\beta=\beta[k-n]} (1 - \mu\gamma_n),
$$

where

$$
\gamma_n = \frac{\gamma_0}{1 + \mu\gamma_0}\left(1 - \mu\sum_{j=1}^{n-1}\gamma_j\right)
$$

is defined recursively with γ_0 given in (G.16) and $\gamma_1 = \frac{\gamma_0}{1+\mu\gamma_0}$. For small μ, this implies that

$$
\frac{d\sigma(\beta[k-n], k-n)}{d\beta[k]} \approx \left.\frac{d\sigma(\beta, k-n)}{d\beta}\right|_{\beta=\beta[k-n]} \tag{G.17}
$$

for each n. Combining these yields the approximation (G.13).

Example G.3

Example G.1 assumes that the phase angle θ is fixed, even though the purpose of the adaptation in a phase tracking algorithm is to allow θ to follow a changing phase. To investigate the time-varying situation, let $\beta[k] = \theta[k]$ and $\sigma(\beta[k], k) = \sigma(\theta[k], k) = \sin(2\pi fkT + \theta[k])$. Suppose also that the dynamics of θ are given by

$$\theta[k] = \theta[k-1] + \mu\gamma(\theta[k-1], k-1).$$

Using the approximation (G.17) yields

$$
\begin{aligned}
\frac{d\text{LPF}\{\sigma(\theta[k], k)\}}{d\theta[k]} &= \frac{d}{d\theta[k]} \sum_{i=0}^{P-1} b_i \sigma(\theta[k-i], k-i) \\
&= \frac{d}{d\theta[k]} \sum_{i=0}^{P-1} b_i \sin(2\pi(k-i)T + \theta[k-i]) \\
&= \sum_{i=0}^{P-1} b_i \frac{d}{d\theta[k]} \sin(2\pi(k-i)T + \theta[k-i]) \\
&\approx \sum_{i=0}^{P-1} b_i \frac{d}{d\theta[k-i]} \sin(2\pi(k-i)T + \theta[k-i]) \\
&= -\sum_{i=0}^{P-1} b_i \cos(2\pi(k-i)T + \theta[k-i]) \\
&= \text{LPF} \left\{ \frac{d\sigma(\theta, k)}{d\theta} \bigg|_{\theta=\theta[k]} \right\}.
\end{aligned}
\tag{G.18}
$$

Example G.4

Example G.2 assumes that the parameter a_1 of the linear filter is fixed, even though the purpose of the adaptation is to allow a_1 to change in response to the behavior of the signal. Let $\sigma(\beta[k], k)$ be formed by filtering a signal $u[k]$ that is independent of $\beta[k]$. To be precise, let $\beta[k] = a_1[k]$ and $\sigma(\beta[k], k) = \sigma(a_1[k], k) = a_0[k]u[k] + a_1[k]u[k-1] + a_2[k]u[k-2]$. Suppose also that the dynamics of $a_1[k]$ are given by

$$a_1[k] = a_1[k-1] + \mu\gamma(a_1[k-1], k-1).$$

Example G.4   (Continued)

Then the approximation (G.17) yields

$$\frac{d\text{LPF}\{\sigma(a_1[k], k)\}}{da_1[k]} = \frac{d}{da_1[k]} \sum_{i=0}^{P-1} b_i \sigma(a_1[k-i], k-i)$$

$$= \frac{d}{da_1[k]} \sum_{i=0}^{P-1} b_i (a_0[k-i]u[k-i] + a_1[k-i]u[k-i-1] + a_2[k-i]u[k-i-2])$$

$$= \sum_{i=0}^{P-1} b_i \frac{d}{da_1[k]} (a_0[k-i]u[k-i] + a_1[k-i]u[k-i-1] + a_2[k-i]u[k-i-2])$$

$$= \sum_{i=0}^{P-1} b_i \frac{da_1[k-i]u[k-i-1]}{da_1[k]}$$

$$\approx \sum_{i=0}^{P-1} b_i \frac{da_1[k-i]u[k-i-1]}{da_1[k-i]}$$

$$= \sum_{i=0}^{P-1} b_i u[k-i-1]$$

$$= \text{LPF}\left\{ \frac{d\sigma(a_1, k)}{da_1}\bigg|_{a_1=a_1[k]} \right\}. \tag{G.19}$$

Index